Power to the People

AUDREY
KURTH
CRONIN

Power to the People

How Open Technological
Innovation Is Arming
Tomorrow's Terrorists

OXFORD
UNIVERSITY PRESS

OXFORD
UNIVERSITY PRESS

Oxford University Press is a department of the University of Oxford. It furthers
the University's objective of excellence in research, scholarship, and education
by publishing worldwide. Oxford is a registered trade mark of Oxford University
Press in the UK and certain other countries.

Published in the United States of America by Oxford University Press
198 Madison Avenue, New York, NY 10016, United States of America.

First issued as an Oxford University Press paperback, 2022

CIP data is on file at the Library of Congress
ISBN 978–0–19–088214–3 (hardback)
ISBN 978–0–19–757893–3 (paperback)

9 8 7 6 5 4 3 2 1

Paperback printed by Marquis, Canada

For Professor Sir Adam Roberts,
who gave power to the pupil.

CONTENTS

T HE TITLE OF Audrey Cronin's book plays on the slogan of 'power to the people', extensively used by radical groups in the 1960s and famously turned into a song by John Lennon. The slogan demanded that all those instruments of influence and decision should be taken away from elites and distributed to the masses. This same slogan was appropriated in the 1970s by computer hobbyists as they showed how using silicon micro-chips they could build affordable machines that would give individuals access to data processing capabilities hitherto reserved for large corporations, universities and the Pentagon. In this case the vital power went beyond speaking freely, and voting, demonstrating and organising, and on to the ability to perform an ever-expanding range of tasks quickly and effectively. Over the past half century we have come to expect access to extraordinary data bases, complex games, navigational aids, satellite imagery, and instantaneous communications, if necessary encrypted. The process began with clunky desktop computers; now almost everything required can be carried on a pocket phone. Powers that were only available to a narrow elite barely a few decades have reached the people.

For most of us this represents fantastic progress. But Audrey Cronin points to a darker side. These powers, so widely distributed and now so taken for granted, so important to our prosperity and way of life, can be used to cause great harm. It is not news that most technologies can be used for multiple purposes, not all benign, or that there is often a close link between new technologies and the military. In the past the resources of the state were vital when turning scientific insights into usable products. This was true with ships and aircraft, chemical weapons and nuclear energy, radar and rockets. Even modern computing and the internet would not have advanced so quickly without military support. Often if a military value had not been demonstrated first the civilian benefits would not have been realised.

Technological developments have become increasingly unmoored from the military, taken forward by multinational companies, motivated by profit, and interested more in commercial prospects than any security implications. The military are now one customer among many. This has not meant, however, that these technologies are now useful solely for benign purposes. What can now be bought off the shelf can support individuals in a variety of menacing undertakings. To help identify which new technologies might support political violence in the future Cronin develops a 'lethal empowerment theory', looking at how capabilities were harnessed by individuals intent on terrorist acts in the past. In her first example she shows how Alfred Nobel's development of dynamite for industrial customers was turned to advantage by 19th Century anarchists appreciated the ease with which it could be acquired and the havoc that could be caused when thrown in the direction of individuals they disliked. In the second example she notes how Mikhail Kalashnikov's design of the AK-47 assault rifle for use during the Second World War might not have been as refined as others but worked for Soviet soldiers because it was easy to carry and to operate and very durable. For the same reason it was sought by the various guerrilla groups and militias that came into being after 1945. These provide salutary examples of the progressive democratisation of violence. It may need a state or a proto-state to put together an army of size and substance but individuals or small groups can acquire a range of disruptive and damaging capabilities.

This happens with the technologies of the digital age in several ways. A smart phone can help terrorists identify, find and track targets, pick up a recipe for a home-made bomb, send orders to attack, and then live-stream the carnage. If they decide not to broadcast their attack or distribute their own images they know that the job will be done for them by witnesses who will not be able to resist the opportunity to do so. Even without resorting to direct violence, social media provides means of making lives miserable by defaming the character of targets and sending packs of angry stalkers to harass them. The more criminally minded can send malicious messages to millions in the hope that a few will provide sufficient feedback to enable their bank accounts to be emptied, or extort bitcoins by disrupting the workings of major corporations and frightening the rich. Is it not necessarily the case that the big cyber threats—hacking into the most sensitive secrets, denial

tworks, and subverting critical in-
overnments. There are many other
e. Drones bought for home use can
d with 3-D printers. Coming down
will be opportunities for malicious
and direct them onto destructive
ning scenarios, but most alarming
es her to back up her analysis with
demonstrating the destructive po-

a present set of security challenges
vere. Now as Cronin rightly notes
ome things might be done it does
re unfortunately many individuals
r a variety of philosophical or quite
reats will still tend to come from
gns to achieve some sort of political
e clues to law enforcement. A large
hreats, including those still relying
rchists, involves good intelligence

t what attitude we should take to
tenment, and the extraordinary re-
to be techno-optimists. The people
ovides us with reasons to take care.
be done with bio-technologies and
ted. Although there is no reason to
gulation in many areas may mean
e and consciences of big companies
ing capabilities. Some of the recent
t hate campaigns and misinforma-
The US and its allies are not the
tems and exploring artificial intel-
e that future dangers can easily be
hard-headed, smart and nimble to
ely Audrey Cronin has provided us
hrough these issues and an agenda

| The Age of Lethal Empowerment

W E LIVE IN an epoch of unprecedented popular empowerment. Increasing access to information, rising global living standards, growing literacy, and improving medical care and longevity are just a handful of the benefits derived from the modern march of technological innovation. Yet the same technologies that are furthering prosperity are creating critical new security vulnerabilities.

The worldwide dispersal of emerging technologies, such as commercial drones, cyber weapons, 3D printing, military robotics, and autonomous systems, is generating gaping fissures in the ability of conventional armed forces to combat lethal capabilities of non-state actors, most notably terrorists, but also rogue lone actors, insurgent groups, and private armies. As these malcontents gain power through emerging technologies, the defenses of the law-abiding are increasingly breached. Traditional rights, such as privacy, property ownership, and freedom of movement, are under unprecedented attack by autocratic regimes, not to mention private companies, which monetize our online behavior. Not only are democratic governments unable to protect innocent citizens from such invasions, they engage in them themselves, reading individuals' emails, monitoring their phone calls, tracking their every move online, and consolidating their behavioral data.

Every day we learn of fresh examples of the new breed of technologies being used for malevolent ends or posing potential new dangers. But while we've become aware that intelligence services may be listening to your phone calls and tracking your emails, and that Apple, Facebook, Google, and other corporations are mining our data and selling it to advertisers, the potential uses of the new technologies for violence are less discussed. Is anyone protecting you from having your Internet-connected Jeep Cherokee hacked and acceler-ated into a wall? The ease of doing so was demonstrated in 2015, when security specialists Charlie Miller and Chris Valasek commandeered a Jeep, with the agreement of its driver, journalist Andy Greenberg, disabling the accelerator as he drove down a highway, just as a tractor-trailer truck was barreling up from behind him. Or, what happens when your Amazon delivery drone carries a bomb-filled package instead of prescription medicine? The Islamic State, Hezbollah, and Jund al-Aqsa (an al-Qaeda-associated group) have deployed explosives this way. In 2017, the Islamic State jerry-rigged commercial drones to carry explosives and filmed nearly a dozen of them dropping small bombs on US and Iraqi tanks, vehicles, and people, eventually killing more than a dozen and injuring many more.[1] Drones and self-driving cars are not only ex-citing advances for consumers; they are also new means of conducting political violence by militant groups and self-radicalized individuals.

Never before have so many had access to such advanced technologies capable of inflicting death and mayhem. Unless we better understand the rapidly developing threats, governments, especially democracies, will be in-creasingly unable to combat them.

This book explores how individuals and groups who engage in political violence have repeatedly made use of emerging technologies to wreak havoc, and how they're likely to do so in the future. Many scholars have argued that such non-state combatants rarely innovate new forms of weaponry; that they are tactically sophisticated but technologically crude.[2] Regarding the danger of more sophisticated weaponry getting into their hands, especially so-called weapons of mass destruction, the focus has been mainly on groups acquiring weapons from states rather than learning to produce their own. That is too lim-ited a purview, conditioned by the experience of the last fifty years. A longer historical view, stretching back to the nineteenth century, reveals that, in fact, terrorists and insurgents have repeatedly seized opportunities presented by new technologies, innovating deadly new uses for them, and sometimes even adopting them ahead of their use by traditional military forces. When clusters of new technologies become widely accessible to the public, their innovations are particularly lethal and fast-spreading. We are in such an era now.

Placing our current technological advances into the historical context of key innovations used to wage politically disruptive violence, this book will first explore why certain kinds of emerging technologies are rapidly adopted by rogue actors. We'll then consider how those emerging today are already changing who is able to muster formidable armies and show up on the battlefield, how individual acts of violence can be perpetrated, and even who may gain the ability to catalyze state-to-state wars. Traditional warfare concepts of the massing of force, logistics, and power projection— the understanding of which has been essentially stable for centuries—must evolve in the face of novel challenges. Imagine a successor to the Islamic State, for example, teaming small numbers of fighters with autonomous suicide attack vehicles, flying unmanned systems through urban sewer systems, and assassinating key government leaders with facial recognition software, all while using 3D printing to reproduce an endless supply of replacement parts, such as drone blades and motors. Technological advancements are heightening global instability in ways that extend far beyond the battlefield. States will still be the dominant form of governance, but tomorrow's question is what *kinds* of states? New technologies enable increasingly powerful non-state actors to affect the answer. Power is shifting away from democratic states, and they must prepare for, and defend against, the potentially seismic consequences.

The United States is especially vulnerable to being caught on the back foot by this surge of technology and weapons diffusion. The country leads the world in the development of high-tech weaponry, and its defense strategy is predicated on this technological superiority. Yet the effectiveness of that formidable arsenal in the face of emerging threats must be thoroughly re-examined. The failure to win decisive victories in the wars in Afghanistan and Iraq has demonstrated this. Meanwhile a recent re-emergence of major-power competition, particularly between the United States and China, is likely to keep the focus of military planners on large-scale, high-end weapons systems rather than on building capabilities and strategy to defend against more pervasive and less obvious emerging threats.

The US defense community is predominantly focused on competing technologically with state rivals at the high end, while slighting centuries of experience in how violent individuals and groups improvise, sometimes unleashing unexpected waves of political disruption that topple regimes and trigger interstate war. Many of the most revolutionary changes in warfare, the most surprising victories and losses, have occurred as a result not of new technologies on the battlefield, but of broader changes in society affecting who used weaponry, where, when, why, and how. With the maturing of the digital revolution and the beginning of the robotics revolution, we are seeing the combination of rapid technological innovation and that kind of widespread *societal* change, especially in how people shape, spread, and evaluate information, then acquire the means to act upon it. The result will be dramatic transformation of the character of war, not just in the form of ongoing cyber warfare or further integration of autonomy into military weapons, already well underway, or the widespread proliferation of high-powered drone weaponry from state to state. The changes will also involve the redistribution of relative power between states and nonstate actors, who will surprise traditional militaries and police forces with innovative uses of the technologies.

Despite the breathless claims of many technology promoters, what is crucial today for national security is not the transformative power of the new technologies, but the transformative power of human beings throughout the world to adapt them to unanticipated purposes. That is where their true revolutionary power lies. Individuals and groups harnessing the emerging innovations may actually be capable of undermining global stability and, as in the past, even sparking interstate war.

Technology is no longer necessarily on the side of the major powers. This book explains why and what to do about it.

The book is not a treatise on technology; it does not go into great detail about the design and capabilities of the emerging technologies, which are discussed in nontechnical terms. The focus, instead, is on the potential uses of the technologies and the need for strategy that anticipates them. The competition that matters most now to the security of the United States and its allies is not in technology development but in strategy development. And in that they have a long way to go. Developing clearer doctrine for how and when to deploy the new breed of weapons, and how and when not to deploy them, is vital. Crafting scenarios of how the technologies may be deployed, and combined, in innovative ways by rogue actors is also crucial.

No anti-technology screed, this book aims to illuminate how to maximize the promise and minimize the risks of emerging technologies. That requires understanding how to distinguish between technologies that will advance global stability and those that will undermine it, widening the scope beyond high-profile arms races and state-to-state competitions. Key questions include: How and why do non-state actors innovate differently from state actors? Why do violent non-state actors adopt certain technologies more readily and widely than others? How have past technological advances driven waves of political violence and destabilized the world, and what does that historical record show us about the threats from today's emerging technologies? Finally, how can that history guide our actions going forward?

Answering those questions requires close examination of the double-edged aspect of technological innovation.

Inno..tedly the engine that drives human progress. No one
wou..hroughout human history, advances in engineering,
me..puter science, and so forth have vastly improved the
qu..s, propelling economic, social, and political develop-
m..gical innovation involves risks as well as promises,
..es present more risks than others. Those risks may
..gies, such as the danger of a nuclear cataclysm, or
..technologies becoming accessible to people who
..ome lethal and destabilizing consequences are
..human error. Understanding the risks is vital

sl.
di.
egy reg.
rs in Iraq
offensive a.
of drones and
ks and reduce
had been un-
tlefield minus
tary to target
ch preemptive
not lead to de-

chnology and
in eradicating
objectives. No
conflicts than
Syria and the

vernance over
gle, Facebook,
their mandate
g the Internet
ccess—the so-
abilities, such
nomic growth
States does'
edge wi'

easibl'
rit?
n

nologies that were developed with good
ia and drones. They were created to solve
expanding public access to information and
nane. But some of them can be fashioned
nd precise weapons. They have also inad-
thical guidelines, such as the protection
e state's monopoly on the use of armed
e and national policies. By not thinking
rging technologies as they've been de-
benefits and profitability and not their
e, neither the companies creating them
have built viable long-term strategies
purposes.

mprehensive strategic approach to the
technologies has been lacking, one his-
many of the innovations have emerged
ave tended to be techno-optimists. The
e the commercialization of technology
er economic, political, cultural, and secu-
century, excitement over new inventions,
elegraph, and the automobile, swept the
ild it into the world's dominant power.

eyday during the Second World War, when
ded its industrial capacity, transforming into
he Allied Powers won the war largely due to
iovations, including radar, the aircraft carrier,
ind, of course, the atomic bomb. Technological
s of US military strategy, with the emphasis on
power, speed, range, stealth, cryptography, and

reconnaissance for warfare waged on traditional battlefields, in
the buildup of a deterrent nuclear arsenal. Over the course of
century, the result was the innovation of a string of sophistic;
technologies, such as ballistic missiles and airborne early war
trol systems (AEW&C). All of these were vital to Cold War con
arguably, to the West's ultimate success and the collapse of the
Yet with the growing predominance of asymmetric warfare, e
the first Persian Gulf War (1990–1991), the US military sav
returns for its massive expenditures on these capabilities.

The second reason for the failure to develop adequate strat
the emerging technologies is that the 9/11 attacks and the w;
Afghanistan led to an urgent focus on rapid deployment of new
defensive capabilities. The United States rushed to make use
autonomous robots in order to curtail further terrorist attac
the ghastly toll of improvised explosive devices (IEDs), which
anticipated. New technologies, many of them sent to the bat
the usual extensive testing and refinement, allowed the mil;
individual terrorists, develop better reconnaissance, and laun
strikes, saving many American and allied lives. But they did
cisive victory.

These wars have shown that even with awe-inspiring t
overwhelming force, military operations have limited effect
unconventional forces and advancing political and security
sooner did the United States begin to extricate itself from these
new fronts of battle emerged, with the outbreak of civil war i
rise of the Islamic State.

The third reason little progress has been made in go
emerging technologies is that private companies, such as Goo
and Microsoft, are driving so much of their development, with
being to generate profits rather than to secure peace. Makin
more accessible and selling products enabled with Internet a
called Internet of Things—and artificial intelligence (AI) caj
as voice recognition, have indeed become core drivers of eco
which is in the national interest. What's more, if the United
keep up a lively pace of innovation, its current competitiv
eclipsed by other major powers, especially China.

Simply clamping down on private innovation is neither
United States and its democratic allies, nor desirable. Auth
have held a tighter grip over their technology innovators
cess to the technologies. For instance, China censors the di

of service attacks against government networks, and subverting critical infrastructure—can only be mounted by governments. There are many other examples of the democratisation of violence. Drones bought for home use can be weaponised. Guns can be manufactured with 3-D printers. Coming down the line with artificial intelligence there will be opportunities for malicious actors to get inside autonomous vehicles and direct them onto destructive paths. Cronin asks to consider some alarming scenarios, but most alarming of all is that her extensive research enables her to back up her analysis with many real-world examples of individuals demonstrating the destructive potential of new technologies.

This is therefore a book that points to a present set of security challenges and warns how they might get more severe. Now as Cronin rightly notes technology is not strategy. Just because some things might be done it does not mean that they will be done. There are unfortunately many individuals who might wish to cause random harm for a variety of philosophical or quite personal reasons, but the most serious threats will still tend to come from those wishing to mount sustained campaigns to achieve some sort of political goals, and in those cases they may provide clues to law enforcement. A large part in dealing with any sort of terrorist threats, including those still relying on the methods of the 19th Century anarchists, involves good intelligence and social resilience.

This book raises larger questions about what attitude we should take to new technologies. The spirit of the enlightenment, and the extraordinary record of scientific progress, encourages us to be techno-optimists. The people still want the extra power. But Cronin provides us with reasons to take care. There are already restrictions on what can be done with bio-technologies and this is an area that remains highly regulated. Although there is no reason to be confident in this, the difficulties of regulation in many areas may mean that we may have to rely on the good sense and consciences of big companies to think through the implications of coming capabilities. Some of the recent efforts by Facebook and Twitter to combat hate campaigns and misinformation may be steps in the right direction. The US and its allies are not the only countries working hard on new systems and exploring artificial intelligence, so it would be unwise to assume that future dangers can easily be contained. Governments will need to be hard-headed, smart and nimble to address the coming challenges. Fortunately Audrey Cronin has provided us with a framework with which to think through these issues and an agenda for action.

information on its national Internet with its Great Firewall and exercises close government oversight of its IT corporations, such as Tencent, Alibaba, Huawei, and Baidu. This autocratic control is unacceptable for democratic regimes, and it's far from certain, at any rate, that autocratic powers will be able to sustain it as the technologies continue to develop, offering citizens increasing means of evading government oversight.

Yet ceding totally free rein to companies is also unwise. They naturally emphasize the benefits and underplay the risks. DNA editing may be able to conquer genetic diseases, but it may also result in unintended consequences, called "off-target effects." Quantum computing may resolve intractable puzzles but also destabilize global security systems by breaking prevalent forms of encryption.[3] Artificial intelligence may be used to protect populations, but it may also become unmoored from human control.[4] A laissez-faire approach has already proven foolish: widespread linked sensors in everything from refrigerators to thermostats provide opportunities for hackers to wreak havoc with vehicles, in households and bank accounts, and with assaults on utilities and other infrastructure. And the vast amount of behavioral information gathered through sensors and aggregated in massive databases threatens privacy and self-sufficiency. A middle way between unfettered innovation and oppressive regulation will be necessary if the United States and its allies are to navigate between the Scylla of anarchy and the Charybdis of authoritarianism.

Finding that optimal approach toward the future of security will require an appreciation of why some new technologies are predictably evolutionary in their effects on conflict, while others unexpectedly and fundamentally alter the equation. The first are sustaining technologies, and the second are disruptive.

Disruptive technologies often democratize access to lethal capabilities, shifting unprecedented power into the hands of civilians. This was the case, for example, with the invention of dynamite, which fueled an outbreak of politically motivated bombings in the nineteenth century, destabilizing several of Europe's autocratic regimes and contributing to the outbreak of the First World War. It's the case again today with the cluster of emerging capabilities.

The word "disruptive" is often used to describe any kind of technological advance that is unanticipated, such as in a 2010 report by the National Academies of Science that defined a disruptive technology as "an innovative technology that triggers sudden and unexpected effects."[5] This misses a critical aspect of the disruption caused—that it shifts power from dominant players, whether in business or the governmental and military domains, to surprise actors. Such a shift is underway today and it presents a bedeviling problem for military and national security forces.

In most military planning, not to mention popular histories of the evolution of military technology, sustaining innovation at the high end draws the lion's share of attention. The balance of tanks, ships, missiles, and aircraft among states is intricately parsed to formulate assessments about likely outcomes in a next major war.[6] The focus on these capabilities is important, but it is too narrow. Major new threats to the international security order are coming from innovations at the low end, developed from technology created for widespread commercial use. While the current wave of disruptive commercial technologies offers a bounty of opportunity for businesses, and a boon of great new products and services for consumers, it poses wicked problems for national security. Anticipating the nature of the threats requires an appreciation of the dynamics by which disruptive technologies emerge and are adopted.

Disruptive technologies meet specific criteria that differentiate them from sustaining evolutionary ones. To consider the differences we can turn to Harvard Business School professor Clayton M. Christensen, who coined the term "disruptive innovation" in a 1995 article reporting on findings from a study he conducted.[7] He wanted to understand why well-managed, customer-sensitive firms that were pioneers in their fields, such as Texas Instruments, Sears, Xerox, and IBM, often suddenly faced crippling competition from smaller firms that took superior advantage of technological developments, many of which the powerhouses had helped innovate. A famous case is Apple's commercialization of computer mouse technology that Xerox had pioneered. Analyzing case studies from two different economic sectors—how progressively smaller disks replaced 8-inch digital disk drive technology and how hydraulic-powered excavators replaced mechanical shovels—Christensen developed a framework he then explored through a range of cases involving other firms. That work led him to a theory of disruptive technological change.[8]

Most new technologies, he pointed out, are sustaining rather than disruptive. Sustaining technologies improve an established product or service along lines current customers value and expect. Those changes can be either incremental or radical—for example, adding global positioning system (GPS) capability to a car dashboard versus developing a fully self-driving Tesla Model S. Performance is enhanced, product design is improved, more sophisticated software is written, or new product features are added. These improvements drive established customers to upgrade to new models, which are typically more expensive, and this often leads to higher company profits.

The dangers are that enhancements may eventually exceed the demands of existing customers, while price increases turn them away. Meanwhile, large numbers of potential customers at lower income levels are priced out of the market from the start—like the millions of Honda or Fiat buyers who would never shop for a Tesla Model S.

Disruptive technologies, on the other hand, shake up existing markets, undercutting the competitive advantage of established industry leaders. These disruptions may involve creating an entirely new market, as with that for commercial drones, or the expansion of an existing one at the low end by making a product cheaper, such as with smartphones or tabletop printers. In the latter case, the lower-priced product often initially offers poorer performance, but succeeds nonetheless because it attracts so many customers, not only due to affordability but to appealing features that make up for performance shortfalls. Christensen writes, "Products based on disruptive technologies are typically cheaper, simpler, smaller, and, frequently, more convenient to use."[9] Often the novelty of disruptive products, or their performance issues, leads to a limited initial market of "early adopters," who are seen by dominant firms as fringe customers whose requirements they do not need to cater to. That opens the door to upstart competitors.

Christensen's theory is not without its critics. Some argue that he chose only successful cases and question the predictive capacity of the theory. Others point to things omitted from his case studies (like widespread economic downturns that affected established companies), the murkiness of some examples ("sustaining" companies bought by other companies that thrived), the vast majority of "disruptive" start-ups that fail, and his strictly profit-based definition of "success." Still others mischaracterize Christensen's original argument, vastly expanding the meaning of "disruption," as we've seen. The theory of disruptive innovation has nonetheless been one of the most influential business concepts of the past two decades and helps explain how lower-performance technologies sometimes overcome better-performing ones.[10]

A classic case of disruptive innovation is Henry Ford's creation of the Model T car in 1908. The automobile was nothing new. Karl Friedrich Benz of Germany had invented the first gasoline-powered automobile in 1886,[11] and other European companies had developed cars incorporating far more advanced technologies, such as Porsche's 1899 four-wheel-drive electric car or Renault's 1902 drum brakes. Compared to these, the Model T was a joke, a crummy little putterer with a maximum speed of around 40 miles per hour that came in only one color, black, and didn't even have doors.[12] Yet while the sleek racing cars and classic roadsters of the period were playthings of the prosperous, the assembly-line-produced Model T was so cheap that masses of Americans bought it. The original price of a "Tin Lizzy" was $850 compared to the Cadillac Model 30, which sold for $1,400 (and up), or the 1908 Packard 30 touring car, whose standard sticker price was $4,000.[13] So popular was the Model T that Ford was able to drive the price continually lower, to under $300 by 1927.

The extraordinary impact on the whole market for automobiles in the United States can be seen in a set of statistics for car ownership by country from 1924. In the United States the count was one car for every seven residents compared to one for every 78 residents in the United Kingdom, 108 in France, 470 in Germany, and 654 in Italy.[14]

Over the past several decades, we have seen this disruption process unfold in many domains. The personal computer disrupted expensive mainframe computers. Smartphones disrupted cell phones. Cheap tabletop printers and scanners disrupted high-volume Xerox machines. Discount retailers like Walmart disrupted dominant department stores such as Sears. Health maintenance organizations disrupted traditional health insurers.

Established firms often struggle with the competition from disruptive technologies not due to a lack of talent, imagination, or entrepreneurship. Often they are actually far more advanced in developing new technologies, and they're frequently alert enough to acquire start-ups with technology that looks promising. Instead they typically fail to take full advantage of disruptive technologies due to their current business incentives, which encourage them to focus on the more predictable profitability of pleasing their existing customers by serving their needs better and better. Meanwhile the entire market morphs. They face what Christensen called "the innovator's dilemma," the problem of meeting the competing mandates: to maintain advantage in an established line of business while also not being overtaken by the innovation commercialized by upstarts. While hindsight can make the decisions about how to navigate the challenges seem obvious, they are generally anything but easy to make as disruption is getting underway.

Leading corporations are pressured by shareholders to produce reliable short-term revenue growth, which in turn generally assures robust stock prices. Ironically, the more successful the firm, the more difficult it is to resist this pressure. The most dependable way to assure strong short-term growth is to focus on incremental improvement to upmarket, higher-performance products with premium prices. So the pursuit of the kinds of initially small and often poorly defined low-end markets that generally constitute first adopters of disruptive technologies are not initially attractive to them. While established firms may not be forced out of business by these upstarts, they are often unable to gain sizeable market share as they seek to catch up, because with disruptive technologies, the early leaders capture a decisive "first mover" advantage.[15]

A vital point for those who wish to prevent malevolent uses of the currently emerging technologies is that despite the mythology that often develops later, technological disruptions never appear out of the blue: Ford could not have built the Model T without the development of the electric utility industry and global supply chains, for example.[16] Clusters of new technologies lead to disruption, and careful analysis can reveal how such clusters yield disruptive potential. That analysis is urgent now, but the governments of both democratic nations and autocracies are still mainly focused on improving sustaining military technologies.

Military technology has typically evolved by making improvements upon existing capabilities, such as accuracy or stealth. Upgrades are most often incremental, but even some radical leaps still qualify as sustaining, such as the development of precision-guided munitions, first used by the Germans in the 1940s, and the jump to stealth aircraft, developed beginning in the 1970s. As is true with businesses offering upgraded products, the constant development of improved military weaponry has increased costs, in this case dramatically. The arsenals of the world's leading powers are stocked with Porsches not Model Ts, the most extreme example being nuclear weapons, which are so politically and financially costly to develop that only nine states have as of yet been able to do so.

Sustaining innovation in major capabilities such as missiles, ships, aircraft, and tanks is indeed important. Staying at the forefront of such high-level technological development is vital to maintaining the perception of a state's relative power in the international arena, and these stores of massively lethal weaponry have deterred direct wars between the major powers for over seventy years (though several proxy wars have been supported by them). Yet history is replete with examples of the dangers to militaries from disruptive technologies.

States that have failed to adapt to these threats have found their power undermined by changes in the "marketplace" for conflict. Such low-tech changes need not compete head-to-head with dominant militaries. For example, AK-47s proved more important than smart bombs for the Vietcong, and IEDs proved more important than Predator drones for the Afghan Taliban. Or sometimes the process unfolds gradually and involves unanticipated uses of beneficial technologies, and unexpected combinations of them, so that the consequences become obvious only in retrospect, when power has diffused and it is too late for status quo actors to put the genie back in the bottle. For example, in the 1990s, the United States assumed that communications connectivity would result in war-winning synchronized attacks, but the Internet also helped al-Qaeda synchronize its operations and played a key role in the global spread of suicide attacks. The rapid development of both the motivation and the means of violence has changed the power equation between states and groups of individuals.

Today's disruptive technologies are following familiar patterns of innovation and diffusion among non-state actors. But the effects of digital technologies will unfold faster and be more extensive than prior disruptive technologies because of the Internet's unprecedented global reach and mobilizing capacity. Never before has a military power as dominant as the United States shared basic technologies so perfectly designed to democratize the global distribution of power.

The result is that we are in an era of open technological innovation that differs from the mainly closed dynamics of the twentieth century, yet we continue to use closed-technology frameworks to explain it.

In a period of closed military technological innovation, states can largely control access to major technological developments such as nuclear weapons, jet fighters, or precision-guided munitions. These technologies are hard to develop, require high levels of expertise, and are carefully protected by things like security classifications or copyrights. Experts speak of "proliferation" of these weapons and use phrases like "dual use," meaning that they sometimes have two categories of users: civilian and military. But they're expensive, rare, and difficult to build.

That model of military technological innovation dominated throughout most of the last century. The iconic image is J. Robert Oppenheimer, working away secretly in a lab with a small team of scientists, developing the nuclear bomb.

At the end of the twentieth century, the United States consciously shifted from closed technological development to open development, spurred by post–Cold War confidence in the triumph of democracy and US dominance. Virtually all of today's digitally based technological developments originally derive from US publicly financed basic and applied research dating to the mid-twentieth century. Today's smartphones, for example, would not exist without US-government funded programs that created key components, including the microchips, touchscreens, and voice activation systems. That's not to take away from the stunning creativity that happened in Silicon Valley—or even in Bill

Gates's parents' garage or Mark Zuckerberg's dorm room at Harvard. But the US government also consciously shared formerly protected basic technologies such as GPS and ARPANET to kick-start what followed.

Thirty years later we are witnessing the maturing of an open technological revolution driven by the commercial sector. In open technological innovation, there is popular access to advanced lethal technologies. You do not have to be a nuclear scientist to use them. Indeed, you don't even have to fully understand them, because many platforms are designed to help people experiment. A much broader range of people is now involved in innovation, including professionals, professional consumers (or pro-sumers), hobbyists, and consumers, so we should speak of "multi-use" technologies not dual use.[17] New potentially lethal technologies now diffuse through commercial processes, not just to states but to a range of non-state actors. Fortunately, we can build more relevant models for this period by examining the history of global diffusion in the nineteenth century, another time of open technological innovation.

Just as important to understanding why some technologies rapidly diffuse is analyzing why others do *not* diffuse. In the case of dynamite, for example, better explosives were created at about the same time, including balastite, guncotton, and gelignite (which dynamite's inventor himself preferred), but dynamite was the one to broadly catch on. Why? Examining the history of diffusion allows for what I'll call a "lethal empowerment theory," helping to anticipate which new technologies hold the greatest potential to become popular tools for political violence in the future. The theory holds that disruptive lethal technologies are

- Accessible
- Cheap
- Simple to use
- Transportable
- Concealable
- Effective (i.e., providing leverage and more "bang for the buck")
- "Multi-use" (i.e., beyond "dual use"; suitable for a wide range of contexts)
- Not cutting-edge—usually in the second or third wave of innovation
- Bought off-the-shelf (or otherwise easily purchased)
- Part of a cluster of other emerging technologies (which are combined to magnify overall effects)
- Symbolically resonant (which makes them more potent than just their tactical effectiveness)
- Given to unexpected *uses*

They give individuals and small groups greater power not because they are superior to the high-end technologies of states, and not because the individuals using them can go toe-to-toe with conventional militaries, but because they help mobilize individuals, extend their reach, and provide them with unprecedented command-and-control abilities. In the pages ahead, we'll see how new technologies have been harnessed for these purposes in the past, and then explore how the emerging disruptive technologies of today are being adopted for the same purposes going forward.

Because the focus here is on violent individuals, terrorist groups, and insurgents who seek political power, criminal syndicates, gangs, drug lords, wildlife poachers, human traffickers, and an enormous range of other criminals who are exploiting the new technologies are not covered. Doing so would require a book of its own, which I hope will be written. Likewise, many positive uses of new technologies by benevolent non-state actors, such as nongovernmental organizations, will not be covered here. Among them are combating corruption, increasing government transparency, and providing a voice for the oppressed in authoritarian states—wonderful technological advancements that have already been discussed well in other books.[18]

This book is divided into three sections. The first introduces predominant ways of thinking about the innovation and diffusion of military technology and demonstrates their shortfalls as regards the current era. It also examines consistent patterns of the diffusion of lethal technology to violent nonstate actors in the modern era.

Part II takes a close look at how two key innovations—dynamite and the AK-47 assault rifle—drove global waves of non-state violence, in both cases culminating in major upheaval in the international order.

That history informs the analysis in Part III of how clusters of emerging technologies are offering non-state actors unprecedented disruptive power. Today's drones, advanced robotics, 3D printing, and autonomous systems have more in common with dynamite and Kalashnikovs than they do with military technologies like the airplane and the tank.

The last chapter presents a set of broad policy prescriptions for managing down risk and preventing democratic governments from falling further behind the current trend favoring more lethal power in the hands of terrorists, insurgents, and rogue actors. It also offers specific advice, based upon in-depth historical studies of the diffusion of key technologies, regarding which ones are more or less likely to spread and be destabilizing.

The power of states has not been eclipsed, and they must not fall into the trap of being manipulated into overreaction to non-state threats, which is a chief aim of terrorism. The history of modern terrorism demonstrates that attacks are generally designed either to provoke state reactions that undermine a regime's legitimacy (as in nineteenth-century tsarist Russia), to polarize targeted populations so that they are ungovernable (as the Armed Islamic Group did in Algeria during the 1990s), or to mobilize supporters who are inspired into siding with the group (as the Palestinian Liberation Organization did with the 1972 Munich Olympics massacre).[19] All of these are strategies of leverage that draw their power from government responses, inducing states to undermine themselves. A self-defeating state response is a particular risk when there is ambiguity about an attack's origins, as is often the case with cyberattacks. It is also a growing challenge with the evolution of autonomous weapons and artificial intelligence. Serious study of the growing risks of new technologies and development of robust strategic approaches to limiting them can guide rational responses and keep rogue actors from needling states into lashing out against each other or undermining themselves.

There is nothing inevitable about how new technologies are used. Technology is neither good nor bad in itself. Technology is a tool, and it is what we do with it that makes all the difference.

PART I | Theory

FIGURE 1.1 (Clockwise from upper left) 27th July 1962.: View of a US Navy Polaris test missile sitting on its launch pad prior to blast off, Cape Canaveral, Florida. Photo by US Navy/Getty Images. The US Navy submarine USS George Washington (SSBN-598) is launched at the Electric Boat Division of General Dynamics, Groton, Connecticut, 9th June 1959. Armed with Polaris missiles, the vessel is the United States' first nuclear ballistic missile submarine. Photo by Paul Popper/Popperfoto/Getty Images. Physicist J. Robert Oppenheimer (R), dir. of the Institute of Advanced Study, discussing theory of matter in terms of space w. famed physicist Dr. Albert Einstein, while sitting at table in his office. Photo by Alfred Eisenstaedt/The LIFE Picture Collection/Getty Images.

| Classic Models of Military
Innovation: Shaped by the
Nuclear Revolution

I do not know how the Third World War will be fought, but I can tell you what
they will use in the Fourth—rocks!

Albert Einstein, in an interview with Alfred Werner, 1949[1]

T ECHNOLOGICAL REVOLUTIONS THAT affect military innovation
can be either open or closed.[2] During a closed revolution, social, political, or economic forces restrict access to emerging technologies. Military
or scientific elites are able to limit their availability, as nuclear weapons,
battleships, or radar. Doing so enables them to direct the development and
track the spread of technologies, preventing them from dispersing to the general public and guiding their evolution through stages of innovation and
adoption. By contrast, during an open revolution, emerging technologies are
accessible to the public, and individuals and private groups are free to not
only buy, use, and distribute them, but also to invent new purposes for them,
new forms of them, and new surprise combinations of them. Neither model
is absolute; both types of innovation generally occur during periods of major
technological invention. The nuclear revolution that predominated for most
of the twentieth century was closed, while the information revolution, which

overlapped with the advent of nuclear technology and is driving the most disruptive innovation today, is open.

The predominant thinking on military innovation has been shaped for decades by the dynamics of the closed nuclear revolution. This robust body of work has a great deal to teach us, but in crucial ways it is out of step with today's technologies. Innovations in open revolutions follow their own patterns of development and adoption that are less examined and understood. Rather than conforming to the well-known stages of military innovation and adoption, their development and adoption more closely resemble commercial processes. Fortunately, there is deep historical precedent for understanding those patterns as well.

To help us understand the differences, this chapter will review some of the classic arguments and rich debates about the evolution of war and technology throughout history, to draw the best insights from them and to determine where we need to move beyond them.

The Historical Relationship between War and Technology

In most study of military technological innovation, the focus is on state use of force against other states, as if the weapons, strategy, and tactics used by non-state individuals and groups, such as al-Qaeda, Hezbollah, or ISIS, as well as the key questions of how, when, and which weaponry they acquire, from where, and why, are virtually irrelevant. This is unsurprising, as the stress on the military power of states has a strong rationale.

Military strategists have placed an emphasis on armed forces gaining the technological edge because in many conflicts military technology has made an enormous difference, especially where there is a large technological gap between the belligerents. Take the case of the Maxim Gun (whose development is further discussed in chapter 5). At the Battle of Shangani during the First Matabele War, 25 October 1893, a 700-man column of the British South Africa Police defended itself from an ambush of 5,000 Matabele warriors. Using Maxim Guns, the British force wiped out 1,500 warriors before the Matabele King called off the attack.[3] At Omdurman in the Sudan, on 2 September 1898, over half of a 50,000-man Mahdist warrior force was wiped out by an Anglo-Egyptian force of a quarter of that size, which suffered roughly 400 casualties. Put another way, the ratio of Mahdist to Anglo-Egyptians killed was 11,000 to 48. In 1915, Churchill wrote in *The*

River War that the battle was "the most signal triumph ever gained by the arms of science over barbarians."[4]

In the twentieth century, advancements in sea power, and then the development of air power and nuclear weapons, were crucial to victories for the United States and its allies in both world wars and in the Cold War. For example, aircraft carriers provided close air support to amphibious landings, aerial reconnaissance for invaluable operational intelligence, standoff torpedoing of enemy ships, and defensive countermeasures to similar aerial advantages for the enemy. In the 1942 Battle of Midway, the Japanese Navy entered with more forces and more experience than its American counterpart, yet Japan lost four carriers and two heavy cruisers, and the United States lost one carrier and a destroyer. The number of casualties was 3,057 Japanese to 307 Americans, a ratio of about 10:1. The battle turned on the cutting-edge naval technology of both sides, American cryptographers' breaking of the Japanese Navy's secret code, and the brilliant leadership of Admirals Chester Nimitz, Frank Jack Fletcher, and Raymond Spruance. The Battle of Midway is widely considered the decisive turning point in the allied Pacific campaign.

Another reason for the focus on states and large-scale weaponry is that the gap between state and non-state actor capabilities has been vast. It has thus far been beyond the means of non-state groups to either build or seize, even as high-end design of weapons systems like the F-35 stealth aircraft and the next-generation Navy frigate has continued apace. In addition, interstate rivalry has resurged. After a brief period of relative international stability following the end of the Cold War, state powers have presented new threats, including efforts by Iran and North Korea to develop long-range missiles and nuclear weaponry, a resurgence of Russian aggression in Ukraine and Syria, and China's military buildup in and around the South China Sea, constructing and fortifying artificial islands, and bullying smaller states like Vietnam and the Philippines.

Yet the focus on big threats and exquisite technologies has led to dangerous oversight in our thinking about how technological innovation more generally affects warfare and confers power. The prevailing notion is that more sophisticated lethal technological capabilities concentrate and enhance state power. That viewpoint can be traced to sociologist Max Weber, who in 1919 argued that the modern state depends upon a monopoly of the use of force.[5] The increasing concentration of lethal power by states has been the dominant development in military affairs for the past five centuries. The process greatly accelerated in the period from the First World War through the end of the Cold War, arguably the golden age of Western military technological innovation. The climax of that modern period was the development of nuclear weapons, dramatically ending the Second World War and setting up a Soviet-American arms race that risked global nuclear obliteration, assuring that massive resources and strategic focus were devoted to the nuclear threat. Surely if any proof of state monopoly of military power and technology were needed, the ability of the United States or Soviet Union to use nuclear intercontinental ballistic missiles (ICBMs) to annihilate the planet in a matter of seconds seemed to fit the bill.

Strategists in the aftermath of the atomic bombing of Hiroshima on 6 August 1945 believed that technology had so changed modern warfare and weaponry that prior concepts, such as air defense and protection of sea lines of communication, no longer applied. Just before the Hiroshima bombing Yale professor Bernard Brodie, a well-respected expert on sea power, had written a paper predicting that the battleship was about to make a comeback. Reading the *New York Times* headline about Hiroshima a few days later, he reportedly told his wife, "Everything that I have written is obsolete."[6] He soon became famous for his book *The Absolute Weapon: Atomic Power and World Order*, which laid the foundations for deterrence theory and all of the related nuclear strategic concepts that followed.[7]

Thereafter, a belief in the decisive role of increasingly advanced, complex, and lethal technologies for use on the battlefield was the predominant driver of twentieth-century studies of the evolution of technology and warfare. How could it be otherwise? Humans had perfected the means to kill each other by the millions. The most famous of the early studies was the 1962 military history book by Brodie and his wife, Fawn (a prominent historian in her own right), entitled *From Crossbow to H-Bomb,* which the two authors described as being "about the history of the application of science to war."[8] The book opened in 481 B.C., when the Greek engineer Harpalus designed pontoon bridges made of hundreds of fifty-oared galley ships across the Dardanelles, thus solving a major logistical problem for King Xerxes I and enabling the Persian army to invade Greece.[9] From there it moved through the centuries at an accelerating clip, explaining the impact of gunpowder, the development of early artillery, the transition from wooden to iron ships, and a sweeping inventory of fascinating, crucial technological developments.

A central assertion in the Brodies' book was that the development of weaponry had been quite slow for most of history, by comparison to the nineteenth century and especially the twentieth, when its pace had sped up.[10] But the authors did not explore in depth instances when new technologies were actually counterproductive in warfare. They also saw the trajectory as having been toward control of increasingly lethal technologies by states and did not deeply investigate periods when power became more widely distributed, such as in ninth- and tenth-century Europe. The book culminated with the development of thermonuclear weapons, arguing that "[N]uclear bombs made it clear . . . that there could be no question among reasonable and objective men about the decisiveness of strategic bombing in future wars."[11] Reality quickly intruded, with nuclear weapons not deployed in the wars in Korea, Algeria, and Vietnam, and caveats were offered. The 1973-updated edition of *From Crossbow to H-Bomb* admitted that the war in Vietnam "probably resulted in a net slowing down in technological development rather than the reverse," and concluded with an insightful discussion about the increasing and worrisome costs of high-end weapons systems, questioning whether such concentrated expenditures could be justified.[12] But the focus on the innovation of large-scale high-technology systems and how they enhanced state power persisted.

The spectre of nuclear obliteration overshadowed attention to the potency of simple and inexpensive weaponry used in places like Algeria and Vietnam, such as IEDs and AK-47s. Strategists such as Brodie, mathematician Albert Wohlstetter, and economist Thomas Schelling, working in places like the RAND Corporation, the Massachusetts Institute of Technology, and Harvard University, devised theoretical models of how nuclear war would unfold and, more important, how to prevent it from happening. Concepts specifically designed for nuclear strategy, such as deterrence and compellence, were then adapted to fit conventional warfare, as with the focus on escalation dominance in Vietnam. Deterrence theory argued that the threat of Mutually Assured Destruction—annihilation of both, or all, sides in the event of large-scale nuclear attack—was a linchpin of preventing such an attack, and that maintaining that deterrent required a tit-for-tat buildup of capabilities, especially of bombers, missile-firing submarines, and land-based ICBMs, which made up the nuclear "triad." Terrorism, insurgency, and civil war continued throughout the Cold War; but nuclear devastation was the nightmare and kept the focus squarely on high-end technology.

Theories of Closed Technological Innovation

The ominous shadow of nuclear weapons, and especially the US-Soviet nuclear contest, shaped theories of military technological innovation throughout the twentieth century because most academics and policymakers saw innovation as a closed loop. In closed innovation processes, state organizations create and control high-end military technologies, and others gradually copy, steal, expand, or build upon them in a process of global proliferation. The state with more and better military technology is advantaged, according to this theory, because there is a direct relationship between military technology and political power. Military technological innovation thus drives warfare and, by extension, the side with the better technology can usually expect to win.

This was the Cold War intellectual backdrop for examining the importance of technology in the long history of warfare. As with nuclear weapons, the impetus for epochal change could come from a single technology. For example, in 1962 one historian traced the history of European feudalism to the invention of the stirrup around the eighth century A.D., arguing that the stirrup enabled the Europeans to anchor themselves on horseback and drive the lance with the full momentum of both horse and rider. Shock cavalry then repulsed invasions by nomadic horsemen, contributing to the ascendancy of armed knights, and playing a key role in the establishment of feudalism in Europe.[13] The medieval longbow was another technology singled out for its impact. It enabled Englishmen and Welshmen to cut down charging Frenchmen in the Hundred Years' War, contributing to major English victories, most famously the battles of Crécy (1346) and Agincourt (1415).[14] (Though, of course, the French nonetheless ultimately prevailed.) Both examples elicited spirited dissent from other historians; but the belief that a single military technology could be the pivot point in history easily gained traction in the nuclear age.[15]

The broader concept of military technological revolutions caught on during the mid-1950s and gradually grew in scope over subsequent decades, culminating with the argument that between 1500 and 1800, improvements in artillery, new types of fortifications, and an ability to deploy larger armies contributed to European dominance of more than one third of the world.[16] The centralizing narrative about military power and the nation-state had a few holes in it: Europeans had outsourced their military power to private contractors during the seventeenth century, after all.[17] But the view that military technological innovation drove the evolution of warfare prevailed.[18] In 1989, historian Martin Van Creveld opened his well-known *Technology and War: From 2000 B.C. to the Present* with "The present volume rests on one very simple premise which serves as its starting point, argument and raison d'être rolled into one. It is that war is completely permeated by technology and governed by it."[19]

This may have been true of nuclear weapons, and aircraft carriers and radar during the Second World War; but throughout the century, states with the most technologically advanced capabilities often lost against lesser-equipped opponents—because they lacked the necessary will or they faced opponents whose technological disadvantage was slight or they could not use new technology effectively, or for a variety of other reasons.[20] A broad range of social, political, and economic factors well beyond military weapons technology affects, sometimes even "governs," outcomes in war, such as how governments mobilize their populations or build infrastructures or tap into their wealth.[21] And the roles of commercial technologies, like the printing press, timepieces, railroads, and telegraphs, well beyond the purview of the military, are often crucial to the outcome.[22]

Still, the predominant narrative of Cold War historical studies of technology and warfare is that a gradual concentration and control over increasingly sophisticated and lethal technologies by the ruling elite and militaries of nation-states consolidated their power, and if the more technologically advanced opponent could build better machines, systems, and doctrines, he could reasonably expect to dominate in battle and win wars.

The view that technological innovation by militaries is essential to strategic advantage is also reinforced by work in political science. Most of this research focuses on the organizational processes by which major military powers develop large-scale weaponry (both in peacetime arms buildups and during wars), how it is adopted by military organizations, and how they deploy it in combat with other major powers.

The field of military innovation studies traces its roots to early twentieth-century organization theory, including scientific management theory and studies of government bureaucracy.[23] Most scholars in this area stress that states must invent, adapt, or integrate successful innovations into their military operations or they will fall behind rivals, drawing a direct line between a state's prowess in military innovation and its military power.

The history of the two world wars again looms large, as does careful analysis of innovation between the wars. Studies of capital-intensive programs including strategic bombing, amphibious warfare, aircraft carrier warfare, and submarines predominate, yielding key insights about why technologies may or may not be used effectively for advantage in battle.[24] For example, the Germans were the most technically advanced of any of the combatants at the outset of the Second World War, yet still failed to use radar effectively. Inter-Service rivalry, resistance to radar in the Luftwaffe (World War I flying ace Gen. Ernst Udet reportedly said, "if you introduce that thing you'll take all the fun out of flying!"), abandonment of shorter wavelength research, and failure to develop effective operational doctrine all contributed to their defeat. The British, who lagged behind the Germans technologically, more than made up for the shortfall by the way they wove radar into every aspect of air defense, partly by necessity as they absorbed withering German air attacks during the 1940 Battle of Britain. According to Churchill, "[I]t was operational efficiency rather than novelty of equipment that was the British achievement."[25]

Other human factors determined how effectively technological advantages were capitalized on by various state belligerents. Sometimes training made the difference. When the war started, the United States already had a robust fleet of submarines capable of long-range cruising, for example, but commanders had been peacetime-trained to attack well-escorted enemy warships, meaning they avoided doing anything risky that could reveal their location, especially not surfacing to periscope depth where hostile destroyers or aircraft could see and destroy them. All the emphasis in interwar exercises was on stealth and the use of sonar, rather than actually espying targets through a scope. That training failed during the war, when the Allied mission changed to attacking fast-moving convoys of Japanese merchant ships. Political scientist Stephen Rosen calculated that only 31 of 4,873 known submarine attacks were directed by sonar.[26] Yet most commanders hewed to their instinct to be invisible, missing target after target, a practice that changed only when more aggressive younger skippers took over during the war. Thirty percent of US submarine commanders were relieved for cause in 1942.[27]

These studies were part of a boom in the field in the United States and United Kingdom that produced many additional important insights about how a range of human elements influence the course of military innovation and how technologies are deployed. Dozens of scholars, some of them serving military officers taking time out to earn doctorates at civilian universities,[28] zoomed in on how individual military Services change,[29] how they evolve their doctrines,[30] how they fight with each other over weapons and resources,[31] how they respond to civilian leaders,[32] and how the cultures of their Services (Army, Navy, Air Force, or Marines) affect their perspectives.[33]

All of this organizational analysis played a crucial role in the unprecedented strength and technological dominance of the United States and its allies during and shortly after the Cold War. But though it revealed qualifying factors in how effective large-scale, high-end military innovations are in warfare, even highlighting a few less well-known examples of individual innovation in the ranks,[34] the view that sophisticated military-controlled technologies were the linchpin of strategic advantage still prevailed.

The Revolution in Military Affairs (RMA) framework that emerged in the United States toward the end of the twentieth century followed directly in this tradition. It focused squarely on the vital role of high-end, large-scale technologies, arguing that future technology, specifically a system of US-dominated information age technologies, including precision-guided munitions, surveillance satellites, battlefield command, control, and communications, networked operations, and other computer-dependent systems, would virtually remove any guesswork from future conflicts.[35] The overwhelming defeat of Iraq in the First Gulf War of 1990–1991 seemed to confirm that wisdom. Some advocates even argued that information technologies had fundamentally transformed the nature of war by making the battlefield transparent and controllable. In the words of US admiral Bill Owens, whose 2000 book *Lifting the Fog of War* was widely read, "When technology is correctly applied to the traditional military functions—to see, to tell, and to act—a powerful synergy is created. . . . Together, these create the three conditions for combat victory: *dominant battlespace knowledge, near-perfect mission assignment*, and *immediate/complete battlespace assessment*."[36] (Emphasis in the original.)

Unfortunately, the subsequent wars in Iraq and Afghanistan, in which insurgents somehow found effective ways to combat American forces as well as politically outmaneuvering efforts to establish stable regimes, proved this argument wrong.

Plentiful research on tactical and operational innovation from the "bottom up" (or at least from the "middle up") in counterinsurgencies had been conducted by the European colonial powers in earlier years, inspired by their struggles to combat rebellions in their colonies. Classic books include ones by Trinquier (French, Indochina), Galula (French, Algeria), Kitson (British, Kenya, Malaya, others), and many others.[37] But most US professional military educational institutions removed counterinsurgency from the curriculum in the 1970s, after the Vietnam War ended. In the post-9/11 campaigns in Iraq and Afghanistan, the US military again began tapping this colonial literature to revise its strategic doctrine, issuing a new Army/Marine Corps Counterinsurgency Field Manual in 2007. Those conflicts then stimulated a flood of excellent new thinking about the need for innovation and learning at the tactical and operational levels in technologically asymmetrical conflicts.[38]

Still, whether they focus on top-down or bottom-up processes, today most military innovation studies continue to look one-sidedly at traditional armies, and the power of high-end systems developed in closed processes of military technological innovation.

What has not been adequately accounted for is that we are in an era of open technological revolution, in which remarkably powerful computers and a wide range of related technologies are easily accessible. Although the resulting change in the character of conflict is now coming into focus, it is developing more gradually than that propelled by the shock of a nuclear detonation. We are in a moment of reckoning similar to the one Bernard Brodie woke up to in August 1945, but the dangers are not as apparent.

We must engage in a sweeping re-examination of the role and impact of technology in war, and political violence generally. Central to this must be recognition that the nation-state is less capable now of achieving dominant control over lethal technologies, and that although the menace of nuclear obliteration still haunts, a cluster of innovations is providing new capabilities that non-state actors can use to challenge the effectiveness of major power armed forces. We must become more aware of, and focus more on, both the risks and opportunities of these emerging technologies with respect to violence. That requires a course correction to the prevailing understanding of the nature of innovation.

Innovation Is Double-Edged

Today most people use the word "innovation" to refer to new and exciting inventions, whether of technology, products, or services; but innovation was not always seen in a purely favorable light. An innovation is an idea, practice, or technological development that is new to its adopter, but not necessarily better.[39] Up to at least the 1930s, innovation was as often considered bad as good. According to Canadian historian Benoît Godin, for centuries being called an "innovator" was an insult. In 1636, the English Puritan Henry Burton actually got his ears chopped off for innovating on church doctrine.[40] Even as industrialization arrived, technological innovation was widely seen as a social plague; manufactured products left craft workers behind, and the cultural changes ushered in led to horrible working conditions for many, as well as to loss of heritage, fragmented families, and political revolutions that destabilized governments and killed millions in revolutionary wars.

The current perspective of innovation as being positive derives largely from the work of Austrian-American economist Joseph Schumpeter (1883–1950), who argued in his 1942 book *Capitalism, Socialism, and Democracy* that technological innovation drives economic progress, causing processes of "creative destruction" that sweep aside dominant players, making room for new ones.[41] This roiling process, pairing pain and gain, is the reason capitalist systems ultimately prevail, he argued.[42] The outcome of the Second World War, especially the dropping of nuclear bombs on Japan, seemed strong confirmation of that view. The opposite of "innovation" became "stagnation."[43]

The study of processes of innovation flourished in the subsequent years, particularly in economics, anthropology, sociology, political science, and history, with each discipline focused on a particular aspect. Economists concentrated on commercial innovation of new products; anthropologists studied processes of cultural change; sociologists investigated how societies evolve; and political scientists mainly studied how organizations innovate, as we have seen. We can draw on some of this work to examine how emerging technologies are likely to be adopted, and adapted, by non-state violent actors in the future.

Lethal individuals and groups who are *not* members of recognized organizations also use standard processes of technological innovation, yet how they do so has been barely considered by those who study military innovation. Individuals and groups share both their military tactics and technologies with each other in established ways, and have done so for centuries.

The Social Nature of Diffusion

While the study of innovation is about how new technologies are developed, diffusion is the study of how they spread. The touch point for studying diffusion is Everett Rogers's 1962 book, *Diffusion of Innovations*, where he defines diffusion as "the process in which an innovation is communicated through certain channels over time among the members of a social system."[44] Rogers is a scholar of communications, which explains his emphasis on that. His work focuses on diffusion of commercial products, laws, regulations, organizations, and ideas—specifically things like snowmobiles, insecticides, contraceptives, teaching practices, medical drugs, and farming practices. Dozens of sociologists, anthropologists, and business scholars have followed him into this field, studying the global diffusion of every imaginable commercial product or government practice from medical drugs to cell phones.[45]

The narrower study of the diffusion of military innovations was kicked off four decades later, with a focus overwhelmingly on the transfer of military technologies, doctrines, and practices from one state to others.[46] European and American scholars considered how other countries were reacting to, assimilating, and exploiting the military technologies, organizations, and doctrines at the heart of Western dominance. Ranging beyond the major powers, they addressed historical adoption of both material technology ("hardware") and military practices ("software") by Arab, Israeli, and South Asian states, notably including the eighteenth-century transfer of military drill and firepower from the British East India Company to Indian sepoy armies that had previously been organized around the use of elephants as "weapon systems."[47]

A few years later, in 2010, the enormous toll taken by suicide terrorism in the wars in Iraq and Afghanistan prompted analysis of the spread of that tactic, as well.[48] Political scientist Michael Horowitz offered an original book, *The Diffusion of Military Power*, that examined the diffusion of four "major military innovations" including carrier warfare, the nuclear revolution, battle fleet warfare and—most important for our purposes—suicide terrorism. This time the research stretched beyond traditional state-to-state military frameworks to consider the global diffusion of non-state actor technologies and practices. During a period of open technological innovation, it was an important early step in the right direction—as we will further explore in the next chapter.[49]

Another advance in military diffusion studies with great relevance to how non-state actors will adopt today's emerging technologies concerns the loss of "first-mover advantage." As is the case in the commercial sector, in military innovation, technologies have often been co-opted by competitors. In the mid-nineteenth century, French engineers were the first to build a steam-powered warship, an ironclad warship, a mechanically powered submarine, and a steel-hulled warship. Yet, employing Britain's superior industrial capacity and greater organizational agility, the Admiralty made better use of French inventions and integrated them into their naval strategy. Despite the submarine's effectiveness in 1917, the Germans neglected U-boats in the interwar years and the Americans moved ahead. In the early development of the airplane the technology evolved so quickly that European competitors often waited until another power had developed a system and committed to producing it first. Competing states could then gather intelligence on that model, make changes (such as mounting more effective machine guns, improving speed or maneuverability), and relatively quickly produce a plane designed to surpass the enemy's version. The British led in developing the aircraft carrier, yet the Americans and the Japanese fully realized the offensive potential of airpower at sea and surpassed them.[50]

This research does not emphasize how state militaries might lose first-mover advantage to non-state actors. The focus remains on inter-state competition and the danger of falling behind competitors. A frequently cited example is the ancient Chinese, who invented gunpowder around A.D. 800 and were dynamic military innovators for centuries, using state-of-the-art weapons to sweep through Central Asia, Vietnam, Burma, and Nepal during the Ming and early Qing dynasties (from 1368 to about 1760). The Great Divergence (as the period when the Chinese fell behind Western military developments and prowess is known) was a time when the Qing dynasty faced only small wars and regional disputes. The Chinese successfully suppressed Russian and Mongol threats, but there was no real danger from Western

European powers until the late eighteenth century, with the rise of maritime trade. As scholar Tonio Andrade puts it, "Compared to earlier periods in China's history, this period [1760–1839] was extraordinarily free of warfare. China's armies atrophied, and military innovation slowed."[51] The Chinese dynastic air of moral and cultural superiority also colored their mindset, arguably similar to European attitudes toward their colonial opponents a century later.[52] With all the attention to the development of European armies and the rise of Western dominance, the fact that the Chinese were such innovators, and that the diffusion of technology moved from East to West, was overshadowed.

While these important steps in advancing the study of diffusion of military technology provided a basis for widening the scope to include more attention to non-state actors, they didn't propel a leap into that terrain. Even Horowitz, whose study of suicide terrorism ventured furthest from Cold War models, wrote that "military power is the measure of how states use organized violence on the battlefield or to coerce enemies."[53] The focus was on arguing that the most effective state adopters of major military innovations are more likely to win the next inter-state war. While important caveats about the fact that command of the most sophisticated and powerful technology does not always lead to victory had been produced, they did not catalyze a sea change in strategic analysis and development.

Despite efforts by social scientists to be objective and scientific, this argument ventures perilously close to the same technological determinism that some military historians have been accused of, because even the "best" technology, meaning the most sophisticated, does not ensure either victory in battle or a viable overall strategy.

Technology Is Not Strategy

More emphasis on the fact that there is no direct line from technological innovation to successful strategy is vital. With an eye trained to that recognition, the wealth of counterexamples is sobering. Take the case of successive efforts to make wars short by delivering a knockout blow to the enemy. In 1861, American inventor Richard Gatling argued that his hand-cranked machine gun would allow Union forces to quickly win the Civil War. A similar logic guided the British, French, and Germans in the use of chemical weapons, including phosgene, chlorine, and mustard gas in World War I.[54] In 1917 British staff officer J. F. C. Fuller argued that a system of mobile armored vehicles could overrun German defenses and bring the

First World War to a rapid close. In 1942 Arthur "Bomber" Harris assured Winston Churchill that long-range bombers would end World War II by 1944.[55] Operation Linebacker during the Vietnam War, a 1972 high-intensity bombing campaign to destroy North Vietnamese infrastructure, did not destroy their will to fight, as President Nixon had initially hoped.[56] From 2013, American policymakers argued that armed UAVs ("drones"), along with Special Operations, could destroy al-Qaeda's leadership and effectively end the war on terrorism.[57] Each weapon had an impact, but they were not decisive.

Depending upon what the problem is, technological innovation may or may not solve it. Strategy is the application of means to ends, and technology is only one in an assortment of means states can use to attain political objectives. In many cases, it may not be the most effective one, and technological "solutions" often fail, also sometimes having revenge effects. In 1914–1917, Peter Strasser of the Kaiser's Navy believed that dropping bombs (essentially hand grenades) from German zeppelins would shock the British and French publics into rapid surrender.[58] Instead of cowing British civilians, the effort outraged them and strengthened their resolve. In the next war, Hitler essentially repeated the same logic with the V1 and V2 rockets of the Third Reich, launched against Britain, France, and Belgium in 1944–1945. They killed or injured some 85,000 Londoners but had no discernible effect on the war's outcome.[59] Hitler was looking for a quick technological fix to offset the strategic blunders he had made, and he failed.

Of course, as said, the United States has learned this truth painfully, first in Vietnam and then in the ongoing wars in Iraq and Afghanistan. In Vietnam, the United States had technological superiority, including superior firepower, airpower, and helicopter mobility, but misjudged the war from the start, failing to discern the nationalist motivations of the North Vietnamese as distinct from their Communist ideology, as well as the weakness and corruption of America's South Vietnamese ally. In Iraq, the toppling of Saddam Hussein with "shock and awe" technology led to sectarian violence and insurgency. In Afghanistan, replacing the Taliban with a centralized nation-state ignored centuries of Afghan history. In both of the latter wars, the most powerful technology has been the IED.[60]

There is no question that technology can be critical to gaining advantage in wars or even be decisive. The point is that historically this is not always the case and definitely not so for every social or political context. Today we are in an open technological context that makes familiar models obsolete. The disconnect between sound strategy and the effective use of new technology is more obvious than ever.

Historical Context Matters

Without an appreciation of the historical context, we cannot ask the right questions about lethal technologies and their impacts on war. Long before a conflict occurs, changes in political ideas, governance, economic distribution, or speed of communication may affect or even shape who is likely to show up brandishing a new weapon or inventing new ways to kill. Sometimes technology causes social change first, and then affects military-technical innovation; sometimes the process works the other way around.[61]

One of the most eminent military theorists, Carl von Clausewitz, was keenly aware of the social factors in military outcomes. He attributed Napoleon's success on the battlefield in the early nineteenth century, for example, as mainly due to the French Revolution having upended the political order and inspired hundreds of thousands of people to follow him into war.[62] Having watched Napoleon's victories unfold from the loser's side, Clausewitz, in his timeless classic *On War*, gave no attention to the role of technology, because he thought that technology was virtually irrelevant to Napoleon's success.[63] A warrior who first faced combat at age thirteen, Clausewitz was struck instead by the social and political factors driving Napoleon's triumphs, particularly the "passions of the people," which were the engine of the powerful French army. Although Clausewitz was a keen advocate for the Prussian Landwehr, or militia, and strongly supported the introduction of conscription, he knew Prussia could never mobilize the teeming, fervent armies that Napoleon led: Prussia had a smaller population, yes, but also, the king feared popular revolution.[64] When Clausewitz wrote, "Wars must vary with the nature of their motives and of the situations which give rise to them," that is what he meant.[65]

Given that Americans are among the biggest fans of Clausewitz's work, it's somewhat strange that such strong faith in military technology as the decider of future wars persists in the US military establishment, and that they tend to take the social setting out of the equation and decontextualize technology in their own war planning. A lack of emphasis on the risks of diffusion of technologies to the broad public may have been a key driver of the current open technology revolution.

Opening Pandora's Box

The shift from closed development to an open revolution was spurred by deliberate US government policy, as part of the post–Cold War euphoria about the US-dominated "new world order." Publicly financed basic and applied research from the 1960s, 1970s, and 1980s drove the technology boom of the 1990s. With federal government support, ARPANET became the Internet. Tax dollars developed the Global Positioning System. The Google founders continued the development of their search engine with funding from a National Science Foundation grant. All of the major components of smartphones derive from US government programs, including the microchips, the touchscreens, and natural language voice activation, such as Apple's Siri system.[66]

The flip is most starkly illustrated by comparing the management of the highly secret Manhattan Project, which resulted in the nuclear bomb in 1945, to the current development of machine learning AI technology. Private US companies, foremost Microsoft, IBM, Facebook, Amazon, Apple, and Alphabet (Google's parent), drive research in the latter. In 2017 alone, these companies booked more than $21 billion in expenses just to acquire AI-related ventures.[67] By comparison, the US Defense Department reported that it spent about $7.4 billion in 2017 on AI, big data, and cloud computing combined.[68] In 1965, about 75 percent of all US research and development came from the US government, but now it's more like 20 percent.[69] Meanwhile the big technology companies have entirely globalized their operations. "They want to be Switzerland, selling everything to everyone," as former US secretary of the navy Richard Danzig commented.[70] In January 2018, for example, Google announced a new AI institute in Beijing, stating that "science has no borders."[71]

There is no question that the United States and its allies must continue to develop high-end weaponry and security technology, and the wealth of research on large-scale military innovation is vital to that project. In the wars in Syria, Iraq, and Afghanistan, as well as in Somalia, Niger, and other conflict zones throughout the world, new systems have afforded important advantages and helped save lives. They have also helped win battles. But, again, they have not won the wars.

As the Information Age barrels along at increasing speed, we are embarking on an era of full automation, autonomy, and, ultimately, artificial intelligence. Yet most analysis of the current and future threats applies concepts developed during the Nuclear Revolution. That is not to argue that history is irrelevant; quite the opposite. The scope of analysis must be further widened, not only beyond formal military organizations but also to earlier periods that were precursors to the current disruptive moment.

The next "big thing" in warfare may well be a bunch of little things used in new ways.

To develop an understanding of how the emerging technologies are likely to be harnessed by non-state actors, in the next chapter we'll uncover a number of patterns in the history of innovation and diffusion of lethal tools and tactics for using them.

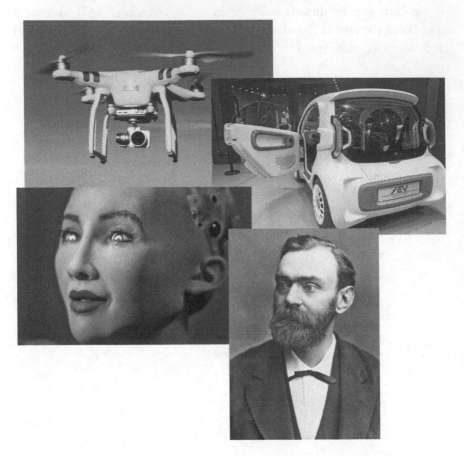

FIGURE 2.1 (Clockwise from upper left) Flying DJI Phantom drone.
© alpinenature/Shutterstock. A 3D-printed low-speed electric vehicle (LSEV) on
display at China 3D-printing Cultural Museum on March 15, 2018 in Shanghai,
China. The LSEV electric car was co-produced by Italy-based X Electrical
Vehicle, or XEV, and China-based 3D printing materials company Polymaker.
All components are 3D-printed, except tires, leather and glass. Photo by VCG/
VCG via Getty Images. "Sophia" an artificially intelligent (AI) human-like robot
developed by Hong Kong-based humanoid robotics company Hanson Robotics
is pictured during the "AI for Good" Global Summit hosted at the International
Telecommunication Union (ITU) on June 7, 2017, in Geneva. The meeting aims to
provide a neutral platform for government officials, UN agencies, NGOs, industry
leaders, and AI experts to discuss the ethical, technical, societal and policy issues
related to AI. Credit: FABRICE COFFRINI/AFP/Getty Images. Alfred Nobel,
1885. © The Nobel Foundation.

CHAPTER 2 | The Arsenal for Anarchy: When and How Violent Individuals and Groups Innovate

SEIF ALLAH HAMMAMI, a twenty-nine-year-old Tunisian man, was living in an apartment in Cologne, Germany, with his forty-three-year-old German wife, Yasmin, and their two children in June 2018. The couple had met online and married after Yasmin converted to Islam, and she was pregnant with their third child.[1] According to German state prosecutors, Hammami was "deeply connected to the Islamist spectrum."[2] The year before, Hammami had twice attempted to travel to Syria, probably to join the Islamic State; but he was turned back in Turkey. So he sought another path. Islamic State video tutorials posted on the encrypted app Telegram demonstrated exactly how to make and weaponize ricin from shells of castor beans. Ricin poisoning is how the Bulgarian Secret Service assassinated dissident writer Georgi Markov as he waited at a London bus stop in 1978, using an umbrella gun that shot a metal pellet the size of a pinhead into his thigh.[3] There is no known antidote.

Hammami ordered a coffee grinder and thousands of castor beans online direct from Amazon Marketplace. He even bought a hamster and started experimenting on the poor creature. Fortunately the British tipped off the Germans about his suspicious online shopping behavior, and concerned neighbors reported his radicalization. Plus, French police had foiled a similar ricin attack a month earlier, in May 2018: two Egyptian brothers in the 18th

Arrondissement of Paris had watched the same Telegram video and tried the same thing.[4]

When police raided Hammami's apartment, they found 83.3 milligrams of already produced ricin and 3,150 castor beans, a vast increase over any prior terrorist cache. They also found 250 small metal balls, two bottles of acetone nail polish remover, light bulbs, cables, and 250 grams of powder removed from fireworks—all materials useful for a remotely detonated, shrapnel-loaded bomb.[5] Hammami had been communicating with Islamic State operatives, and one of Hammami's cell phones had jihadist propaganda, plus more than 9,000 chat messages, 11,000 contacts, and several bomb-making manuals.[6] The attack appeared to be imminent.

What is notable about this case is not the explosives or the ricin-manufacturing attempt: numerous terrorist groups and individuals have tried to weaponize ricin.[7] It was even considered for military use in the First World War, either by coating bullets or shooting it in a cloud of dust, except that ricin could not withstand heat, and there was no antidote if one's own troops accidentally inhaled it.[8] What's new today is fast online commerce and globally accessible video, providing virtually instant access to deadly materials, motivation, and know-how—sometimes even personalized coaching—to virtually anyone interested in smaller-scale or personal attacks. The lethal knowledge that enabled a state-sponsored Bulgarian agent to fire a ricin-laced pellet into a dissident writer has become part of the public domain. In an era of open technological innovation, individual experimentation and surprise outcomes will be the norm.

An Open Technological Revolution

Unlike technologies arising from closed revolutions, those emerging in open revolutions follow commercial patterns, being bought or bartered for by the broad public. They can typically be altered by their users, such as hobbyists customizing their Apple I and II computers in the late 1970s and 1980s.[9] Surprise applications often arise, innovated by private individuals or small groups, including lethal ones. They can reshape the "marketplace" of conflict worldwide, causing broad societal effects, more in the manner of the printing press, electricity, or the steam engine, rather than the aircraft carrier, tank, or nuclear weapons.

Open technological revolutions redistribute lethal power, shifting the balance more toward the public vis-à-vis states. This can be good or bad. Private armies, small groups, and individuals gain the ability to carry out new or more destabilizing terrorist attacks or other political violence and that, in turn, forces major powers to react.[10] The increase of power in the hands of non-state actors has been quite substantial in the past, particularly, as we'll see, with the inventions of dynamite and the AK-47, and today's emerging technologies are allowing non-state actors to become more powerful than they ever have been. They will surely be exploiting this power, just as they have throughout history, and while states can often be deterred from using force, some small groups and individuals are virtually impossible to deter. What's more, states often consider it in their interests to provide support and safe harbor to terrorist groups and insurgencies.

In the fifty years prior to the shock of the 9/11 attacks, the nature of non-state political violence led to a false impression that these individuals and groups will always be conservative in their use of weaponry and are not innovators. Analysis drawn from the best available incident database shows that more than 88 percent of all terrorist attacks from 1970 to 2015 have used firearms and explosives.[11] And the vast majority of insurgent groups have likewise relied only on small arms and explosives.[12] The sophisticated orchestration of the 9/11 attacks, using fuel-laden civilian passenger airlines as conventional missiles, led to intense scrutiny of al-Qaeda's deadly tactics, though no new technology was used.[13] But a longer historical view reveals periods of dramatic innovation and rapid adoption of new technologies, followed by intense global diffusion of their use.

To develop an understanding of why periodic bursts of lethal innovation from below occur, with predictive power regarding currently emerging threats, we need to consider their history. We also need to look beyond the history of military innovation to the patterns of adoption of new commercial technologies.

The voluminous body of research on military innovation in weaponry and the tactics for its deployment is of little value in assessing how new twenty-first-century technologies such as the UAV, the autonomous robot, or the 3D printer will be used by malicious non-state actors. And analysis of how state armies adapted to counterinsurgency missions in Vietnam or Malaya, lessons the United States then applied to Iraq and Afghanistan, does not help predict how they will be forced to adapt to technological innovations emerging from Apple, Google, or Facebook, not to mention China's Alibaba, Baidu, or Tencent.[14]

Violent non-state actors do not focus on the innovation of major systems or on creating new complex technologies themselves because they don't have the necessary resources. They also don't need to. Most achieve their ends by some combination of buying or stealing weapons; gaining financial and logistical help, military training, intelligence, and weapons support from states; using established technologies in new ways, as with the hijacking of airplanes in the 9/11 attacks; or using newly available technologies in surprising ways. Individuals, terrorist groups, and insurgents are opportunistic: they exploit whatever technologies they can get their hands on.

Fortunately, in seeking to discern how various types of violent actors will embrace emerging technologies we can draw on two key frameworks of analysis. One has established a clear pattern in the history of the bursts in their innovation, and the underlying forces driving those bursts. The other has identified the distinguishing features of new commercial sector technologies that are rapidly adopted. The history of new technology adoption by malevolent nonstate actors shows that they are just as eager as any other consumers to buy, build, and use them—they just use them differently.

The Historical Relationship between Political Violence and Technology

Like war, political violence dates to earliest recorded human history. The world's first documented terrorists were the Jewish Zealots of Judea, who sought to incite a mass uprising against Roman rule. Insurgency as a concept is at least as old as the reign of Roman ruler Augustus (43 B.C. to A.D. 14), with the first known use of the Latin word *insurgere*, meaning to rise up or be rebellious, appearing in Book 9 of Ovid's epic poem *Metamorphoses* (A.D. 8).[15] Today an insurgency means a group engaged in a political and military campaign against an established order, using irregular tactics against military forces, and seizing and holding territory.[16] Any tourist who has climbed to Israel's ancient fortress of Masada in the Judean desert knows of the first-century revolts by the Jews against Roman rule; but the Jews also rebelled against the ancient Babylonians (597 B.C.) and the Hellenistic Seleucid Empire (167 B.C.). Historians have documented insurgencies by members of the Persian Empire against Alexander the Great (334–331 B.C.) and by the Egyptians against the Hellenistic Ptolemaic Kingdom (206–186 B.C.).[17]

The technology employed by the earliest guerrillas, rebels, and insurgents was rudimentary. Ancient terrorism typically relied on knives, swords, or strangulation, though sometimes leaders were poisoned. The Sicarii, a splinter group of the Jewish Zealots, were named for their weapon. "Sicarii" translates to "dagger men." They knifed Jewish officials in broad daylight for optimal public notoriety.[18] The Muslim Assassins (also known as

Ismailis-Nazari, A.D. 1090–1275) also used daggers, seeking both martyrdom and publicity by attacking prominent people in mosques or markets, while scores of people watched.[19]

Another group, the Hindu Thugs (from the Hindu "thuggee"), operating on the Indian subcontinent for more than 600 years (fourteenth to nineteenth century), killed up to a million people (estimates vary) using only strips of cloth. Thug cult members strangled random travelers in the service of Kali, the goddess of death. Mark Twain familiarized his readers with their bizarre behavior in two chapters of his 1897 travelogue *Following the Equator: A Journey around the World*, calling them a "bloody terror" and describing the indoctrination and training of new members in their gruesome techniques. "No half-educated strangler could choke a man to death quickly enough to keep him from uttering a sound—a muffled scream, gurgle, gasp, moan, or something of the sort; but the expert's work was instantaneous: the cloth was whipped around the victim's neck, there was a sudden twist, and the head fell silently forward, the eyes staring from the sockets; and all was over."[20] Surprise and secret training were the Thugs' most important weapons.

Terrorists and insurgents have, however, sometimes been quite innovative, whether in their use of existing technologies or seizing on the violent capabilities of new ones. We learn much about the nature of their innovation, and which sorts of it have had the most disruptive impact, by examining the history of terrorism from the late nineteenth century up to today, which I'll call the modern era of terrorism. Political scientist David C. Rapoport has delineated four overlapping "waves" of activity during this time span, in which violent campaigns of a new kind spiked. Each lasted about a generation (forty years).[21] He describes a wave, more specifically, as an international spread of violence in which "similar activities occur in several countries, driven by a common predominant energy that shapes the participating groups' characteristics and mutual relationships."[22] Two critical factors facilitated each upsurge. First, changing communication and transportation patterns provided new avenues for sharing information and new means of travel. Second, compelling ideologies or doctrines attracted adherents, which were, in order: anarchism, anti-colonialism, Marxism/socialism, and religious fundamentalism. The goal of each wave was revolution, Rapoport argues, having different meanings in different waves, that is, popular suffrage, national self-determination, radically reconstructing authority, and a return to sacred texts as sources of legitimacy.[23]

The anarchist wave stretched from the 1870s well into the 1900s and was propelled by the ideas of Russian intellectuals, such as Sergei Nechayev, who advocated revolt "by any means necessary," very much including terrorism, to catalyze popular revolution.[24] The wave grew to encompass a wide array of violent groups, including Russian Nihilists; anarchists (Russian, Italian, French, British, American, and many others); Irish nationalists (the Fenians, Clan na Gael); and the Russian group Narodnaya Volya, which assassinated Tsar Alexander II in 1881.[25] Bombings were the preferred tactic, although guns and knives were also used. This wave climaxed in the assassination of Archduke Franz Ferdinand and the outbreak of the First World War, though anarchist violence continued, and was even ratcheted up, in some countries through the 1920s and even into the early 1930s.[26]

The second period was of ethnonationalist-separatist violence, which Rapoport calls the Anti-Colonial wave, stretching from the 1920s to the 1960s. It was initiated at the end of the First World War but gained momentum after the Second World War. A populist push for national self-determination led to the liquidation of European empires. This was by far the most robust wave in terms of its political effects and the numbers of casualties, and many terrorist movements from this time developed into vigorous insurgencies.[27] Groups formed during this wave had longer lifespans than most terrorist groups; some transitioned into political parties or founded new regimes. They engaged primarily in bombings, but also hit-and-run attacks with small arms, especially assault rifles. The conflicts that erupted included those in Northern Ireland (PIRA), Cyprus (EOKA), Algeria (FLN), and Palestine/Israel (Irgun).[28]

The third wave was that of the New Left, driven by Marxist/socialist ideology. It overlapped with the second wave, emerging in the 1960s and lasting into the 1980s. Rapoport characterizes this surge as primarily a reaction to the Vietnam War. The players include the West German Red Army Faction, the Italian Red Brigades, the American Weather Underground, and the Japanese Red Army. Deeply intertwined with the Cold War, these groups saw themselves as advocates for the Third World against the capitalist West, and the Soviet Union offered them ideological support, training, and weapons. Airline hijackings and kidnappings were the chief tactics. State sponsorship of terrorism also peaked during this period, arming and training groups or using them as proxies, including by Iran (Hezbollah, Hamas, Palestine Islamic Jihad, Popular Front for the Liberation of Palestine-General Command), Libya (PLO, Abu Nidal Organization), and North Korea (New People's Army, Japanese Red Army). After a twenty-year dormancy in the aftermath of the Cold War, state sponsorship, notably by Iran and Russia, is in resurgence now and will likely grow in the future.

The religious fundamentalist wave gained momentum in the 1970s and is still playing out, incorporating Sunni Islamist groups like al-Qaeda, but also the Shi'ite Hezbollah, the Jewish extremist group Kahane Kai, and Japan's Aum Shinrikyo, among others. They actually represent a return to the first-century "spiritualist" roots of terrorism. The distinctive tactic of this wave is suicide attacks, which also have ancient roots but have been made more lethal due to the use of high explosives.

Rapoport laid out his waves shortly after the 9/11 attacks, and they have held up well under academic scrutiny.[29] Scholars debate the exact beginning and end points of each wave, and which groups should be included, and some have proposed adding or subtracting certain waves, such as shoehorning in a German wave between 1920 and 1940.[30] But these are largely quibbles, and no other analysis has described broad global patterns of the last two centuries of political violence better.

An examination of the technologies used in each wave, and the tactics for their use, reveals how avidly accessible new lethal inventions are embraced, and also that the more accessible a lethal technology is, the more widespread and disruptive the resulting political violence is.

How Technologies Were Harnessed

The anarchist wave was sparked by Alfred Nobel's invention of dynamite in 1867. The political arguments that motivated the anarchists were at least seventy years old, but because dynamite was commercially available, and also relatively easy to make, and was the first safe explosive for attackers to make use of, it provided would-be revolutionaries an alluring new means of *acting* on those long-standing arguments. They conducted shocking attacks that garnered international attention, spread a fascination with anarchism, as well as other revolutionary ideas, and inspired many subsequent attacks. The widely publicized attacks also stirred public horror and beguilement with dynamite itself, a key element in its spread. The first wave, labeled "anarchist" but also incorporating nationalists and social revolutionaries, killed thousands of innocent civilians, assassinated presidents and prime ministers, destabilized governments, and played a role in the coming of the First World War.

The anti-colonialist wave also had major destabilizing effects, and it was also propelled by a newly available and easily obtained technology: the AK-47 assault rifle.[31] Like dynamite, the rifle also developed symbolic resonance that helped it spread. Pressures for decolonization had complex roots, including the dashing of colonial people's expectations of national self-determination after World War I through the Versailles peace settlement. In cynical haggling over the spoils of war, with little or no involvement of indigenous inhabitants, the victorious powers at Versailles transferred German and Ottoman territories into mandates under their own control, including Syria and Lebanon (France), Iraq and parts of Iran (Britain), Rwanda and Burundi (Belgium), and Palestine (Britain).[32] Momentum built during the Second World War, with the 1941 Atlantic Charter promising to "respect the right of all peoples to choose the form of government under which they will live."[33] But it was after the war that violence against the regimes surged. It was fueled, in part, by how cheap, easy to operate, and widely distributed the AK-47 was. Invented by Soviet tank commander Mikhail Kalashnikov in 1947, the rifle, officially known as the Avtomat Kalashnikova model 1947, was deliberately disseminated globally by the Soviet Union to Communist allies, proxy clients, and anti-West terrorist groups. So effective was the weapon against state military forces that it was ironically a factor, along with Stinger missiles, in the USSR's demise in the 1980s. The Afghan Mujahideen used the gun to blistering effect against Soviet troops, bogging them down in a quagmire. The resultant deep economic and political pressures in the USSR ultimately drove Soviet forces out of Afghanistan, one domino in the process of collapse that followed.

With the aid of the rifle, the rate of success in insurgencies increased in the aftermath of the Second World War, going from less than 25 percent between 1775 and 1945, to more than 40 percent from 1945 to the present, according to one researcher.[34] (We will examine insurgency success rates in greater depth in chapter 6.) In the same period, the number of member states in the United Nations went from 51 to 193.[35] The gun became a powerful symbol for "freedom fighters," every bit as compelling as dynamite was for the anarchists, and today, some newly independent states, and militant groups within them, still proudly display images of the classic AK-47-shaped weapon on their flags and in political posters.[36]

In the third wave, terrorists did not make use of an accessible new technology they got their hands on; they harnessed two sustaining commercial technologies hijacked for new uses: television and the airplane. The watershed event was the 1972 Munich Olympics massacre, which demonstrated the powerful potential of international broadcasting to draw attention (some 800 million viewers, more than 1 in 5 people on the planet) and sympathizers to a cause, in that case Palestinian nationalism. Even though the operation was tactically a failure, yielding none of the Palestinians' ransom demands and resulting in the deaths of all eleven Israeli athletes and five of the eight Black September operatives, the global networked television coverage propelled the PLO to prominence (Figure 2.2) and inspired other aspirational terrorist groups to copy their techniques.[37]

FIGURE 2.2 Munich Olympics Hostage Crisis, 1972. Photo by Russell Mcphedran/The Sydney Morning Herald/Fairfax Media via Getty Images.

The third wave's innovation with airplanes was first the hijacking and then the bombing of commercial planes. Between 1968 and 1972 a range of groups carried out 396 aircraft hijackings, and at the outset, the goal was to draw attention to the cause, not to kill the crew and passengers. The hijackers ordered planes to fly to a particular destination, such as Cuba, or demanded money in exchange for hostages. This changed dramatically during the 1980s, when four horrific attacks on airliners were executed. In 1983, a bomb hidden in the baggage hold blew a hole in Gulf Air Flight #771, killing 107 people. In 1988, a radio cassette player concealing Semtex with a timed detonator destroyed Pan Am 103 over Lockerbie, Scotland, killing 259 people in the plane and 11 on the ground. In 1989, a dynamite bomb detonated by a Sanyo tuner destroyed Air India flight #182, killing 329 people. That same year, a PETN bomb destroyed the French flight UTA #189 (Union de Transports Aeriens), killing 179 people aboard.[38] In each of these cases, the device was hidden in an unaccompanied checked suitcase.

While these terrorist activities garnered major worldwide notice, the third wave of terrorism had more limited global effects than the first two, as no regimes were overthrown or new states established as a result. This is in part due to a clampdown on the ability to bring explosives onto planes. The research and development budget of the Federal Aviation Administration for aviation security more than tripled, going from $9.9 million to $30.3 million between 1989 and 1991. In addition to instituting new ways to screen baggage for the nitrogen in high explosives, a key low-tech innovation was baggage reconciliation, meaning that no baggage was allowed to stay on a plane without its accompanying owner also onboard.[39] Military and police departments also refined and expanded their ability to rescue hostages, such as Germany's 1973 establishment of Grenzchutzgruppe 9 (GSG 9), an elite tactical response unit modeled after the British SAS and Israeli special operations.

Another key factor affecting the third wave was incremental professionalization of the media, especially the editorial guidelines and self-restraint major traditional media outlets gradually developed.[40] In the United States, where a free press remained sacrosanct, responsible media outlets engaged in varying degrees of self-censorship, becoming more aware of the downsides of completely unfiltered reporting of live acts of violence or detailed after-action reports of government operations and tactics. In most countries, a range of government measures were implemented to deny, delay, or dilute the real-time coverage, sometimes imposed ex post facto—in the form of injunctions on materials deemed deleterious to national security.

In addition, many of the third wave terrorist groups were ideologically aligned with the Soviet Union, China, or Eastern Europe, and supported by them both financially and through the provision of weapons. When the Cold War ended, not only did Marxist ideology lose its luster, but the state sponsors, no longer fighting proxy battles against their ideological enemies, pulled back their support for groups.

The current fourth wave of terror has been characterized by a resurgence of suicide attacks and the harnessing of the propaganda power of social media, a blend of an old tactic made more deadly and a new technology being embraced. Suicide attacks have been documented going back to as early as the first century, and they were by no means confined to the Middle East. During the eighteenth and nineteenth centuries, unsuccessful suicide campaigns against European and American colonial powers were common among Asian Muslim communities of the Indian Ocean region, including the Malabar Coast of southwest India, Aceh on the northwest tip of Sumatra, and Sulu in the southern Philippines.[41]

The fourth wave has been driven largely by religious fundamentalism, and suicide attacks are made a more viable tactic when such zealotry is involved.[42] Due to the new airline security measures, pre-placed explosive attacks involving planes became difficult, if not impossible, but they could be carried out if the operatives boarded planes and then either detonated a bomb while aboard, or commandeered the plane and used it as a weapon.

New digital communications tools were also important to the planning and coordination of the 9/11 attacks, and they have fueled continuing terrorist activity since. The proliferation of suicide bombings by the LTTE

(Liberation Tigers of Tamil Eelam, or "Tamil Tigers"), al-Qaeda, ISIS, and many other recent groups (both secular and religious) has been encouraged by the guaranteed widespread publicity via an unprecedentedly globalized new media.

The new digital tools of the Internet and mobile technology have given anyone with connectivity direct access to an audience of potential billions, now including detailed profile information about them. This is the kind of direct marketing that earlier terrorist groups could only fantasize about, and it has enabled a range of actors as disparate as ISIS, the Neo-Nazi National Alliance, and Boko Haram to raise the profiles of their cause, disseminate carefully curated information, attract followers, and mobilize armies. Access to new tools of communication has been a key factor in making this wave so consequential.

How can these insights help us anticipate the nature of the next terrorist wave?[43] In assessing this, we have to take a closer look at the adoption of technologies by violent non-state actors, and their innovation with them. We will come away with a core set of criteria for which new technologies are the most likely to be adopted in the future and in what ways.

How Lethal Non-state Actors Innovate

Those who study how non-state actors such as terrorists and insurgents innovate rarely take a broad strategic perspective about long-term trends or patterns.[44] Scholars have tended to conduct in-depth case studies of individual groups, conflicts, or campaigns,[45] and there has been little tying together of those findings into broad characterizations of how groups adopt technologies and innovate with them.[46] The narrow focus has also limited understanding of how the larger global context of sweeping dislocations accompanying the shift from industrial manufacturing to autonomy and robotics are likely to affect their technological choices now and in the future.

Another limiting factor in analyzing the adoption of technology for political violence is that most terrorism scholars don't focus on the technologies used. Whereas many of those who study the history of military innovation emphasize the hardware, like tanks and battleships, and neglect the software, like military doctrine and organizational learning, terrorism experts do the opposite: they tend to focus primarily on the *software*—ideologies, group strategies, organizational structures, decision-making processes, and leadership—and very little on the hardware.[47]

When weaponry is considered, the prevailing argument is that guns and bombs will continue to be the key technologies of terrorism.[48] In

1999, distinguished scholar Ariel Merari observed, "At the end of the century, terrorists still use the same weapons that they used in its beginning, namely: pistols, rifles, and improvised explosive devices. This stagnation is astonishing, considering the fact that in the late nineteenth century, Russian terrorists adopted new explosives as soon as they were invented."[49] Even after 9/11, in 2015, terrorism scholar Adam Dolnik wrote, "[W]hen one surveys the list of terrorist operations case by case, very few incidents strike the observer as creative *in any way*."[50]

Yet, as we've just seen, some groups have in fact been avid early adopters of new technology, and some have been quite innovative in making use of it. They've done so both opportunistically, jumping on the means that become easily available for carrying out more lethal attacks, and defensively, in efforts to evade counterterrorism measures enacted by states. A number of key insights about how they're likely to innovate will emerge from an examination of the activities of the most creative groups.

We'll first consider the Provisional Irish Republican Army (PIRA), considered one of the most imaginative revolutionary groups of the twentieth century, emulated by numerous others and responsible for training members of groups as far-flung as the Basque group Euskadi Ta Askatasuna (Basque Homeland and Liberty, or ETA) and the Colombian Fuerzas Armadas Revolucionarieas de Colombia (Revolutionary Armed Forces of Colombia, or FARC).[51] During its almost thirty years of operations (1969–1998), the PIRA amassed more than thirty varieties of weapons, including mortars and rockets, Armalite rifles, advanced explosive devices, and military-grade machine guns, notably funded by major contributions from the United States and Libya.[52]

Two things drove the group's technological innovation: the high rate of fatality of its members, as they tried to build homemade bombs (in 1973 alone, they lost thirty-two members to premature detonation); and the need to respond to technological countermeasures by the British security forces.[53] The first pushed them to refine their explosives expertise and develop more pipelines for finished products; the second drove them to be more creative in designing and adapting their timers and explosive detonating devices.

The group's ingenuity in obtaining explosives was impressive. In the beginning they stole them from mining or industrial sites and received large quantities of weapons and explosive materials in frequent small shipments from the United States, which were concealed in clever ways.[54] Between 1970 and 1975, for example, Irish-American ex-pats skimmed off a large quantity of dynamite from the huge New York City Water Tunnel construction project and shipped it to PIRA.[55] The many means they used for smuggling arms and explosives included nestling them in hollowed-out electric transformers, stuffing them in golf bags, and putting them in the coffins of Irish Americans headed back to the old country for burial.[56] They also developed their own explosives, building tons of fertilizer bombs for car bombs in small three-person factories. In the 1980s, when Libya shipped the PIRA large quantities of Semtex, the plastic explosive (along with Kalashnikovs and other weapons), the group was able to shift from production of explosives to the refinement of devices.[57]

The group's technical ingenuity mainly concerned the detonators and explosive devices it devised, designed to outwit British forces who nipped at their heels in a deadly game of cat-and-mouse. In the beginning, PIRA explosive devices were little more than hand-thrown coffee jars filled with nails and explosives, but those were ineffective, and the hurlers ended up dead or incarcerated. The PIRA soon switched to using long command wires to set off bombs, but the British used chase cars or helicopters to go after the operators. So the PIRA found alternative methods of detonation, using model airplane or toy boat radio transmitters, for example. When the British then jammed those signals, the group switched to the Weather Alert radio frequency—and so on, in a race up and down the electromagnetic spectrum. Then they shifted into direct types of remote detonation, from police anti-speeding radar guns, to garage door openers, to the kinds of mechanical pressure plates later used in IEDs in Iraq and Afghanistan. They also refined the use of delayed timers, including watches, clocks, parking meter timers, and eventually infrared and light-sensor systems.[58]

The Japanese Aum Shinrikyo was another highly innovative group, and its attacks in the 1990s were an early shift from predominant Cold War patterns. An apocalyptic cult founded in 1984, Aum included some twenty university-trained scientists from Russia (the former Soviet Union) and Japan. Aum Shinrikyo was well funded, with more than $1 billion in its coffers, and it controlled two dozen properties. Their goal was to kill a large number of people using high technology, especially chemical and biological weapons, and they are best known for their 1995 attack on the Tokyo subway with the chemical nerve agent sarin, in which 13 people died and about 6,000 were injured. This attack was the outcome of years of trial and error beginning in 1990, including testing, developing, or using anthrax (*B. anthracis*), VX nerve gas, chlorine gas, and hydrogen cyanide in very advanced scientific laboratories. Aum developed and used biological weapons, including spraying *Clostridium botulinum* at two US Naval bases, Narita airport, the Japanese Diet, the Imperial Palace, and the headquarters of a rival religious group.[59]

The story of Aum Shinrikyo powerfully demonstrates the dangers of allowing a group the freedom to gather resources, build extensive facilities, attract experienced high-level talent, and engage in years of experimentation. Were it not for eventual Japanese police pressure and a good deal of luck, the death toll would likely have been much higher. The only good news from this case is that, even with extraordinary resources, Aum Shinrikyo repeatedly failed, and responsible governments have learned to be more vigilant— although most agree that the risk of biological threats is growing.[60] And the immediate post-9/11 anthrax attacks in the United States only strengthened that view.

New communications capabilities have also been crucial to the new lethality of another old tactic, "Fedayeen" or "wolf pack" attacks, meaning commando operations by coordinated small groups against urban civilian targets, involving shooting, bombing, arson, and suicide attacks, often over a series of days. The origins of the approach go back many years. In 1947, for example, well-trained, machine-gun armed operatives from Irgun Zvai Le'Umi (Irgun) engaged and overwhelmed the guards as they deployed a large truck bomb against Goldsmith House, the British officers' club in Jerusalem, killing seventeen and wounding twenty-seven.[61] The 1972 Munich Olympics massacre was a commando operation and multiday urban barricade attack. The 2006 attacks on numerous hotels in the city of Mumbai, India, with operatives arriving by sea and supported in real time by handlers in Pakistan are the newer model, with the innovative elements being the use of cell phones and the simultaneity of attacks in several places throughout the city.

Perhaps the most striking recent case of how innovative lethal non-state groups can be is that of the American experience with insurgents' use of improvised explosive devices (IEDs) in the wars in Iraq and Afghanistan.[62] Between 2001 and 2007, IEDs caused more than 70 percent of all American combat casualties in Iraq and more than half of those in Afghanistan.[63] Alarmed by the toll, in 2003, Gen. John Abizaid, head of US Central Command, sent a memo to the secretary of defense calling for a "Manhattan Project–like" effort to reduce the threat.

In response, a small 12-person Army task force on homemade bombs morphed into a very well funded, 1,900-person Joint Improvised Explosive Device Defeat Organization (JIEDDO).[64] The US government appropriated more than $21 billion to the anti-IED effort, which included a wide range of countermeasures such as electronic jammers to disrupt detonators, robotic explosive ordnance disposal equipment, 3D cameras to detect disturbed soil, increased armor for vehicles and personnel, airships to observe insurgent activity, and explosives detectors that sent out neutrons to cause explosive substances to emit back gamma rays—among hundreds of other projects.[65] The technological IED countermeasures by the British security services against the Provisional IRA come to mind, although the US effort dwarfed that one in its financial heft, the size and remoteness of the territories it covered, and the number of enemies it was trying to counter: at its most active, the PIRA had membership in the low thousands, while insurgents in Afghanistan and Iraq easily numbered in the tens of thousands.[66]

The outcome reduced risk but there was a huge asymmetry in cost, accompanied by disappointing overall results. IEDs are made from a wide range of materials, from gunpowder to stolen artillery rounds to plastic explosives to fertilizer, so the cost of each device ranges from essentially free to about $300–$400. The cost of various US responses was billions of dollars, and usually the solutions didn't work for long. The insurgents adapted with low-end, simple innovations that were accessible on the open market.

A few basic examples demonstrate the dynamic. Early in the war, insurgents buried explosives in the road, with hidden pressure plates to set them off when US and allied vehicles drove over them. So the United States and its allies attached long extension arms, like mini steamrollers sticking out in front, to hit the pressure plates at a distance and set off the bombs without harming the passengers. The insurgents estimated the length of the arms, and buried the plates down the road that distance ahead of where the explosives were buried. A similar device called Rhino was a hot metal box hung in front of vehicles to set off roadside bombs that were triggered by infrared heat signatures: insurgents reoriented the aim of the bombs so they still hit the vehicle when they went off.[67]

The United States threw every possible technology at the IED problem, and the insurgents adapted. A high-powered microwave emitter called Blow Torch was supposed to damage the devices' electronic circuitry; the insurgents figured out how to shield them. The notorious "Joint IED Neutralizer" (JIN) was a kind of lightning gun on the nose of a truck, meant to use short-pulse lasers to carve conductive channels in the air and zap roadside bombs. Its Mississippi-based maker, Ionatron, got a $30 million contract, and yet the unwieldy thing never worked.[68] Hundreds of $25,000 "Fido" machines that were developed by an Oklahoma company were supposed to detect explosives better than a dog's nose could (when the sensors were heated to 200 degrees Fahrenheit, to find explosives with very low vapor pressures); in the end, actual trained dogs worked out much better.[69] At times technology displaced common sense.

When the counter-IED program was launched in 2006, retired Army general Montgomery C. Meigs, who has a PhD in history from the University of Wisconsin–Madison, was put in charge. Meigs, who had been teaching at the Air War College in Montgomery, Alabama, had written a book about World War II's Battle of the Atlantic, between German U-Boats and the so-called 10th Fleet, the legendary interagency group that stopped them. The success of US anti-submarine warfare depended on brilliant data analysis of attack patterns and the breaking of the German enigma code, which told Americans roughly where the subs were. Technology and science helped the United States defeat the Germans. So a key part of JIEDDO's initial concept was to crack the nut by using science and data analysis of IED attack patterns.

But this was a completely different challenge. After two years of struggling to neutralize homemade bombs, Gen. Meigs told a reporter from *Wired* magazine:

> [I]t's much more complicated. You have many more degrees of freedom in the IED problem than you did in the U-Boat problem. There's a three trillion dollar a year investment in information technology. . . . And our opponents can go to the world marketplace in information technology and get literally for free off of the Internet very robust codes, cryptographic means, instant communications. And they can buy all kinds of things that can be transformed into ways of arming and initiating IEDs. And sensors. It's all out there. It's in the marketplace. You couldn't do that with U-Boats.[70]

Another key problem with the big-data approach was that soldiers in the field collected IED data inconsistently, so JIEDDO was never able to distinguish the conditions that determined its successes from its failures.[71] Unexploded IEDs that Marines or soldiers discovered got plenty of write-ups; when bombs went off, however, the last thing they wanted to do was take out their notebooks.[72] In short, the better an explosion worked, the less likely it was to be reported on.

What's especially discouraging about all this is that it suggests that insurgents are keeping up technologically and perhaps even winning. A rigorous 2018 quantitative study of declassified IED data gathered in Iraq and Afghanistan between 2006 and 2014 provides strong support for that conclusion. It demonstrates that insurgents kept pace with expensive technological countermeasures, showing that IEDs were just as likely to detonate and kill or maim in 2014 as they had been in 2006.[73] That is in part due to the sheer numbers of IEDs deployed as the conflicts progressed. The total number of IEDs had increased during that period, from 1,952 in 2006 to 5,616 in 2009 in Afghanistan.[74] The role of outside support, especially training, logistics, and supplies from Iran and Pakistan, was also crucial. Still, the effectiveness of IEDs seemed to be increasing.

In short, even in the midst of dramatic technological changes at the high end, insurgents are getting better at innovating low-tech solutions to stymie the most advanced armed forces in the world. Given access to cheap materials and online instructions about exactly how to build bombs, they are dramatically expanding the use of IEDs not only in Afghanistan but also in dozens of other countries, including Colombia, the Democratic Republic of Congo, India, Iraq, Nigeria, Pakistan, Russia, Somalia, Syria, and Thailand, not to mention the United Kingdom, France, and the United States.[75] Insurgent adaptation, self-curated global publicity, the sharing of techniques, and the sheer quantity of simple IEDs can change the political dynamic of a conflict and help the underdog succeed, often by waiting out the stronger power and imposing greater costs than it is willing to bear.

A few key points about the nature of technological innovation for purposes of political violence come out of these cases. One is the powerful role of accessible new communications capabilities. For most of history, insurgents and terrorists tended not to learn about each other, either because there were no global communications or because the government had the upper hand and controlled the means of communication. Manuals by revolutionary leaders began to emerge in the nineteenth century, when leaders were more likely to be literate and to gain access to a printing press.[76] In the twentieth century, it became common for groups to share books of advice about revolutionary tactics, from Mao's *Little Red Book* to Ernesto "Che" Guevara's *Guerrilla Warfare* to North Vietnamese general Vo Nguyen Giap's *How We Won the War*. Smartphones and social media technology have provided communications capabilities on an entirely new level. Violent groups and individuals have already harnessed them in many ways in addition to using them to coordinate attacks. They're targeting certain profiles of individuals to recruit, sharing or mischaracterizing enemy actions, and live-streaming choreographed attacks, with direct global reach. They will surely continue to innovate with these tools and with any new communications tools that become widely available.

A special point to make about how communications technology has and will continue to spread violence concerns the phenomenon of "contagion."[77] Regarding political violence, this is the tendency of well-publicized crimes (mass murders, assassinations, bomb threats) to trigger copycat activities, especially when the crime was thought to be "successful" in some way.[78] Contagion has also been documented regarding suicides, especially highly publicized ones, such as that of Marilyn Monroe. This pattern became known as the "Werther Effect," after the protagonist in Goethe's 1774 novel *The*

Sorrows of the Young Werther who shoots himself at the end. In the late eighteenth century, the author was widely condemned, because many feared that the book's publication would set off a wave of suicides, and a number of people did kill themselves as a result.[79]

As the effect pertains to political violence, it was beginning in the 1970s that researchers noticed an increase in violent crime following well-publicized assassinations (such as the 1963 killing of John F. Kennedy) and murders in the United States.[80] Other research found that widely publicized bombings and kidnappings inspired copycat attacks.[81] Careful analysis of aircraft hijackings between 1968 and 1972 also demonstrated a clear relationship between successful (but not unsuccessful) aircraft hijackings in the United States.[82]

A key example was the 1970 hijacking by the Popular Front for the Liberation of Palestine (PFLP) of three planes (one American, one Swiss, and one British) with more than 300 hostages to a former British airstrip near Amman, Jordan, Dawson's Field. (A fourth plane, an American Boeing 747, landed in Cairo, Egypt.) Expecting a commando raid, the PFLP disembarked the hostages and blew up the empty planes on the tarmac, a spectacle carefully choreographed for the news media. Shortly thereafter, the hijackers' demands for the release of Leila Khaled and six other PFLP members from prison were met. A sharp spike in hijackings for ransom followed, both internationally and in the United States, including a September 1971 hijacking to gain the release of several Black Panther prisoners, as well as at least twelve hijackings for money. In 1972, the total number of aircraft hijacking attempts of foreign-boarded planes (as opposed to those boarded in the United States) peaked at more than sixty.[83] Researchers found that the best predictor for a hijacking was a prior hijacking, well publicized and broadly perceived to be successful.[84]

That kind of contagion seems to be resurging today. The journalistic practices of ethical self-restraint that responsible newspaper and television organizations developed in the late twentieth century have lost ground in the past decade, especially because of the rise of social media. Barriers to entry are virtually nonexistent; almost everything can be photographed or videoed on a smartphone, then captured on servers and web pages out of the reach of most governments—or, for that matter, of responsible editors. With truly globalized media and social media, millions of individuals instantly know about both the means and the ends of lethal technologies.

A second key insight from the cases of terrorist innovation we've looked at is that while many factors may be involved in spurring individuals to move from sympathy with violent ideologies/perspectives to actually participating in violence, an absolutely decisive one is access to the means. Focusing on radicalization processes can provide insight into motivations; but the question of whether or not someone acts depends on whether or not he or she has the means to act. Accessibility will make new technologies much more disruptive. Dynamite and the AK-47 were both easily obtained; airliners have been quite successfully placed off limits. Accessibility to newly emerging technologies must be a major concern. Whereas during the last century, most concern about advanced technologies, like nuclear, biological, and chemical weapons, getting into the hands of nefarious actors focused on the problem of "dual use"—meaning that they were of use in both military and civilian contexts, and the civilian uses might make them accessible to the public—we've gone way beyond dual use now. But just as concern about "proliferation" (implying a transfer from states to other states or non-state actors) should be replaced by consideration of "diffusion" (meaning a global spread unconstrained by the types of actors involved), "dual use" has been made obsolete by "multi-use."

As I mentioned in the Introduction, today's new and emerging technologies are useful to four categories of users: the professional, the prosumer (professional consumer or "pro-sumer"), the amateur (or hobbyist), and the consumer. Of these four, the concept of "prosumer" is particularly important but it is not new: it originated in Alvin Toffler's 1980 book, *The Third Wave*. It defines someone who blurs the distinction between "consumer" and "producer." Prosumers use accessible current technologies to produce innovative *new* technologies that combine them in new ways.

Our current era is characterized by clusters of technologies that are perfectly suited to this. For example, in the field of biology, gene editing, automation, and cloud-computing are all intersecting to make tools like CRISPR/Cas9 accessible to the public, and at low cost. Already companies produce very simplified kits that use the enzyme Cas9, which can edit, add or delete DNA sequences, and sell them for use in homes or even middle school classrooms at about $150 a kit.[85] We must take seriously the risk of terrorists, or simply unethical individuals, using more advanced versions of this technology to alter genetic code in ways that could be horrifying.[86]

Another key insight regarding past terrorist and insurgent innovation is that we should not underestimate the willingness of violent players to learn about even quite sophisticated technology, enough to be able to use it for their own purposes, as with the PIRA and its inventive detonators or Aum Shinrikyo and its chemical and biological experimentation. Key factors are whether groups have access to, or the ability to develop, the necessary training and expertise, along with the time and space in which to tinker with new technologies. But we can no longer rely upon the sophisticated know-how that building mass lethal weapons during the nuclear era required. There are few barriers to entry. The dedicated prosumer today is operating on advanced technological platforms, such as computers and gene-editing kits, that they may not even understand nor do they need to: their tools are developed by others and specifically designed to facilitate innovation.

Along with new aspects of non-state actor innovation, we must anticipate that violent actors will keep innovating in response to our efforts to crack down on them, just as the PIRA and insurgents in Iraq and Afghanistan have done. Comparatively low-tech countermeasures to high-tech innovations can be remarkably effective because, especially in democracies, the question of effectiveness is not just whether cutting-edge technology trumps jury-rigged explosives, not just who is spending the most money on defenses and who is broke, but also how the political dynamic of the contest unfolds before the public.

Everett Rogers's Theory of Commercial Diffusion Revisited

To take a systematic approach to assessing whether emerging technologies will be used for political violence, by whom, and in what possible ways, we can apply the lessons from Everett Rogers's theory of diffusion of commercial products. Recall that Rogers defined diffusion as "the process by which an innovation is communicated through certain channels over time among the members of a social system."[87]

Through all of his study of commercial products and ideas, from cellular telephones to nitrogen soils tests for farmers, Rogers identified five key characteristics of innovations that determine how widely and quickly they diffuse: First, relative advantage, meaning the degree to which the innovation is better than what it replaces, especially lower cost and higher social status. Second, compatibility, meaning its alignment with values, needs, and past experiences. Third, complexity: the simpler an innovation is to use, the more likely it will be adopted. Fourth, trialability, or the degree to which an innovation can be tried out and experimented on. And fifth, observability, or the degree to which the innovation is visible to others.[88] The more of these criteria any given potentially lethal technological innovation meets, the more disruptive it is likely to be.

Rogers did not include a single lethal innovation in his analysis. In the following chapters, we will apply it to a wide range of them, from the relatively rudimentary dynamite and AK-47, to the most sophisticated of today's emerging technologies, such as autonomous robotics and quantum computing.[89]

Before turning to that, there is one last important point to make about the adoption of innovations: they are frequently employed for unexpected purposes. In his Pulitzer Prize–winning book, *Guns, Germs, and Steel*, Jared Diamond examines an enormous range of inventions and shows that many inventors have not initially identified the use their creation is ultimately put to, often attempting to spur its adoption for a range of uses before it catches on for some other purpose. Many inventors are driven by a love of tinkering, or progress, or curiosity, and then have to figure out a use for their devices afterwards—or, even more commonly, other people figure it out. "It may come as a surprise," he writes, "to learn that these inventions in search of a use include most of the major technical breakthroughs of modern times, ranging from the airplane and automobile, through the internal combustion engine and electric light bulb, to the phonograph and transistor."[90]

One of Diamond's examples is that Thomas Edison envisioned the phonograph as a high-minded tool to preserve the last words of the dying, teach the alphabet, record books for the blind, take office dictation—not do something as frivolous as play music. When phonographs were installed in jukeboxes, Edison was initially offended (until he sold so many that he began to get rich). Or consider the first petroleum gas engine. Built by Nikolaus Otto in 1866, it was a seven-foot-tall curiosity, vastly inferior to horses, very heavy, and apparently useless. It took decades before Gottfried Daimler thought to put a gasoline engine on a bicycle (or motorcycle, 1885), and then Benz built the first gasoline-powered automobile (1886). In short, a new "thing" often appears first, and then its uses emerge more gradually, usually helped along by many hands. "Thus, invention is often the mother of necessity, rather than vice versa," writes Diamond.[91] Today's inventions, from self-driving cars to the iPhone, will be used for different purposes from those their inventors envisioned.[92]

As we'll see in chapter 3, Alfred Nobel was another inventor living at a time of open innovation who created a wonderful new product, dynamite, but failed to anticipate how it would be used for political violence. Why did anarchists mainly adopt dynamite instead of better explosives, like blasting gelatin, cordite, or ballistite? Why did groups prefer the AK-47 instead of the M-14 or M-16 rifles, which were technologically superior? Why, despite the hype in the early 2000s, did "cyber-terrorism" fail to materialize while non-state actors used digital technologies in more effective ways instead? The answers involved a combination of the nature of the technologies and the nature of human beings and our societies. We'll go in pursuit of them in the remainder of the book.

PART II | History

CHAPTER 3 | Dynamite and the Birth
of Modern Terrorism

THE DEADLIEST ACT of terrorism on US soil before the Oklahoma City
bombing of 1995 was the 16 September 1920 anarchist bombing of
the Morgan Bank at 23 Wall Street, which, as it happens, is less than half
a mile from the World Trade Center, attacked eighty-one years later. The
bomber, Mario Buda, parked a red horse-drawn wagon opposite the bank,
at the corner of Wall and Broad Streets. The wagon was packed with dynamite nestled under a pile of cast-iron window sash weights. Used in the
pulley system for double-hung window frames, these weights were shaped
like candles, and they functioned as bullet-like projectiles.[1] The explosion
killed 40 people and seriously injured more than 200 bystanders on the
streets, mainly messengers, clerks, stenographers, and drivers, but also including four teenagers and five women.[2] J. P. Morgan's son Junius was among
the injured, and his chief clerk, Thomas Joyed, was killed. The power of the
detonation left a scene of horror (Figure 3.1). The head of one of the women
killed was found stuck to a wall, her hat still in place. The hooves of the horse
pulling the wagon were propelled blocks away in four directions. This early
"truck bomb" (or "Vehicle-Borne IED" today) caused $2 million in damage.
Sharp-eyed New York City tourists can still see pockmarks on the granite
façade of the iconic corner building.[3]

This was just one attack of many perpetrated in a global terrorist movement that was propelled by the invention of dynamite. The violence gripped
the world for more than fifty years, from 1867 to 1934. Alfred Nobel had

FIGURE 3.1 Aftermath of the 1920 dynamite explosion on Wall Street, New York. Everett Collection Inc/Alamy Stock Photo.

not anticipated the outbreak of violence his invention spurred. His intention in creating dynamite was to correct for the drawbacks of gunpowder, serving the need of the mining and construction industries for a better explosive. In that, he succeeded brilliantly. Dynamite facilitated enormous infrastructure projects during the Second Industrial Revolution, which erected the skeleton of the world we know today. It cut the pathways for the rail lines and tunnels that carried the goods that fueled the economic boom.

But its explosive power also captured the imagination of small groups and individuals who were inspired to violence as much by the accessibility of such an effective, affordable, and attention-grabbing means of carrying out attacks as by other deeply analyzed sources of the contagion. Dramatic and widespread economic and social changes caused by the expansion of manufacturing, such as job instability and miserable labor conditions, provided a fertile context for violent protest. The leadership of fiery revolutionary leaders stirred support and was a necessary factor. But these were not sufficient to propel the activity. Great economic inequality and appalling factory conditions predated the onset of attacks by decades, and revolutionary ideologies had also been espoused for many years without triggering widespread violence. Dynamite sparked action because it provided the perfect means.

Attacks were carried out in every major country, and the magnitude of the political and social effects of the wave of violence is difficult to appreciate today. It's a forgotten first outbreak of modern terrorism, remembered more for ideological fervor than for technological adaptation. Much of the story has been forgotten, overshadowed or reinterpreted in light of subsequent events, especially the devastating world wars and the Cold War that followed. A close examination of the course of events reveals that non-state terrorist violence was a precursor to, a catalyst even, of those conflicts: a destabilizing threat that roiled nations at a time of unsettling social, economic, and political transition in some ways comparable to our own. While there are surely differences between the late nineteenth century and today, commonalities include rapid changes in the economy and society brought by industrialization, dislocation of workers, growing income inequality, intensifying nationalism, and fast-paced globalization of both communication and trade.

The story of dynamite's diffusion tells us much about the nature of violent movements and about the sorts of threats we can expect to face from the new breed of democratized technologies. In particular, the story helps anticipate which of them will be most readily utilized for terrorism. More advanced and efficient explosives introduced during the time dynamite became available, such as guncotton, ballistite, cordite, and gelignite, were *not* as widely adopted by attackers. Their use was largely contained in one region or confined to specialized industries or military arsenals. So, we'll follow the dynamite! We'll examine what characteristics of dynamite influenced its rapid global diffusion, and why it diffused globally while those other new explosives did not.

To appreciate dynamite's transfixing allure, however, we must begin with a consideration of the forebear of all explosives, gunpowder.

The Advent of Gunpowder

Also known as black powder, gunpowder was massively disruptive. In making knights and knaves equally vulnerable on the battlefield, it undercut the power of landholding nobility and was critical to ending feudalism. It also gradually drove the professionalization of state armies and the consolidation of European state power in the modern era. The story begins more than a millennium ago.

The Chinese invented gunpowder during the Tang dynasty, in the ninth century, and devoted it to military purposes from the beginning. The basic formula was charcoal combined with sulfur and potassium nitrate, commonly referred to as saltpeter. It was called "fire drug," initially used in a weak formulation that powered incendiary bombs, and was also used in noisemakers to frighten the enemy. The Chinese kept refining it, increasing the proportion of saltpeter to make it more deadly, and during the thirteenth and fourteenth centuries, they devised increasingly large artillery weapons that utilized large quantities of gunpowder, with names like "Dropping from Heaven Bomb," "Match for Ten Thousand Enemies Bomb," and "Bandit-Burning Vision-Confusing Magic Fire-Ball." They also built rockets, which were long tubes containing gunpowder. The powder was packed into a hole in the middle, so that when it was ignited, gases could build up, shoot out the back, and propel the rocket upwards. Rocket-launching hand tubes and primitive guns followed. Cannons that shot two-inch balls or lead pellets joined the arsenal during the Ming dynasty, in 1412.[4]

Most Western accounts credit Oxford academic Roger Bacon with the transmission of the formula to Europe in the thirteenth century. The true story is more complex, with likely transmission by the Arabs, who learned about saltpeter, which they called "Chinese snow," around 1240. But Bacon was the first European to refer specifically to gunpowder and suggest its potential in weaponry, in a work he scribed in 1267 summarizing knowledge of the natural universe for Pope Clement IV.[5] The science of physics outpaced chemistry after that, with the creation of ingenious machines such as catapults and pulleys for use in siege warfare, while the formula for gunpowder changed very little. It took a century for Europeans to employ gunpowder as a propellant in weapons, and even then, because the weapons were unreliable, they were used alongside traditional arms, such as the longbow, crossbow, pike, and poleax. The effective employment of gunpowder in European warfare evolved gradually, over hundreds of years.

In the fifteenth and sixteenth centuries, European armies adopted primitive gunpowder bombs that were simple hollow metal containers with a wick, weighing some three or more pounds each. They came to be called grenades, and men chosen for their height and throwing strength were formed into elite groups of infantrymen called "grenadiers." The earliest references to these soldiers are from Austria and Spain, and the word "grenade" comes from the Spanish word "granado," which means pomegranate—about the size and shape of the earliest grenades. By the late seventeenth century most European armies deployed specialized companies of grenadiers, but grenades proved to be unreliable weapons in the more dynamic types of warfare that evolved, better suited to a siege than to battle between armies on the field. Grenades blew up unpredictably, were difficult to deliver effectively by hand, and tended to be more dangerous to the thrower than to the target. So while regiments of grenadiers were retained, gunpowder grenades were abandoned in the mid-eighteenth century.[6]

A more successful early use of gunpowder in warfare was as the propellant in cannons. The first version of a cannon was probably the Arabian *madfaa*, built in the fourteenth century, a deep wooden bowl that held gunpowder and a ball balanced on the top that "popped off" when the gunpowder exploded.[7] Smaller cast-iron or bronze cannons soon replaced it, used by the British and French during the Hundred Years' War (1339–1453). Adopted quickly throughout Europe, the cannon's bombardments spelled the end of the fortified medieval castle, whose high walls ironically increased their vulnerability by offering easy targets. The cannon dramatically shifted advantage in battle from defense to offense, with the watershed event being the devastating siege and sacking of Constantinople by the Ottomans in the mid-fifteenth century, widely considered to mark the end of the European Middle Ages.

Gunpowder was rapidly adopted for use in other forms of artillery as well as in firearms, which called for increasingly professionalized armies trained to deploy them. Diffusion was limited, for the most part, to these forces as the powder and weaponry were expensive and difficult to manufacture, maintain, and operate. Over subsequent centuries, control of these arms by princes and then kings fueled the consolidation of power into what eventually became the modern Western nation-state.

Early Explosive Violence from Below

Hiccups occurred, however, along the way in this centralization of power. Among them was the Gunpowder Plot of 1605, an infamous attempt by Catholic conspirators to overturn the Protestant government of England by blowing up the Parliament building and killing King James I. Gunpowder bombs were difficult to ignite and had to be large to have effect. Guy Fawkes was part of a group of thirteen collaborators who dragged thirty-six 100-pound barrels of gunpowder into a cellar under the House of Lords, directly below where King James was to preside over the opening of Parliament. Regicide was hardly a novel concept, but the notion of sparking a revolution without mobilizing an army—by simply blowing up the leaders of the government—seized the imagination of the English public, despite the fact that the plot failed.

While not the leader of the group, Fawkes had military experience and knew how to acquire and ignite gunpowder, making him the linchpin of the plot. The English government theoretically had a monopoly on gunpowder production, and laws controlled it, but every soldier, militia member, and merchant vessel had it, and in the dozens of powder mills around London storage was lax. In addition, a recent peace treaty with Spain had led to a glut of stores, and as the powder's firepower deteriorated over time, sellers turned to a thriving black market, making ample amounts readily accessible to the plotters.[8] Due to a letter anonymously sent to a Member of Parliament revealing the plot, Fawkes was captured in the cellar in the morning of 5 November 1605, just a few hours before the explosion would have occurred. He was tortured, and then hanged, drawn, and quartered, as were his co-conspirators. Fawkes became a symbol of Catholic extremism and an icon of rebellion against state power, and the plot foreshadowed the rise of modern terrorism more than 250 years later.[9]

While gunpowder transformed warfare, it was not well suited to attacks by individuals or small groups. It was unwieldy, difficult to ignite, and of variable strength. Targeting explosions precisely was impractical.[10] A notorious case in point is the attempted assassination of Napoleon Bonaparte on Christmas Eve in 1800. A horse-drawn carriage laden with gunpowder-filled barrels blew up just after he passed it on the way to the opera. That gunpowder was of poor quality and blew up late. The explosion was powerful, however, wounding or killing some thirty-five or forty Parisian bystanders.[11] These included the fourteen-year-old Marianne Peusol, whose mother sold vegetables and baked goods nearby. The perpetrators paid Marianne to hold the bridle of the carthorse while they lit the fuse and then ran away. The attempt backfired. Poor Marianne's coldly calculated demise, and that of the unlucky waiters, bakers, and shopkeepers who died with her, pushed those on the fence about support of Napoleon further toward sympathy with him.

In subsequent decades, numerous inventors tried to make gunpowder less volatile and more accessible for widespread use, but the devices they designed were still bulky, heavy, and unreliable. One of these inventors was Italian nationalist Felice Orsini, whose "Orsini bomb" was a spherical grenade-like steel ball filled with four pounds of gunpowder, studded by a series of "horns," which were percussion caps containing fulminate of mercury. The device looked vaguely like a porcupine. Orsini used the bomb in a spectacular 1858 attempt to kill French Emperor Napoleon III in Paris. The target escaped; but the attack killed eight and wounded more than 140 other bystanders, and the widely publicized episode both heightened public interest in explosives and increased Napoleon III's popularity.

Gunpowder Helps Build the Modern World

Gunpowder was used more effectively in great infrastructure projects that furthered the consolidation of modern state power, especially in building canals and railways. First used in mines in 1627, gunpowder was for hundreds of years the only means apart from manual labor and horsepower of excavating solid rock.[12] It helped dig canals during the seventeenth to the nineteenth centuries, starting with the Languedoc Canal across southwest France, finished in 1681. By the nineteenth century, building railroads was a top priority in every industrializing country, and gunpowder facilitated faster construction. Building rail lines was painstaking work, especially in challenging terrain, and the ability to blast through rock not only sped up the process but provided gravel for ballast on the railroad beds and boulders for building retaining walls along tracks, as well as opening up mines for iron ore and other metals for spikes, bolts, rails, cars, steam engines, and bridges.

With rapid expansion of lines, railroads connected shipping lanes in vast transportation networks linking buyers and sellers initially across states and then whole continents—not to mention connecting armies and battlefields.[13] By 1850 almost 6,700 miles of track had been laid in Britain. The United States reached 9,000 miles the same year.[14] Ten years later, in 1860, that number had more than tripled, and at the start of the American Civil War the following year, the country boasted 30,000 miles of track, greater than the rest of the world combined, with most of it controlled by the North.[15]

Railroads were essential to the logistics of the war, the first large-scale conflict in which their value was showcased. They magnified the impact of the North's stronger industrial base, allowing faster troop movement as well as resupply, and greatly expanding the strategic and operational reach of the Union army. Indeed, a large number of battles were fought over control of railroads and railheads, notably the First and Second Battles of Bull Run and the Battle of Vicksburg. Sherman's March to Atlanta, hub of Confederate rail traffic in the South, was aimed largely at tearing up railroad tracks all along the way.[16]

Even before the war's end, construction of the Transcontinental Railroad commenced. Beginning in 1863, veteran officers from the US Army Corps of Engineers supervised the work, overseeing former soldiers, Irish immigrants, and Chinese laborers, who did virtually all of the work by hand, using pickaxes, hammers, and crowbars. Dirt and rock were carried away in baskets by hand and on mule carts. Irishmen were drawn mainly from the overcrowded cities of the Eastern seaboard, while Chinese laborers were brought in from the West

Coast by the Central Pacific railroad. They faced the grueling challenge of leveling beds spanning 1,912 miles, including through prodigious mountain ranges such as the Rocky Mountains and the Sierra Nevada. The pace of progress was excruciating and the human toll immense, with hundreds of workers dying in avalanches, rockslides, falls, explosions, and other accidents.[17]

Gunpowder was responsible for many of the deaths. It was hardly ideal for the job. As what's called a low explosive, gunpowder deflagrates rather than detonating, meaning that its explosion creates shockwaves that travel at only hundreds of meters per second (m/s) by comparison to many thousands of meters per second for high explosives. Remember, gunpowder does not have to shatter someone to harm him. It's a different story when digging mines or tunnels. With very little brisance, or shattering power, gunpowder couldn't pulverize a boulder on its own, and it was used as a propellant to push energy into rock to split it. Workers had to hand-drill boreholes into solid granite, for example, put the gunpowder in the hole, tamp it in to keep the energy from escaping—sometimes with clay or sand on top to concentrate the explosion—and light a fuse.[18] If the powder was not well tamped it would sometimes shoot back out of the boreholes at the workers. When it did work, it tended to break off huge chunks of rock that had to be broken up with a sledgehammer.[19] While much faster than hand-splitting, the excavation work moved forward less than a foot a day, and it could only proceed when it was not raining, snowing, or overly humid, as gunpowder was useless when wet. Hence the phrase "Keep your powder dry."[20]

Where gunpowder was more efficient was for explosions in small, enclosed spaces, such as a gun barrel. But even then it was filthy and produced an enormous cloud of smoke that wafted into soldiers' faces, making it impossible to see—leading to the phrase "fog of war"—not to mention disclosing soldiers' locations to the enemy. It also dirtied the bores of guns, so that they had to be cleaned frequently lest they clog or jam, and it was hard to store and fired irregularly.[21] The saltpeter that accounted for gunpowder's explosive power was also in short supply, so that all of the colonial powers suffered from shortages, which drove Dutch, English, French, Danish, Swedish, and Austrian competition for saltpeter in East India from the seventeenth to the nineteenth centuries.[22] For both construction and military purposes, a new explosive was badly needed.

Invented in 1866, dynamite was the first widely accessible, commoditized, inexpensive, and portable high explosive. Compared to gunpowder, dynamite was more concentrated, reliable, and powerful, and was very easy and safe to use. It was immediately embraced by the construction and mining industries.

Alfred Nobel's Vision

One of the primary uses of gunpowder was mining iron ore, for which there was a limitless hunger as industrialization sped forward. As with rail construction, it was a dangerous tool for the job, killing thousands of miners in badly controlled explosions or by poisoning them in its toxic gaseous clouds. Dynamite was designed primarily to replace it.

Alfred Nobel was an eccentric Swedish inventor, determined not only to solve the problem of mining deaths, but to save his family from financial ruin.[23] Born in 1833 to Immanuel Nobel and Andriette Ahlsell Nobel in Stockholm, Sweden, Alfred made an inauspicious start in life, arriving just as his father declared bankruptcy. His mother Andriette bore eight children, only three of whom survived beyond age twenty-one. Throughout Alfred's youth, the family barely eked out a living and Alfred often went hungry. This likely fueled his passion for wealth, but he was also driven by a love of invention inherited from his father. Alfred's father, Immanuel, went to work developing weapons for the tsar in 1837. Four years later, when Alfred was nine, Immanuel moved the family to St. Petersburg, where he designed the first effective sea mines, filled with gunpowder. The Russians used them during the Crimean War of 1853–1856 to keep the British at bay. When Alfred came of age, he took up study at the Technical Institute in St. Petersburg (Figure 3.2).

The third of four sons, Alfred was a pensive and sickly child. As an adult he complained of migraines, rheumatism, and poor digestion, and throughout his life he walked with a stoop. He called himself a "misanthrope" and shunned publicity. In his older years he declined to have a ship christened *Alfred Nobel* in his honor, remarking drily, "[I]t would seem like bad luck to christen her after an old wreck."[24] But he had a lively mind and excelled in academics, especially chemistry. Never marrying, he devoted his life entirely to his twin passions of invention and entrepreneurship. Receiving his first patent at age thirty, he made his greatest scientific and industrial contributions before reaching forty.

His particular fixation was with creating an explosive that was powerful and effective, yet practical and accessible for mining and construction workers' use. Like most great inventers, he did not start completely from scratch, but he managed to overcome obstacles that had stymied his predecessors. Those included Nikolai Zinin, a Russian professor of chemistry who was one of Nobel's teachers in St. Petersburg, as well as three prominent French chemists and another Swede. Nobel was also influenced by the work of Ascanio Sobrero, a professor at the University of Turin. In 1846 he had

FIGURE 3.2 Picture of young Alfred Nobel. Album/Alamy Stock Photo.

combined glycerin (used for soap), nitric acid, and sulfuric acid, creating a yellow oil he called "pyroglycerine," and later "nitroglycerine." But Sobrero missed the potential of the discovery, finding the mixture unstable and with too violent a blasting effect to be practical. Indeed, while he was conducting an experiment, a test tube exploded, badly scarring Sobrero's face. Fed up with the dangerous brew, the professor published the details of the work and abandoned it.

Despite much effort, no one could figure out how to stabilize nitroglycerine for use in explosives. In 1858 it began to be used instead as medicine for headaches and toothaches, eventually becoming an effective treatment for angina pectoris (heart pain).

Some fifteen years after Sobrero's efforts, Nobel took up the research and sought a way to release the explosive energy in nitroglycerine in a more controlled manner. Nobel had met Sobrero and learned of his nitroglycerine experiments in the late 1850s, when they both worked at the Paris laboratory of their mutual mentor, well-known chemist Thomas J Pelouze.[25] Unlike the professor, Nobel was willing or desperate enough to take enormous risks. The key was not to rely on a fuse, he realized, but to use a small amount of gunpowder to detonate the liquid—one explosive setting off another. He first succeeded with this method of detonation in 1860, and he patented "blasting oil" in 1863.[26]

Returning from St. Petersburg to Stockholm, with a cobbled-together combination of loans and small grants the Nobel family rented part of a run-down estate called Heleneborg, with a backyard shed that Alfred fashioned into a laboratory to manufacture nitroglycerine oil. The family was to learn tragically about the risks. In September 1864 an explosion at the laboratory killed Alfred's youngest brother, Emil, and five other people. For the remainder of his life, Alfred never spoke of the incident, and the trauma of this event may help explain why Alfred was a loner.[27] He was undeterred, however, in his experimentation, desperately needing money to support his parents.

Stockholm soon passed a law forbidding experiments with explosives within the city limits, so Alfred moved his laboratory onto a barge and did his experimenting in the middle of Lake Mälaren. Pressing forward with production, he sold "Nobel's Patented Explosive Oil" at 2.5 Swedish crowns (or about $0.50) per pound.[28]

Nitroglycerine is a high explosive that detonates with blast wave speeds of 7,000 or 8,000 m/s—thus, some twelve times faster than gunpowder. But it remained highly volatile and difficult to transport, especially if it got too hot or too cold. Even if it arrived safely, it was unreliable. Sometimes it did not explode at all, burning like black pitch, fizzling out. The oil also tended to ferment and leak when it was stored, becoming even more unstable, dangerous, and unpredictable over time.

The danger of nitroglycerine was well understood, but the need for a more effective explosive was so great that Nobel's shaky startup took off anyway. Over the next few years, he established nitroglycerine factories in Winter Bay, Sweden, and on the shores of the Elbe River near Hamburg, Germany. This German plant, the Krummel works, had to be rebuilt twice, first after an explosion in 1866 (of nitroglycerine) and another in 1870 (of dynamite).[29]

Nobel traveled all over the world to market and sell his blasting oil. Hearing of the power of the explosive—and equally horrified and tantalized by the publicized family tragedy—mining companies soon placed large orders. Many deadly accidents ensued. On one occasion, the Central Pacific Railroad ordered three crates of nitroglycerine, and one of the crates exploded, destroying the Wells Fargo office in San Francisco and killing fifteen people.[30] On another occasion, a German guest staying in a New York hotel asked the receptionist to store a crate for him and went off to conduct his business. The clerk put the box in a corner of the bar, where one of the customers kicked it over. Noticing the crate was starting to smell bad and emit reddish smoke, the clerk asked a porter to bring it outside. Seconds later it exploded, injuring nineteen people and ripping a hole a yard deep in the middle of Greenwich Street—all of which was luridly described in the next day's *New York Times*.[31] Not long after, a steamship carrying a large quantity of nitroglycerine exploded near Panama's Atlantic coast, killing forty-seven people.[32]

Nobel tried to assure the public that the oil was safe, engaging in highly publicized stunts such as pouring a small quantity of oil on a flat piece of iron and striking it with a hammer, lighting a small vial of cool oil with a match (it just fizzled), or passing packages of oil around for attendees to handle.[33] But the mounting death toll belied his assurances, and he became vilified in the press as a "devil in the guise of a man," a "traveling salesman in death," and a "mass murderer."[34] Country after country passed laws regulating or banning the transport of nitroglycerine, and Nobel returned to the laboratory.[35]

To avoid financial ruin, he had to find a way to stabilize nitroglycerine. Working alone, again on a barge, this time in the middle of the Elbe River, Nobel decided to mix it with something inert, to dilute and steady it. After months of trial and error, he thought of using kieselguhr, a silicon clay common in Germany. Combining three parts nitroglycerin with one part clay produced a malleable putty that could be shaped into rods that were stable. But they were hard to ignite. Even in direct contact with flame, the rods just burned. To solve this problem, Nobel invented a detonator, which consisted of a wooden plug filled with gunpowder that was lit by a fuse. He soon replaced the powder with a few milligrams of mercury fulminate, and he changed the plug to a copper capsule.[36] When ignited, this blasting cap, as he called it, triggered a shock wave of high pressure that reliably exploded the nitroglycerine putty.

The blasting cap was Nobel's greatest innovation. In tackling the ignition problem, he was the first person to think of using one explosive to detonate another—a stroke of genius. Indeed, many scientists believe that the invention of the blasting cap overshadowed all of Nobel's other inventions as it solved a wide range of problems in the evolution of explosives, with this detonation method later used in everything from rocket engines to atom bombs. But one problem with nitroglycerine still remained to be solved; the oil had a tendency to seep out of the clay over time. So Nobel wrapped the clay in wax paper, and in 1866, the classic sticks of dynamite, so familiar from their frequent use in Road Runner cartoons, were born.

Nobel's original mixture, which became known as "kieselguhr dynamite," or Dynamite No. 1, set the pattern for more than a hundred copycat mixtures combining nitroglycerine with various substances, including saltpeter, charcoal, chalk, sawdust, and even sugar, all generically known as "dynamite." Its explosive power greatly accelerated the building of major infrastructure, blasting efficiently through mountains, cutting holes in riverbeds to anchor bridge foundations, leveling land for laying railroad tracks, and crafting roads straight through rough terrain. Construction of the Lehigh Valley Railroad's Musconetcong tunnel, almost a mile long, which was the first major US project built using dynamite, begun in 1872, required some 17,000 pounds of dynamite per month.[37] The St. Gotthard rail tunnel through the Swiss Alps, which opened in 1882, was dug with dynamite.[38] The Brooklyn Bridge, built between 1870 and 1883, relied on dynamite to blast through the bedrock at the bottom of the East River, as did the Williamsburg Bridge and dozens of other vital bridge projects. Construction time for the Grand Tunnel project in Rio de Janeiro, Brazil, was cut from a projected seven years down to eleven months by using dynamite.[39] Over 60 million pounds of dynamite helped build the Panama Canal between 1904 and 1914.

Construction workers still had to exercise caution, especially if the dynamite was old, wet, frozen, or the chemicals had begun to "bleed" out from the wax casing. If stored, it had to be turned periodically to keep the oil from pooling. Accidents persisted. But it was much safer and easier to transport than pure nitroglycerine and improved the lives of masses of workers. It also facilitated rapid globalization at the end of the nineteenth century by opening up so many new travel and shipping routes, and catalyzed industrialization and its accompanying economic development in Europe, North America, and Latin America.

Dynamite Becomes the People's Weapon

The safety and explosive power of dynamite also riveted the attention of civilians wishing to perpetrate political violence. It is no coincidence that the first wave of modern terrorism followed in the wake of its invention. Sticks of dynamite could be bought cheaply and safely secreted in a coat or small bag. While dynamite was hard for laymen to make, it was easy to buy or steal. It captured the attention of people throughout the world because it was so accessible, cheap, easy to use, easy to transport, and easy to hide. It could be bought off the shelf and stuffed in a pocket. Even as other explosives overtook it in sophistication, power, and effectiveness, it only became more popular.

By placing enormous power in the hands of individuals, dynamite helped to destabilize the international order, shifting power from governments to popular movements revolting against sharp economic inequality and social discrimination, as well as the humiliations and disorientation of the massive changes brought about by industrialization, which left so many feeling left behind. A glacial pace of political reforms in response to public demands further stoked anger. In the years following the invention of dynamite, citizens of many stripes, from hardened political activists to lone-wolf discontents, took violent actions with it that further polarized domestic populations, inspired copycat attacks, and destabilized governments, whose reactions were often counterproductive and played a role in igniting the First World War. Armed with just a few sticks of dynamite, anyone could conduct a terrorist attack of then unprecedented power.

The global wave of terrorism that dynamite unleashed involved bombing incidents in fifty-two countries (Figure 3.3). The United States had by far the highest number (216), most of which (139) occurred between 1905 and 1920 (64 percent). In France, 145 bombings were documented, with almost half (47 percent) concentrated in the three years spanning 1892 to 1895. In Spain 153 documented bombings occurred, and in Italy 102. The Russian Empire, comprising Russia (99), as well as parts of modern-day Poland (46), Georgia (16), Ukraine (15), Latvia (8), Lithuania (3), and Belarus (2), saw a surge of attacks during the Revolution of 1905–1906 with 107 of 189 bombings occurring during this period.

Three groups of agitators were particularly influential in this first global wave of terrorism: Russian revolutionaries, Irish nationalists, and anarchists.

The Narodnaya Volya and the Killing of the Tsar

The most infamous early use of dynamite was the killing of Tsar Alexander II of Russia in 1881, fifteen years after Nobel invented the explosive. The members of Narodnaya Volya, or People's Will,[40] believed in replacing the monarchy with a more democratized governmental structure. The group's successful assassination of the tsar became a model for the unprecedentedly violent wave of anarchist terrorist attacks that followed.

People's Will was a group of young Russian intellectuals who were excited by the economic and political transformations sweeping through Europe due to industrialization, and the attendant progressive ideas brewing, yet frustrated by the slower pace of change in Russia. When Russia was defeated in the Crimean War (1853–1856) by an alliance of France, Britain, and the

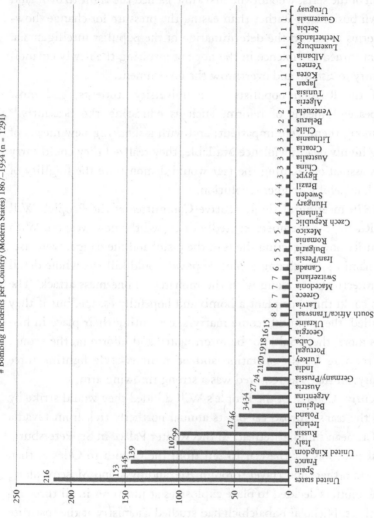

Bombing Incidents per Country (Modern States) 1867–1934 (n = 1291)

Country	Incidents
United States	216
Spain	153
France	145
United Kingdom	139
Italy	102
Russia	99
Poland	47
Ireland	46
Belgium	34
Argentina	34
Austria	27
Germany/Prussia	24
India	21
Turkey	20
Portugal	19
Cuba	18
Georgia	16
Ukraine	15
South Africa/Transvaal	8
Latvia	8
Greece	7
Macedonia	7
Switzerland	7
Canada	6
Iran/Persia	5
Bulgaria	5
Romania	5
Mexico	5
Czech Republic	5
Finland	4
Hungary	4
Sweden	4
Brazil	4
Egypt	3
China	3
Australia	3
Croatia	3
Lithuania	3
Chile	3
Belarus	2
Venezuela	2
Algeria	1
Tunisia	1
Japan	1
Korea	1
Yemen	1
Albania	1
Luxembourg	1
Netherlands	1
Serbia	1
Guatemala	1
Uruguay	1

FIGURE 3.3 Number of Bombing Incidents per Country (1867–1934)

This chart was created based on data we compiled for the purpose of analyzing patterns of anarchist and other terrorist group use of dynamite and other explosives for the period 1867–1934. A detailed explanation of the data and methodology may be found in Appendix B and downloaded from the forthcoming Power to the People (P2P) website at www.audreykurthcronin.com.

Ottoman Empire, the outcome seemed to confirm that the country was backward. Tsar Alexander II, later called the Tsar Liberator, instituted a wide range of reforms, including strengthening the court system, introducing compulsory military service for all classes (not just peasants), and granting limited self-governance in provincial assemblies called *zemstva* (nobility were appointed and peasants elected).[41] The most remarkable reform was the emancipation of the serfs, who for the first time gained the right to own land and their own businesses. Rather than easing the pressure for change, however, the reforms increased the determination of the populist intelligentsia. Some of them turned to violence in the hope of inspiring the newly emancipated peasantry to rise up and overthrow the government.[42]

Many of the Russian populists were university students, and most advocated peaceful means of reform, such as educating the peasantry.[43] But some among them were impatient, and with a shocking new means of perpetrating highly deadly violence available, they realized they could carry off a bold assassination. Killing the tsar would demonstrate the fragility of the state and, hopefully, trigger revolution.

At the 1879 meeting of the Executive Committee of the People's Will in Lipetsk, Russia, the members excitedly discussed the new weapon. With layers of security now protecting the tsar, the pistol and the dagger were useless. But dynamite's tantalizing explosive power could kill the whole detail of guards protecting him along with the monarch in one mass attack. The perpetrators could throw or plant a bomb and hopefully escape, but if they were also killed, they would become martyrs, cementing their place in history. What's more, they could not be interrogated and inform on the group. No special training for the operation and no military-style fighting force were necessary; all that was required was a strong throwing arm.

The Executive Council of the People's Will decided they would strike by blowing up the tsar's railroad car on its annual northerly trek from Livadia Palace, his Crimean summer retreat, to the Winter Palace in St. Petersburg. The original plan was for the tsar to sail from the Crimea to Odessa, then pick up the train there and travel through Alexandrovsk and Moscow along the way. The plotters decided to place explosives at junctures in all three intermediate cities. Nicholai Kibalchich had studied chemistry at the Institute of Transportation Engineers in St. Petersburg for two years and knew how to make crude explosive devices. Working secretly in a rented Odessan apartment, he built dynamite bombs, which could be remotely detonated by connecting two electric wires.[44] One plotter, Mikhail Frolenko, got a job as a guard on the Odessa railroad so he could place one bomb on the tracks there. But bad weather caused rough seas, so the tsar abandoned the Black Sea route and entrained at Simferopol, bypassing Odessa.[45] In Alexandrovsk, Andrei Zhelyabov bought land next to the railroad tracks and dug a tunnel in the frozen earth in the side of a ravine the train would travel along. He was in position with a bomb just as planned when the tsar's train passed above, but when Zhelyabov connected the wires, nothing happened. He had incorrectly wired the device.

The Moscow bomb was their last hope. The group had rented a house 150 meters from the Moscow-Kursk railroad and dug a tunnel from under the house to the tracks. While that work progressed, group member Grigory Goldenberg went to Odessa to retrieve the unused dynamite from that bomb. He stashed it in a suitcase and conspicuously lugged the heavy load through the railway station, refusing offers of help. A suspicious porter fingered him to the police, and he was caught at the next station.

Goldenberg was jailed with a cellmate named Kuritsin, who posed as a political prisoner but was actually a police informant and passed along all that Goldenberg told him about the plot. Under skillful interrogation, and having gained good background knowledge, an officer named Dobrinski convinced Goldenberg that the tsar wanted to negotiate, if only the terrorist activity would end, and that Goldenberg could be the dealmaker. Naively accepting assurances that his compatriots would not be harmed, Goldenberg revealed the names, addresses, and intimate details of 143 "noble members of the People's Will." Having gotten everything he needed, Dobrinski handed Goldenberg off to another policeman, who dropped the pretense. "Remember, not a hair on any of my friends' heads is to be touched," Goldenberg told him. "We don't bother about their hair," the new officer replied. "We want their necks."[46] Upon realizing that he'd been induced to betray the group, Goldenberg tore his prison towel into strips and hanged himself in his cell. The members of the Executive Council went on the run. The tsar's train had still not passed through Moscow, though, and the plotters waiting there persisted.

On 19 November 1879 two members remained at the house, one to signal that the tsar's car was passing and the other to detonate the bomb. The operation went off without a hitch. The villagers of Rogozhskaya Zastava were stunned by an enormous explosion, which tossed one railroad car into the air, derailing the others. Unfortunately for the bombers, their intelligence was faulty: the tsar's train had already passed. Their spectacular blast had blown up a baggage car full of Crimean oranges, producing, in the words of one government minister, nothing but marmalade.[47]

The People's Will wasn't finished yet, though. An expert carpenter named Stepan Khalturin who sympathized with the populist cause approached the group with a plan. He would plant a bomb under the dining room of the Winter Palace if they supplied the dynamite. They agreed, and Khalturin got a job as a peasant carpenter named "Batishkov," working first on the imperial yacht, and then at the palace. Gaining the guards' trust as a genial simpleton, Khalturin could come and go freely (especially through the back door) and soon learned the layout of the building.

Khalturin secreted the dynamite into the palace in small portions over the course of the next three months. Living in the basement carpenters' quarters, he at first slept with the dynamite under his pillow. But the noxious fumes forced him to shift it to a trunk ostensibly containing a dowry for his future bride. On 5 February 1880, having squirreled away a total of seventy pounds of dynamite, Khalturin lit a long fuse, giving himself fifteen minutes to escape. The explosion killed eleven sentries who lived in quarters between the basement and the second-floor dining room, but the dining room's granite floor held firm. The explosion would not have killed the tsar at any rate, as he was not in the room due to the late arrival of his dinner guest.[48]

A little over a year later, on 13 March 1881, Tsar Alexander II's luck ran out. The director of the operation that killed him was Sophia Perovskaya, the daughter of the former governor general of St. Petersburg.[49] Every Sunday Alexander visited the Mikhailovsky Manège (military riding academy) for roll call, traveling in a bulletproof carriage given to him by Napoleon III. The carriage was further protected by an entourage of Cossacks, soldiers on horseback, as well as foot soldiers and policemen. This impressive show of force proved defenseless against the bombers. Perovskaya sent four operatives armed with oval-shaped dynamite bombs.[50] One of them, Nicholas Rysakov, threw a bomb under the carriage, which exploded, injuring one of the Cossacks and a small boy, who lay on the road screaming in agony.[51] The police immediately captured Ryasakov.

The tsar left his carriage to console the injured and a second bomber, Ignati "Kotik" Grinevitski, attacked, throwing a bomb directly at the monarch's feet. The explosion shattered Alexander's legs and injured some twenty other people, including Grinevistski, who died later that day.[52] In a bizarre twist, the third operative, nineteen-year-old Ivan Emelianov, hastened to help Alexander into a sleigh, his unused bomb still clutched under his arm.[53] The fourth bomber, Timothy Mihailov, lost his nerve and fled. Alexander was hastily transported to the Winter Palace, where he bled to death an hour later.[54]

The Executive Committee of the People's Will comprised fewer than thirty members, yet the attacks orchestrated by this small cadre drove the Russian government into a panic.[55] In the aftermath of the tsar's killing, the imperial Russian police were granted expanded powers, including emergency provisions to forbid all social, public, and private gatherings. They crushed the People's Will, arresting 150 people, including virtually all of its known members, six of whom were executed, including Sophia Perovskaya.[56]

The assassination of the "Tsar Liberator" led to a halt of government reforms in the years before the October Revolution of 1905. The new tsar, Alexander's son Alexander III, was determined to, in his words "put an end to the lousy liberals," returning Russia to the brutal ways of his ancestors.[57] The government retreated toward autocracy, conservatism, xenophobia, and Russian nationalism, declining to make incremental changes that might have held off the rebellion. In coming decades no moderate middle developed between a growing cadre of increasingly ardent socialists and a backward government unwilling to modernize. Much later, the Bolsheviks took advantage of this governmental torpor, killing a much weaker tsar and his family, using a different and much broader campaign of terror to incite and then protect a national revolution.

Though the dream of the People's Will to spark a peasant uprising was crushed, group members were viewed as valiant martyrs by the anarchists who followed their lead. The members of People's Will were no anarchists; they aspired to build a new type of national government that would institute universal suffrage and political liberty and provide land for all the Russian people. But the image of the doomed bombers, nobly awaiting their executions, sacrificing themselves to serve the public, formed part of the mythology of the anarchist movement.[58]

Alfred Nobel was horrified that his invention had facilitated such potent attacks that struck at the heart of state power.[59] Suddenly, a formerly powerless small group could launch highly effective assaults that breached even the toughest security. The lesson was that high explosives used in well-orchestrated acts of violence targeted at key individuals could unhinge a state and divert the flow of history.

The Skirmishers and Clan na Gael

It's understandable that the early Russian terrorists get most of the attention in consideration of the emergence of modern terrorism—they killed a tsar and laid the groundwork for the Russian revolution. But extreme Irish republicans were brilliant tactical innovators and exploiters of dynamite, and their role in the crescendo of terrorism should not be overlooked.

Leading the way were the founders of the Fenian Brotherhood. They were mainly Irishmen exiled to the United States after an 1848 attempt to carry out an armed uprising against the British government failed.[60] The poorly armed Irish rebels had been easily defeated, but in exile their passion only grew. They opposed British supremacy in the Old Country and fiercely advocated Irish nationalism. Many were battle hardened from fighting in the US Civil War and learned innovative gunpowder bombing techniques from their service. Time bombs had been used on both sides, notably for the Confederate sabotage of General Grant's Union headquarters in 1864, where a "clockwork torpedo" was planted on an ammunition barge loaded with 30,000 artillery shells. The device was a box filled with 12 pounds of gunpowder and detonated by a timer. Its explosion killed 43 Union soldiers and injured about 126.[61]

Some Fenian Brotherhood members sailed back across the Atlantic after the war to lead an uprising in Ireland, joining with members of the Irish Revolutionary Brotherhood (as they were called in Ireland). In 1867 they launched a raid on the arms cache at Chester Castle in England, planning to transport the guns and ammunition to Ireland and spark a rebellion.[62] As they were again outgunned and badly organized, the plan failed. But they were unbowed. One of their members, Robert Burke, who had been working to procure more arms, was captured and held in Clerkenwell prison in London, and the others decided to blast him out with a barrel of gunpowder. The attack, on 13 December 1867, caused a massive explosion that killed twelve innocent people and injured at least 120, including children living in nearby houses.[63] Despite all of that damage, the operation failed miserably. Instead of tearing open the cells, the explosion destroyed the wall around the prisoners' exercise yard, and Burke was not freed.[64] It also fixed the attention of the English government on the Irish unrest, with Prime Minister Gladstone commenting that the effect of the Clerkenwell bombing was "to bring the Irish question within the range of practical politics."[65]

In the aftermath of the Clerkenwell bombing, six people were tried. Only Michael Barrett was convicted. Sentenced to death, Barrett was the last person publicly hanged in Britain, before a jeering crowd of thousands outside London's Newgate prison on 26 May 1868—three days before the Capital Punishment Amendment Act of 1868 ended the practice. It also emerged that the British police had been warned of the bombing in advance by the government in Dublin yet had failed to respond. "They know as little how to discharge duty in connection with Fenianism as I do about translating Hebrew," a member of the Dublin police force dryly remarked.[66] The government responded by increasing the number of special constables and establishing a secret services department at Scotland Yard, dedicated to the Fenian threat.

Meanwhile, the Fenian movement was thrown into disarray, split between moderates who sought political reforms, traditional military resisters, and extremists. The Irish Revolutionary Brotherhood were appalled by the civilian deaths, condemned the use of explosives at Clerkenwell, and even dropped the word "Revolutionary" from their name, becoming the Irish Republican Brotherhood (IRB). They cleaved to a military approach and decades later orchestrated the 1916 Easter Rising armed rebellion against the British. The minority of extremists rejected the idea of slowly building toward a general insurrection. By 1875, they had embraced the concept of small group violence "by a little band of heroes" instead.[67]

These extreme Irish nationalists applied their experience with explosives to make more effective dynamite bombs for their campaign of 1881–1885, perpetrating better executed attacks through a strategy referred to as "skirmishing." The successes and failures of that campaign drove technological innovation among terrorist groups for decades to come.

The British Home Office anticipated the violence. In late 1881, Chief Inspector of Explosives Colonel Vivian Majendie conducted a series of controlled experiments, blowing up various structures so as to assess the destructive capacity of explosives that "could be conveniently carried secretly by a single individual."[68] His report foreshadowed the increasingly sophisticated attacks that followed.

Two groups, the Clan na Gael and the Skirmishers, were remarkably pioneering in their use of dynamite, as well as improvised explosive devices, especially between 1881 and 1885.[69] Both groups remained closely connected to members of the Irish diaspora in the United States. Dynamite was a crucial inspiration for them, heralded by their US supporters. Some Irish Americans even used prominent elective positions to tout revolution through this new technological means. In 1883, John F. Finerty, the Irish American Congressman from Illinois, put it this way: "In this struggle, this vendetta, which England has now distinctly challenged, SCIENCE . . . must match itself against STRENGTH. . . . In this our battle for vengeance and for liberty, one skilled scientist is worth an army."[70]

The Irish Americans provided support for operations, including ample funding and weapons. Irish American Jeremiah O'Donovan Rossa even established the Brooklyn Dynamite School, to teach explosive manufacturing and raise money for the Irish nationalists from the tuition he charged. He openly bragged in the Irish American newspaper *The Irish World* about the skills of its graduates, including men such as John Francis Kearney and

Thomas J. Mooney, who were leaders in the dynamite campaign.[71] Another Irish nationalist magazine, published in New York at the same time, was called *The Dynamite Monthly*.[72]

The methods and materials the groups used evolved in sophistication as the campaign proceeded. The Skirmishers' first operation was actually conducted with a crude gunpowder explosive, but it incorporated an inventive time-delayed fuse for detonation. Delayed fuses became a trademark of these two groups and their successors. The first dynamite they used they made themselves, but they soon switched to commercial dynamite, which was much more powerful.

The combination of an explosive, a detonator, and a time delay unit became their signature.[73] Scotland Yard received warning of the importation of this new kind of device from an informant in the United States, and searches of ships from the United States were ordered. In June 1881, the manifest of the SS *Malta*, a steamer arriving in Liverpool from New York, showed it had a cargo of cement; but the hold also contained six casks filled with explosive devices, each one equipped with a clockwork timing device and an 11-ounce cartridge of dynamite.[74] A few days later, another ship arriving from New York, the *SS Bavaria*, was found to be transporting two more devices. Subsequent innovations for detonation included a brass tube filled with acid. Clan na Gael initially adopted the same brass tube setup, but then rigged up a clock linked to a pistol as a time-delay igniter.

Another distinguishing characteristic of the nationalists' campaign was disregard for civilian lives. Unlike the Russians, these extreme Irish nationals deliberately targeted the public. The Clan na Gael planted devices in enclosed public places, aware that explosions in contained settings maximized lethality. These attacks included detonating three devices in the London Underground and four more in railway terminals. The three bombs set off in the Underground were planted in or near cars full of passengers, and those detonated in the railway terminals each consisted of more than 20 pounds of commercial dynamite, meant to inflict mass loss of life.[75] The Clan na Gael also pioneered the idea of executing a sustained campaign of explosions in public venues, designed to maximize public anxiety and political effect.[76]

The most spectacular attacks of the Clan na Gael bombers were orchestrated by James Gilbert Cunningham, who sailed from New York to Liverpool on the steamship Adriatic on 10 December 1884. He brought with him a large brown trunk containing almost 60 pounds of Atlas Powder "A" (a specific type of lignine dynamite) plus traditional dynamite, fuses, and detonators.[77] His first attack, carried out with a bomb concealed in a workman's basket, was on the Hammersmith train out of Aldgate Station in London, on 2

January 1885. The attack injured a few people but caused little damage. He and his co-conspirators launched much more ambitious attacks on three icons of British power—the Tower of London, the Palace of Westminster, and the House of Commons—all on the same day, Saturday, 24 January 1885, which came to be known as "Dynamite Saturday."[78] They hid bombs under their clothing and surreptitiously deposited them and set the timers.[79]

This period, which became known as the Dynamite War, ended two years later, in the summer of 1887. More than 100 people had been killed or wounded—not a large number by today's standards but shocking at the time.[80] The dynamite campaign had been largely funded in the United States, but by 1885 American supporters had lost enthusiasm and were pushing instead for the broad armed insurrection favored by the IRB.[81] The IRB feared a loss of sanctuary in the United States, and still believed that, rather than forcing the British leadership to address the Irish Question, the bombings had dissipated resources, hurt the cause, and harmed the Irish working population in Great Britain.

It was also becoming more difficult to carry out the attacks. Changes in British intelligence-gathering and policing were crucial, especially instituting the widespread use of paid informers and dispatching fifty-four additional police officers to ports and continental entry points, which sharply reduced the importation of dynamite.[82] During Queen Victoria's July 1887 Golden Jubilee, for example, Scotland Yard's Special Branch foiled attempts by Clan na Gael to bomb both Houses of Parliament and Westminster Abbey.[83]

Meanwhile the IRB steadily focused on building toward the civil war they had begun to feel was inevitable. Moderates had repeatedly sought to pass Home Rule bills in the British Parliament, but the first two (1886 and 1893) were defeated. A third bill passed in 1914 but was never enacted, and a fourth in 1920 resulted in the partitioning of Ireland between North and South. The Irish War of Independence (1919–1921) led to the establishment of the Republic of Ireland in the South (1922), while Northern Ireland remained part of the United Kingdom. And as we discussed in chapter 2, Northern Irish Republican bombings later continued throughout much of the twentieth century, finally ending with the 1998 signing of the Good Friday Accords.

The International Anarchist Movement

Anarchism rejected all forms of centralized power. It was one stream of thought in a global argument over the nature of industrial capitalism and what kind of state should replace the old monarchial order. Widespread disillusionment with autocratic regimes, fueled in part by endemic famines, led to attempts at revolution in 1848 in France, Germany, Poland, Italy, the Netherlands, and the Hapsburg Empire, aiming to establish democratic republics with universal male suffrage and freedom of the press. They all failed and stoked a resurgence of autocratic power. Anarchism gained ground thereafter, as an alternative, more radical movement of protest.

The movement drew on a body of writing that criticized the very nature of government authority. Influential in its evolution was William Godwin's *Political Justice*, published in 1793, during the French Revolution.[84] Godwin argued (among other things) that governments come into being to control violence and injustice; but in their inevitable disposition toward war and despotism, the same governments end up perpetuating even greater forms of violence and injustice. The best way to achieve justice and equality, he concluded, is to replace fraught concepts like private property, laws, and punishment with individual standards of judgment, reason, and morality. In other words, government is not better than a state of anarchy. In subsequent decades, other philosophical anarchists wrote influential works, most notably Pierre-Joseph Proudhon, whose *What Is Property? An Inquiry into the Principle of Right and of Government* was published in 1840.[85] Proudhon was the first to declare, "I am an anarchist."

Anarchism was embraced by Russian nihilist revolutionaries, such as Peter Kropotkin, Sergey Nechayev, Mikhail Bakunin, Nicholas Mozorov, and Sergey Stepniak-Kravchinsky, starting in the 1860s. They called for the "creative destruction" of the old regimes and the triumph of "liberty" against "despotism," distributing their credos through pamphlets and leaflets.[86] While famous Prince Kropotkin, with his royal Russian blood and courtly manners, did not directly engage in the violent deeds he endorsed, and though most philosophical anarchists actually rejected violence, a host of devotees took up the call for attacks. The movement subsequently gained ground in Europe, the United States, and Latin America, embraced with particular enthusiasm by socialist workers' movements. Some attacks were attributed to anarchists without definitive evidence, or claimed by them retrospectively, which only added to the mystique of the movement, and eventually a multifarious range of agitators, from true anarchists to nihilists, nationalists, social revolutionaries, and populists, came to be loosely referred to as "anarchists," and that is what we will do here. A few avidly endorsed terrorism, arguing that it was the only way to destabilize the structure of the states and propel the masses into an inevitable violent revolution. The invention of dynamite in 1866 spurred Italian, French, German, and Russian anarchists to act more vigorously on this call for violence.

The number of casualties inflicted by attacks attributed to anarchists was small compared to the bloodshed perpetrated by terrorist groups today. Our careful review of historical newspaper accounts and other primary and secondary sources suggests that between 1867 (when dynamite was patented) and 1934, excluding casualties in Russia, where the movement was much more deadly, more than 1,500 people died from terrorist bombing attacks and more than 5,000 were injured, including bombers who accidentally blew themselves up. (See explanation of database and methodology in Appendix B.)[87] In Russia the violence was more intense. In the years from 1901 to 1916 an estimated 17,000 people were killed or wounded in incidents of insurrection. Historian Anna Geifman estimates that a majority of those victims were killed in anarchist attacks, with more than 2,000 murdered in the immediate aftermath of the 1905 Revolution alone.[88] And of course, the number of people killed in government crackdowns is unknown.

Though generally inflicting comparatively fewer casualties than today's terrorism, anarchist wave attacks punched above their weight. They triggered the destabilization of fragile regimes, as the philosophical anarchists had advocated, but by spreading fear and xenophobia, and spurring government repression, their actions ultimately led to outcomes opposed to those the philosophical anarchists had sought. The violence fueled the rise of extremist factions on both the Left and Right, contributing to the rise of both Bolshevism and nationalism, for example. Regimes struggled to find the right response to public fear and anger, and many overreacted and lost their bearings. Vigorous state crackdowns in France, Italy, and Spain drew still more people to violence, and deep domestic polarization emerged, as in the United States, where whole classes of immigrants were profiled as bomb-throwing revolutionaries. In Russia anarchists contributed directly to the success of the 1917 Revolution, by providing the Bolsheviks with a pretext for consolidating their power in a massive bloody crackdown. In short, the global political effects of anarchist terrorism vastly exceeded the body count from attacks. In a 1908 address before both houses of Congress, President Theodore Roosevelt declared, "[W]hen compared with the suppression of anarchy, every other question sinks into insignificance."[89]

The violence eventually touched every continent except Antarctica (see Figure 3.4). Egypt, China, Japan, and Australia all gave rise to significant anarchist groups, and anarchists punished by exile from Europe carried the movement to the new world, such as from the United Kingdom to the United

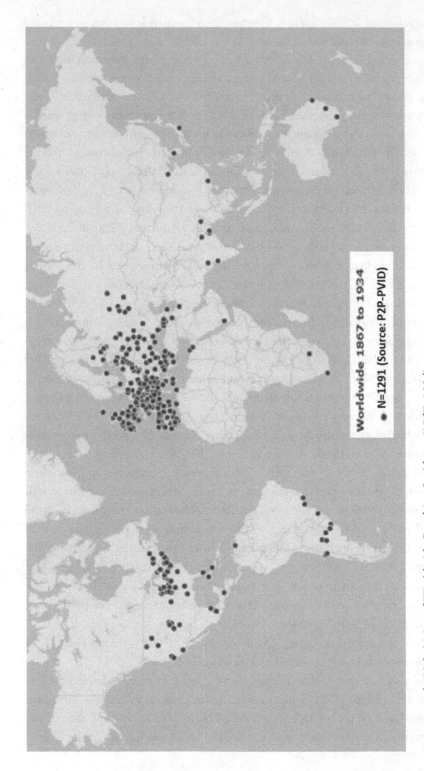

Figure 3.4 Global Map of Worldwide Bombing Incidents (1867–1934)

States and from Spain to Cuba. Even countries that experienced no violent incidents were seized by fascination about and fear of anarchists.

From the outset, the anarchist wave terrorists were best known for highly publicized plots against monarchs and senior officials, though they would soon begin indiscriminately killing civilians as well. Throughout 1878, they engaged in a string of assassination attempts. In January, Russian revolutionary Vera Zasulich shot and wounded the chief of the St. Petersburg police, setting off a wave of attempts throughout Russia, which spread across Europe. Operations against the Prussian kaiser Wilhelm I were launched in May and June of that year, and attempts were made against the Spanish king Alfonso XII in October and the Italian king Umberto I in November, all unsuccessful.[90] In Russia, following the 1881 assassination of the tsar, small decentralized groups calling themselves Younger People's Will and Terrorist Section of the People's Will carried out operations, sometimes using poisoned daggers and firearms, but bombs made with homemade dynamite were most popular, some with added pellets of strychnine for good measure.[91]

From early on, as with the Irish Nationalists, the anarchist terrorists targeted civilians in addition to official representatives of the state. In February and November 1878, alleged anarchists threw bombs at crowds in Florence and Pisa, for example. The historical evidence does not support a sharp distinction between terrorism and assassination, even for the anarchist movement, which was best known for the latter.[92] Verified anarchist killings from 1882 to 1884 included a shoemaker, a policeman, a moneychanger and his two young sons in Austria; a nun in France; and a pharmacist and a banker in Germany.[93]

Anarchism became more closely associated with anti-capitalist ideology and the fight against oppression of workers in the wake of the Chicago Haymarket massacre of 1886. The labor movement was gaining momentum at this time, protesting child labor, long working hours, and the lack of safety precautions in factories. During a peaceful rally in support of the eight-hour workday near Haymarket Square, which was heavily policed, someone threw a dynamite bomb that killed a policeman. The bomber has never been indisputably identified—historians are still arguing. The weapon was a homemade mine, with dynamite packed into a spherical casing of two brittle lead hemispheres, soldered together and ignited by a fuse.[94] Some policemen panicked and started shooting into the crowd, and some of the demonstrators shot back. In the end, three protestors died, with about three dozen injured. Seven policemen were killed and some sixty more were injured.[95] In the aftermath, eight prominent Chicago anarchists were swept up, tried, and convicted of killing police.[96]

The global press corps heavily covered the trial, especially passionate speeches from the accused that resonated with supporters of workers' movements throughout the world. Well beyond the United States, photographs of the condemned men became symbols of oppression and judicial murder, especially in Cuba, the United Kingdom, Mexico, Spain, and Italy. On 11 November 1887, four of the eight accused were hanged wearing white shrouds, and they became known as the Chicago martyrs.[97] Thereafter the labor movement became more closely linked to anarchism, with bombings and other violence occurring at many strikes and labor disputes, sometimes instigated by anarchists, at other times by police spies. Indeed, the use of *agents provocateurs* to encourage anarchists to commit crimes as a pretext for a police crackdown was well known in autocratic regimes like Russia, Germany, and Austria, but it also happened in Britain, France, Italy, and the United States.[98] Sometimes provocations crossed borders and masked state involvement: for example, German spies who infiltrated among German-speaking anarchists in Chicago between 1880 and 1904 tried unsuccessfully to persuade them to kill American leaders.[99]

As the anarchist wave gained momentum, attacks on heads of state also intensified, concentrated in major urban areas such as Paris, Madrid, Lisbon, and Rome. From 1892 to 1901, anarchists carried out an unprecedented number of assassinations of presidents, prime ministers, and monarchs. These included the killings of French president Marie François Sadi Carnot in 1894, Spanish prime minister Antonio Canovas del Castillo in 1897, the Empress Elisabeth of Austria in 1898, Italian king Umberto I of Italy in 1900, and US president William McKinley in 1901.[100] No other terrorist movement has killed as many state rulers.[101]

State countermeasures were initially largely ineffectual. When security details were beefed up in response, making high-profile targets harder to hit, dynamiters focused more on attacks on soft targets at large public gatherings, hurling bombs into restaurants, opera houses, and religious processions. Between 1880 and 1914, anti-anarchist legislation passed in virtually every European country, instituting prohibitions on the use of explosives, penalizing incitement to commit anarchist attacks, and prohibiting membership in an anarchist association. In the United States, the states of New Jersey, New York, and Wisconsin adopted laws prohibiting support for "criminal anarchy," and two federal laws passed in 1903 and 1907 banned anarchists from immigrating and deported any who did.[102] Yet these measures also proved largely ineffective in cracking down on attacks, which further stoked public fear.

Assassination attempts continued. In Spain, three serious attempts were made on King Alfonso (1905, 1906, and 1913). All spared the monarch but killed or mutilated dozens of bystanders. The 1906 bomb, concealed in a bouquet of flowers, was tossed down from a balcony toward the king and his new bride, Victoria Eugenie, as they drove through the streets of Madrid in royal procession after their wedding. Between 23 and 33 soldiers and innocent bystanders died, and more than 100 were injured. The queen's bloodstained wedding gown was later deposited in the church of Almadena, standing opposite the explosion.[103]

Attacks in Spain became particularly frequent, with so many explosions in Barcelona between 1905 and 1908 that it became known as the "City of Bombs." A 1909 labor strike and anti-draft riot in the city led to five days of street fighting and then heavy-handed repression, which saw the slaughtering of 200 people, as well as torturing of many prisoners and the execution of 5 of them, including the well-known anarchist Francisco Ferrer. In 1912, following a vicious crackdown on the anarchist press, Spanish prime minister Canalejas was assassinated.

The wave of violence continued to spread, with attacks throughout continental Europe, the United Kingdom, and both North and Latin America. This growing violence was spurred in part by Russian anarchists and revolutionaries who fled the country in the wake of the 1905 Russian Revolution. Sweden experienced its first terrorist attacks in 1907, Argentina faced a wave of anarchist explosions and assassinations between 1905 and 1910, and then again in 1920–1928. Nationalists from India, labeled "anarchists" by the British, launched numerous attacks in London and on the subcontinent starting in 1908,[104] and a string of attacks occurred in China between 1905 and 1912.[105]

Anarchist attacks ebbed during World War I, which erupted due to a cascade of events triggered by a terrorist assassination, though not one perpetrated by avowed anarchists. The origins of the First World War are complex, the subject of mammoth histories and lively debates that will not be engaged here. But the fact that the assassination of Austria-Hungary's Archduke Franz Ferdinand was the trigger is not in dispute. It was carried out by a team of three young Bosnians—Gavril Princip, Nedeljko Cabrinovic, and Trifko Gabrez—who, though not identifying themselves as anarchists, may have been inspired in part by anarchist writings they are known to have read, such as Peter Kropotkin's *Universal Wealth* and Mikhail Bakunin's *Statism and Anarchy*.[106] The assassins' motivations were primarily nationalist, however, and they had a murky connection to the Black Hand, a nationalist group consisting mainly of Serbian Army officers who opposed Austrian-Hungarian rule and apparently thought the assassination of heir-apparent Franz Ferdinand (who was actually trying to pacify the Serbs) might disrupt the Empire's hold on power. They armed the boys with four revolvers and six small 2.5-pound bombs, from the Serbian State Arsenal at Kragujevac.[107] No evidence was ever found, however, that the Serbian government was aware of or directly involved in the plot. Gavril Princip evidently believed that killing Franz Ferdinand would spark a popular uprising, leading to the independence and unification of southern European Slavs. The events he triggered did disrupt the Empire's hold on power, but not with the results he had anticipated.

Many anarchists thought that the war did, though, offer an opportunity to catalyze widespread opposition to centralized state authority, as the regimes were responsible for the deaths of millions in the trenches. With the eruption of the 1917 Bolshevik Revolution, Russia was hailed as the promised land, but the anarchist fight there also did not go to plan. It was led by Nestor I. Makhno, a native Ukrainian who was released from prison in 1918, having served nine years for murdering a policeman. Returning to Ukraine, he began to build a guerrilla army. Between 1918 and 1921, Makhno gathered more than 80,000 into the Revolutionary Insurrectionary Army, fighting in the countryside under the black anarchist flag. As in the Spanish civil war fifteen years later, however, anarchism's ideology of opposing all forms of centralized authority made organizing anarchists into a coherent military force a challenge: there were no officers, commanders were elected, troops were to rely on individual self-control, and justice was dispensed by consensus. Although Makhno was a brilliant guerrilla leader, in the chaos of the Russian civil war, his ragtag army, after some difficulties, was easily overmatched by the treachery of the Bolsheviks and the size of their forces. Following a temporary truce so that Makhno's Black Army and the Red Army could combine forces to defeat the White Army in the south, the larger and better-equipped Red Army destroyed the Makhnovist forces. Makhno fled and died as a miserable, bitter, forgotten man exiled in Paris.[108]

In the cities, Leon Trotsky, Vladimir Lenin, and Josef Stalin used the newly established Cheka (counterrevolutionary secret police) to crush all political competitors, including anarchists, arresting, imprisoning or executing hundreds of them in Moscow, Petrograd, Kharkov, and Odessa.[109] Lenin had witnessed his older brother's execution for his part in a failed assassination attempt against Tsar Alexander III, and, twenty years later, he was determined to avenge his brother and make the Bolsheviks a more effective, ideologically pure, and brutal, anti-tsarist force.[110] Becoming disillusioned, underground anarchists based in Moscow declared war against the Bolshevik government, which they considered a new form of autocracy. On 25 September 1919, in an effort to disrupt the concentration of power, the anarchists threw a dynamite bomb into the headquarters of the Moscow Committee of the Communist Party during a plenary session. The explosion killed twelve committee members and wounded fifty-five others, including the brilliant theoretician Nikolai Bukharin.[111] But instead of turning the tide on the Bolsheviks, the attack strengthened their hand. They boldly branded all anarchists criminals, rounding up and executing anyone, anarchist or not, who challenged their right to rule.

For the remainder of the 1920s, a brutal campaign to eliminate every trace of anarchism from Soviet Russia unfolded. Many prominent anarchists who had initially supported the Bolsheviks fled Russia and wrote damning exposés from the Left about the new Soviet Union, including Emma Goldman (*My Disillusionment with Russia*, 1923) and Alexander Berkman (*The Bolshevik Myth*, 1925).[112] The anarchist flag appeared in Russia for the last time in February 1921 at Peter Kropotkin's funeral. His passionate writings about collectivism, the evils of the state, the role of cooperation, and the inherent goodness of the individual had helped launch the movement fifty years earlier, and terrorist bombings and assassinations had helped destabilize the tsarist state, as he had advocated. But they had also paved the path to power for the even more brutal Soviet state that took its place.[113]

In the aftermath of the Russian Revolution, terrorist attacks surged again in capitals throughout the world.[114] These included the 1920 attack on Wall Street (40 killed and 200+ wounded), mentioned at the outset of this chapter, a March 1921 bombing of the Diana Theatre in Milan (31 killed and 100 wounded), and a June 1921 bomb explosion in Belgrade that injured 10 people.[115] In Spain, which had remained neutral throughout the First World War, the Primo de Rivera dictatorship squelched anarchist activities for a while (1923–1930), but they resurged in the period leading up to the Spanish Civil War (1936–1939), ending when the brutal Fascist government that took power in the war crushed all domestic opposition.

The other major area of anarchist activity after the First World War was the United States, where violence increased markedly. In late April 1919, Italian anarchists in the United States sent thirty package bombs to senior officials including the attorney general, postmaster general, judges, mayors, police officials, businessmen, and members of Congress. Each contained a number of sticks of dynamite surrounded by metal slugs. Two months later, on 2 June 1919, the same group of perpetrators set off bombs less than two hours apart in seven US cities: Boston (two bombings); New York;

Paterson, New Jersey; Philadelphia; Pittsburgh (two bombings); Cleveland; and Washington, DC. One of the targets in Washington was the home of Attorney General A. Mitchell Palmer. The whole front of the house was sheared off.[116] The devices used represented a step up in explosive capacity, with at least 20 pounds of dynamite in each.[117]

The coordinated June bombings triggered public fear that the government would be overthrown, setting off the first major "Red Scare" in the United States, fanned by officials who warned that a campaign had been launched to start "a reign of terror."[118] Communism and anarchism merged in the public mind as a single radical threat, and one closely associated with dynamite. The press called the plotters "dynamitards."

Congressmen demanded action. Seattle mayor Ole Hanson announced, "I trust Washington will buck up and . . . hang or incarcerate for life all the anarchists. If the Government doesn't clear them up I will."[119] On 7 November 1919, Attorney General Palmer ordered a series of deportation raids in twelve cities, known as the Palmer Raids, beginning with the Union of Russian Workers, an anarchist political organization headquartered in New York, composed mainly of Russian immigrants. Raids in thirty-three American cities and towns in twenty-three states followed in the next few months, with thousands of immigrants arrested and held in sometimes deplorable conditions for months. Hundreds were deported with the enthusiastic support of the public and cheered by newspapers. One headline read, "ALL ABOARD FOR THE NEXT SOVIET ARK."[120]

In May 1920, Nicola Sacco and Bartolomeo Vanzetti, both Italian immigrants and avowed anarchists, were arrested for robbing and killing a shoe company paymaster in South Braintree, Massachusetts. Though the evidence against them was weak, in the highly charged political atmosphere they were both convicted and sentenced to death. Their fate became a cause célèbre for many prominent people in the United States and internationally, including H. G. Wells, George Bernard Shaw, and Albert Einstein, who all pressed for a review of the case. Despite this, on 22 August 1927, Sacco and Vanzetti were executed, a fate now widely believed to be due to their alien status and their political views rather than their guilt.[121]

Bombings continued in the United States until the 1930s, with a last pulse of twenty-four incidents around New York and Chicago between 1927 and 1932. With the stock market crash of 1929 and the Great Depression, the US unemployment rate shot up, escalating from 3.2 percent that year to 24.9 percent in 1933. Many violent clashes between unemployed workers, strikers, and both public and private police forces ensued.[122] In May 1934, Senator Robert M. La Follette of Wisconsin warned of "this impending crisis . . . which will bring about open industrial warfare in the United States."[123] The end of dynamitings in the United States coincided with the New Deal policies of the Franklin D. Roosevelt administration, especially the passage of National Labor Relations Act of 1935, which for the first time guaranteed employees' rights to form or join unions.

Why Dynamite Diffused

Without dynamite, this wave of violence might not have built into a globally disruptive force. Given how horrifying the attacks were, and how much was at stake for the ruling regimes, a crucial question is why dynamite was so accessible and cheap; why didn't governments clamp down on the ability of violent actors to obtain it? As we seek lessons for today from the stories of these leading movements that propelled the violence, we must dig more deeply into why the use of dynamite diffused so rapidly and widely, and why it remained the weapon of choice, despite superior explosives becoming available, such as ballistite and guncotton. As we'll see in the next chapter, dynamite arrived during a period of explosive technological innovation, which was also characterized by mounting social discontent, a period that was quite similar, in those regards, to the present era. Understanding why dynamite had such devastating appeal and was accessible requires delving into a set of key economic, cultural, informational, and political factors that shaped that roiling era and aided its diffusion. The insights gleaned will help us understand whether and how new technological innovations might be similarly harnessed for lethal purposes.

CHAPTER 4 | How Dynamite Diffused

DYNAMITE WAS A commercial product that appeared during an era of innovation and optimism in the private sector. By making possible deeper mines, new infrastructure, and shorter transportation routes, it formed the sinews of the world we know today. Nobel's creation arrived at a time of enthusiasm for science and invention when it was widely believed that almost anyone could invent a new product and become rich; but it also had a darker side, which might easily have been abated if its handling had been tempered with a modicum of common sense.

Looking back at the wave of violence perpetrated with dynamite, hindsight tells us that its sale should have been controlled. How could such a lethal explosive be so widely accessible to the public? Outside of Europe (where it was more likely to be smuggled or stolen), dynamite was very easy for people to purchase, especially in the United States, where it could be bought in hardware stores, sporting goods stores, and even ordered from catalogues. For example, the Watkins-Cottrell company of Richmond, Virginia, advertised "Cutlery, Guns, Sporting Goods, Railroad Supplies, also Dynamite and Fuse," in its 1900 catalog.[1] Explosives were marketed directly to farmers, who learned about them at trade booths and local country fairs. Not everyone thought this was rational. In 1885, one M. Bennett remarked in the *Columbian* newspaper of St. Helens Oregon (US), "The sale of opium and poisons are restricted, but dynamite, the greatest and most terrible destructive engine of the 19th century, may be bought by anyone at 36 cents per pound."[2]

Why, after bombings had begun, wasn't rapid action taken to get dynamite out of the hands of the general public? An important part of the answer is that the invention of dynamite was one breakthrough in a great flourishing of innovation in the commercial sector in the latter third of the nineteenth century, much of which was conducted by tinkering amateurs, working in the equivalent of Silicon Valley's legendary garages. During periods of rapid technological innovation in the commercial sector, with clusters of new technologies emerging, the power of nontechnical enthusiasts to take part in the invention, and in the development of new uses for the emerging technologies, is greatly enhanced.

An Era of Open Technological Innovation

The United States and Europe were aflame with excitement, and widespread optimism, about the string of multi-use technological innovations emerging during the late nineteenth century. Almost anyone who wanted them could get easy access to both information and the materials with which to experiment with all sorts of devices and chemicals. A remarkable range of innovations emerged from the work of prosumers, tinkerers, and hobbyists. In 1884, George Eastman was tinkering at home when he developed photography film in roll form. Italian electrician Guglielmo Marconi invented the radio in the late 1890s using homemade equipment in the attic of his home. Orville Wright, a bicycle manufacturer, along with his brother Wilbur, made the first powered, sustained, and controlled flight in 1903, flying a craft that was designed and made at their home in Dayton, Ohio.[3]

For much of the nineteenth century, there was no clear dividing line between amateur and professional scientific communities, with authors, editors, and readers seeing themselves as co-participants in a great scientific venture—much as do those participating today in crowd-sourced data experiments, do-it-yourself biology labs, and open-sourced AI advances, as well as contributing to and reading open-source journals and popular science blogs.[4] Periodicals explaining science in nontechnical terms started to appear in force to a newly literate public in the mid-nineteenth century.[5] Just as today members of the public are free to purchase drones, build robots, or buy primitive gene-splicing kits, and can download instructions for 3D printing firearms or targeting individuals with facial recognition tools, during this period, the public could get instructions about how to use all sorts of new technologies from the new magazines. In Britain, these included *Popular Science Review* (founded in 1862), *Quarterly Journal of Science* (founded in 1864), and *Nature* (founded in 1869 with the mission "to place before the public the grand results of Scientific Work and Scientific discovery").[6] In France, the *Revue Générale des Sciences Pures et Appliquées* was founded in 1890 for the nonspecialist reader. In the United States, the widely circulated magazine *Scientific American* (founded 1845) closely followed the activities of the US Patent Office and dedicated itself to serving inventors of all classes.[7] It was joined by *The Popular Science Monthly* (now known as *Popular Science*), which was founded in 1872 to serve the interests of the dedicated layman.

The invention of dynamite was widely covered. In 1868, *Scientific American* reported on it by detailing in nontechnical language the paper given by Nobel before the British Association for the Advancement of Science's annual meeting, in Norwich, England.[8] Two years later, in 1874, *Popular Science Monthly* published an article carefully explaining, in nontechnical terms, the actions and properties of both nitroglycerine and dynamite.[9] Lots of people experimented with explosives in their homes, bought dynamite through local stores, or ordered articles about "infernal machines" from the *Scientific American Catalogue*.

Not only scientists and engineers but hobbyists experimented with dynamite uses of all sorts, from the practical to the bizarre. Beginning in Scotland and Austria, and followed by the Americans and the Australians, landowners tried plowing, clearing land, and dislodging tree stumps with dynamite.[10] In 1877, a farmer named Thomas Johnson of Dudley, England, sought to slaughter his cattle quickly by detonating a small amount of dynamite on their foreheads.[11] A few years later, fishermen in Key West, Florida, tried using dynamite to fish in deep water: when the stunned fish surfaced, they scooped them up.[12] (This practice, called blast fishing, continues today in places like Myanmar and Tanzania, damaging reefs and harming sea life, despite longstanding efforts to stamp it out.)[13] A businessman in Ceylon used dynamite to keep sharks away from his pearl divers.[14] In dry regions of both Texas and Southern New Zealand, locals tested whether a dynamite explosion could unsettle the atmosphere and make it rain. When it then rained spontaneously in New Zealand, some observers believed the dynamite had worked.[15]

Popular publications actively encouraged experimenting with dynamite, reporting with enthusiasm and optimism on the efforts of amateur tinkerers. Edited by Columbia University chemistry professor Charles Joy, *Scientific American* carried an easy-to-understand regular feature called "New Inventions, Scientific Principles, and Curious Work," with articles such as "The Dynamite Gun," "Novel Applications of Dynamite," and "Experiments with Explosives."[16] The periodicals also gave specific practical advice about dynamite's pitfalls and possibilities, how to thaw, handle, and use the explosive, and such instructions were globally dispersed, appearing in the *Indian Engineer* newspaper of Calcutta, India, in 1888, for example.[17] At the end of an account of how dynamite had freed two ships run aground on the rocks near Whitby, England, *Scientific American* declared, "So powerful and effective an agent should be better understood and applied to more purposes than it is."[18]

Odd as a public appeal to experiment with dynamite may seem today, for the most part, such easy-to-understand, detailed information about weapons and explosives continues to be freely and legally available in most democracies, although access to commercially sold explosives themselves is usually controlled.[19]

Innovation Was Not Driven by the Military

Underscoring the fact that dynamite was seen as an exciting new invention, rather than as a dangerous tool to be cautiously used or controlled, is that innovation with dynamite as a weapon was driven not by military organizations, but from below. Terrorists leapt ahead of military tacticians in creating sophisticated small explosive devices that used dynamite.[20]

As we saw in the last chapter, gunpowder-filled military grenades had not become a major element of military combat. In the US Civil War, while both armies used grenades, they were ad hoc and ineffective. In the North, Ketchum grenades, which were patented by William Ketchum in 1861, were shaped like big darts, or iron mangoes sporting four fins, with cylindrical plungers at the nose designed to jab into the tamped gunpowder within. When thrown in an arc, these explosive darts had to hit something hard, square on the nose where the percussion cap was, if they were to explode. That was almost impossible to accomplish, especially in the midst of a battle, so they rarely worked.[21] Plus there were easy countermeasures. Confederate soldiers stretched out sheets of fabric over trenches and earthworks so the Ketchum grenades would have a soft landing, fail to explode, and could then be thrown back even harder.[22]

Still, as often happens in war, innovation spurred reciprocal innovation. Copying the Ketchum grenade, the Confederacy developed the Raines, which was named after its inventor, Confederate general Gabriel Raines, head of the Torpedo Bureau. The Raines grenade had a long tail streamer, making its flight a little more stable over further distances, but it was no more effective. Grenades didn't reappear in a significant way in the military repertoire until the 1904–1905 Russo-Japanese War, and these were still crude devices made by stuffing metal cans with guncotton and gunpowder, lit by a simple fuse with a match.[23] The invention of dynamite did not trigger a burst of grenade innovation; the focus of the military was elsewhere.

The military did experiment with dynamite, but it did so mostly with new forms of guns. The military priority was on long-range artillery and dynamite was found to be too shattering and unstable for that purpose. Between 1883 and 1900, the US Army and Navy experimented with pneumatic guns that used compressed air (rather than gunpowder or another explosive) to shoot a shell containing dynamite.[24] The shells kept bursting sideways: in one test of a massive 12-ton gun, for example, the barrel shattered into huge fragments, nearly killing onlookers.[25] In 1888, the Navy launched a cruiser called the USS *Vesuvius*, known as a "dynamite gun cruiser," equipped with three 15-inch pneumatic guns. Despite the name it never actually used dynamite, ultimately being kitted with more stable guncotton shells instead.[26]

The First World War naturally pushed the evolution of all military explosives along, but the emphasis was on big, devastating artillery attacks that used ballistite (which Nobel invented specifically for this purpose in 1887), gunpowder, guncotton, white powder, cordite, and other explosives that could launch or be launched long distances and kill in large numbers. By contrast, military hand-held bombs were ad hoc and primitive, first used in large numbers by all sides to flush the enemy out of the trenches. For example, the British developed a device with a 16-inch throwing handle, which was very heavy and awkward to use, and the Germans innovated a similar design.[27] The Australians contributed the cruder "Jam bombs," which were literally made from jam cans filled with dynamite or gun cotton. The first reliable, mass-produced infantry grenade was the Mills bomb, invented by Englishman William Mills and used in battle in 1915. The Mills bomb was filled with amatol (a mixture of ammonium nitrate and TNT) or ammonal (three parts ammonium nitrate to one part aluminum) and either detonated on impact or with a time-delayed fuse. Its classic pineapple shaped grooves remained standard for the next fifty-five years, until 1970.

The remarkable ingenuity that terrorists brought to their innovation of devices using dynamite is in striking contrast. For them, dynamite was plenty stable, and its explosive power was plenty lethal. They could also obtain it just as easily as anyone else in the public could. Due to how much more effective dynamite was in mining and construction, the market for it was highly profitable. Dynamite was very big business, and that was a major factor in how rapidly and widely it diffused. In fact, the dynamite industry grew faster than any other business had before.[28]

The Global Production of Dynamite

Alfred Nobel had shown inventive genius in creating dynamite, but especially in the early years, his business genius was crucial to its global diffusion. Although the recipe for making safe dynamite had been hard to discover and perfect, it was very easily copied. The competition forced Nobel to keep the price of his dynamite lower than it would have been if he had more exclusively dominated the market, and he tried to gain control in several ways, including patents, rapid expansion, and local partners, but also taking advantage of war.

Obtaining patents was a tactic that he deployed with great acumen. Their acquisition, sale, and control was a linchpin of the global growth of his business empire. He received his first patent on nitroglycerine blasting oil in 1863. Following the invention of dynamite, he quickly secured patents in Sweden (1867), Great Britain (1867), Germany (1868), and the United States (1868).

But while patents helped him hold off some competition in the early years, they were a double-edged sword, as he was in constant litigation to protect them. The mighty struggle he engaged in is a testament to what a juggernaut the growth of the global dynamite business was, and how supportive of its rapid growth governments generally were; but not all allowed unbridled competition. In Britain, for example, a man named E. Jones of Caerphilly (in South Wales) claimed that he should be allowed to produce something he called "Dyna-magnite," which mixed nitroglycerine with magnesia alba (white mineral crystals from magnesium), and he received permission to manufacture it in the United Kingdom in 1879. Nobel fought him, and the House of Lords came down on Nobel's side, deciding that his patent in Britain included "all explosives of the dynamite class."[29] This important finding contributed to the dramatic success of his Scottish plant. But the same explosive later showed up in the United States, where Nobel's patents were virtually ignored, under the name Hercules Powder, and he had no success at all defending his patents in Germany during the 1870s.[30]

The British patent case was a rare victory for Nobel with respect to dynamite. As word about dynamite spread, dozens of knockoff products that mixed nitroglycerine with other stabilizers quickly emerged. Dynamite with kieselguhr clay became known as Dynamite No. 1, with 75 parts (by weight) of nitroglycerine mixed with 25 parts kieselguhr. Because the clay had a little iron in it, the putty was tinged with red. This dynamite was also informally referred to as "guhr dynamite." In 1887, the home secretary of Britain allowed a range of other stabilizers to substitute for kieselguhr and still be valid as Nobel's patented product, including carbonate of sodium, sulphate of barium, mica, talc, and ochre. Other forms of Dynamite No. 1 included Dynamite blanche de Paulilles (mixed with white siliceous earth), Dynamite de Boghead (mixed with ashes of Boghead coal), and E.C. or S1 Dynamite (mixed with carbonate of soda).[31]

The key to the explosive power of all forms of dynamite is how much nitroglycerine there is compared to the various fillers, bases, or stabilizers. Dynamite No. 2 was slower and milder than No. 1, for example, because the amount of nitroglycerine was decreased. Dynamite No. 2 was to be used in places where the huge shattering effect of No. 1 was too much, such as in coal mines or slate and granite quarries. No. 2 was black, and it had only 18 parts by weight of nitroglycerine, mixed with 82 parts of potassium nitrate, paraffin, and charcoal—the reason for the black color. Dynamite au Charbon (mixed with nitrate of baryta, charcoal, and resin), Dynamite grises de Paulilles (mixed with nitrate of soda, resin, and charcoal), and Dynamite de Krummel (could have either 48 to 50 parts nitroglycerine, or 30 to 35 parts), were all Dynamite No. 2 preparations produced in Nobel's factories. Sometimes the same brand could be a No. 1 or 2, depending on how it was mixed: Dynamites des Vonges could have from 30 to 90 percent nitroglycerine, for example. All of these were called "dynamite."

When manufacturers wanted to avoid awkward questions about Nobel's patents, they just gave the product another name: Jupiter Powder, Vulcan Powder, Neptune Powder, and Lithofracteur were all versions of No. 2 Dynamite. Lignin-dynamite was the generic name for mixtures of dynamite that had sawdust or wood-pulp, also known by chemists as Dynamite No. 0 (75 percent nitro mixed with woodmeal). In the United States, there was Hercules Powder, Aetna Powder, and Extra Hercules Powder, among other dynamites. Indeed there were as many as 125 different dynamite varieties, according to Lt. Col. J. P. Cundhill, a British explosives inspector who compiled all of them into a *Dictionary of Explosives* in 1889.[32]

When he couldn't win in court, or his patents were unenforceable, Nobel tried to buy up other companies or make a deal with them to share the market. The fear of losing control of dynamite to locals in other countries was a key factor driving Nobel to build companies so that he could get there first, try to validate the "uniqueness" of his patents, preempt rivals, and dominate the global market.[33]

There was a particularly vicious fight over patents in Australia in the 1880s. The demand for dynamite to dig gold mines in Victoria soared because Australian mines went very deep, requiring a lot of explosives to excavate them.[34] Nobel fought there with the Germans (Krebs & Co.) and with local Australian companies producing dynamite with names like Dynamo[35] and Jones' Dynamite.[36]

France was another special case. France had a state monopoly on the manufacture and sale of explosives, so gunpowder was three times as expensive there as in neighboring Prussia. Since dynamite was cheaper and better than gunpowder anyway, Nobel knew it would do well in France—if he could just sell it there.[37] He searched for a French partner, eventually finding a wealthy (but somewhat reckless) thirty-two-year-old named Paul Émile Barbe, whose family put up the capital and with whom he agreed to share the profits 50/50.[38] Still they struggled to get French government permission to sell.

When the Franco-Prussian war broke out in 1870, the Prussians defeated the French, and suddenly the losing side wanted better, and more, explosives. In 1871, the French Ministry of War allowed Nobel to build a factory at Paulliles (on the Mediterranean, near the Spanish border) and sell dynamite throughout France, even throwing in 60,000 francs in government assistance.[39]

Finding local partners in other countries was how Nobel expanded other parts of his empire so quickly. Being stretched for resources, especially in the early years, he relied on the capital of wealthy merchants, lawyers, and entrepreneurs, on top of the cash he could earn from selling his patents, to get his enterprises going.[40] His first firm, in Sweden, relied mostly on the capital of a Stockholm merchant, and in Germany his partner was a doctor of law.

Nobel always structured the partnerships so that he put up as little of his own money as possible, retaining company shares in return for his patent. The only exception was his Norwegian nitroglycerine firm, where he divested his shares and received cash for his patents, a mistake he never repeated. Nobel held 25 percent of the shares in the Swedish and American companies, and initially more than half of the German one.[41] This large personal stake may have been one reason Nobel wanted the German firm to be a major exporter; however, long-distance shipments from Hamburg were also responsible for many early nitroglycerine accidents. Ultimately Nobel built a large number of local plants around the world so he wouldn't have to transport dynamite very far.[42]

The growth rate of his global business was incredible: between 1871 and 1874 Nobel's manufacture of dynamite almost quadrupled, and his Hamburg plant was the most profitable, exporting about a third of its production.[43] (See Table 4.1, a list of Nobel's earliest explosives factories.)

TABLE 4.1 Nobel's Earliest Explosives Factories

Vinterviken (Winter Bay, Sweden)	1865
Krummel (near Hamburg, Germany)	1865
Lysaker (near Oslo, Norway)	1865
Little Ferry, New Jersey (USA)	1866
Zamky (near Prague)	1868
Rock Canyon (near San Francisco, California, USA)	1868
Hango, Finland	1870
Ardeer (near Glasgow, Scotland)	1871
Paulilles (near Port-Vendres, France)	1871
McCainsville (now Kenvil, an hour west of New York City, USA)	1871
Schleebuch (near Cologne, Germany)	1872
Galdacano, Spain	1872
Giant Powder Works (near New York City)	1873
Isleten, Switzerland	1873
Avigliana (near Turin, Italy)	1873
Trafaria (near Lisbon, Portugal)	1873
Pressburg (now Bratislava, the capital of Slovakia)	1873

This Table was compiled with information mainly from Fant (1991), p. 135, and Lundström (1986). Note that the earliest four factories were established for the production of nitroglycerine oil, before the invention and patenting of dynamite. Note that the Little Ferry factory was destroyed in 1870.

Growth despite Danger

The voracious demand for dynamite is also underscored by the fact that for many years, the manufacture and transport of the explosive was still quite dangerous. One hazard was in transporting it in wet or cold conditions. Water would soak into the putty, forcing the nitroglycerine oil to seep out, making it subject to explosion if jarred or heated.[44] Dynamite also froze at around 40 degrees Fahrenheit and required a complex and difficult process of "tempering," or thawing, to be usable. Detailed instructions for how to thaw dynamite were often included in the box, and miners or railroad workers became good at it; but accidents happened, especially early on. In Parma, Italy, in 1878, for example, a lieutenant colonel in the cavalry set a can containing 1 kilogram of frozen dynamite onto a charcoal stove to thaw. It exploded, and eighty people were killed or injured.[45]

The manufacture of dynamite also continued to be dangerous. Workers at the factories were exposed to its key ingredient, nitroglycerine, in full force. Nitroglycerine oil was prepared in huge vats, usually supervised by men, sitting on one-legged stools to keep them alert, carefully watching a temperature gauge to ensure the mixture didn't get too hot. Women did the hand work and cartridge packing (Figure 4.1). They kneaded the silicon kieselguhr and nitroglycerine putty mixture with their bare hands, as if making bread. Then they put the dough into what looked like a sausage machine, where it was extruded as rods, cut in measured lengths, wrapped in wax paper, packed in tubes, and finally loaded into crates, thirty cartridges in each. It was dangerous work yet, except for the chemists and factory management, the pay was low.[46]

Major accidents included numerous devastating explosions at the Vinterviken factory in Sweden; the complete destruction of the factory at Little Ferry, New Jersey, in March 1870 (it was never rebuilt);[47] and the total destruction of the Nobel Krummel factory twice (1866 and 1870). In 1867, Nobel licensed dynamite to San Francisco businessman Julius Bandmann's Giant Powder Company, which had major accidents in 1869 (two killed, nine injured), 1879 (four killed), followed by three more deadly explosions over the next thirteen years, killing sixty-six workers.[48]

Tight restrictions on the manufacture and transport of dynamite might seem to have been in order, but even quite rigorous regulation of the business could, with sufficient persistence and political pressure, be accommodated or worked around. In this regard, the biggest challenge Nobel had faced soon after he invented dynamite was finding a way to sell it in Britain, which he was intensely keen to do. In his private papers, he wrote:

Great Britain is in my opinion by far the most important of all for this branch of the business, not only on account of the vast internal trade, but also the great facility for export to the colonies. . . . Now the colonial trade is of the greatest importance. Think of all the Indian railways, Australian mining and so forth.[49]

Yet, selling his products in England proved especially problematic due to another tragic accident.

On 30 June 1869, a consignment of nitroglycerine was shipped from Hamburg, Germany, first to the densely populated port city of Liverpool, then southward to Wales, where a part of it blew up near the town of Caernarvon, Wales.[50] The oil was in two carts driven by quarry-men returning from work. They had just passed the small Welsh village of Cwm-y-glo when the nitroglycerine exploded, killing five men outright and injuring twenty-four people (including village children), twelve of them seriously. One person had to have his leg amputated, and two lost arms. The explosion

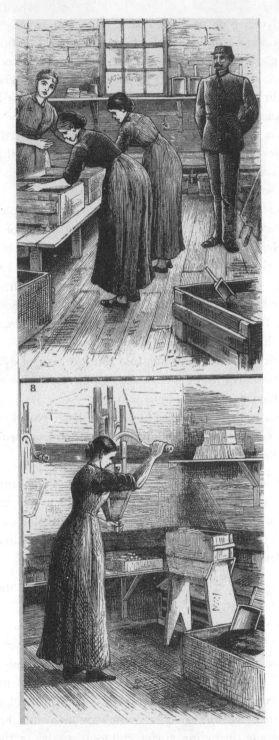

FIGURE 4.1 Female workers mixing dynamite with their bare hands (above) and using dynamite extruding machine (below), Nobel dynamite factory, Ardeer Ayrshire Scotland 1884. The Print Collector/Alamy Stock Photo.

cut two 8-by-8-foot holes in the road, and structures as far as 2 miles away were damaged by the blast. Those who died were pulverized. According to one newspaper account, "A foot, a man's chin covered with beard, and a man's heart were found together about eight yards from the spot."[51]

As a direct result of the Caernarvon accident, on 11 August 1869 the British Parliament hastily passed the Nitro-Glycerine Act, requiring a license for handling any form of nitroglycerine and banning all importing of the oil. Sweden and Belgium also strictly prohibited its use.[52] In subsequent months, demands for licenses from eager British mining and construction businesses still came flooding in; however, the Home Office decided that both nitroglycerine and, by extension, dynamite were too dangerous to transport, and must be made and used in one place. Without either a license or a local factory, Nobel found it impossible to sell dynamite in Britain.[53] Under steady pressure—not least from Nobel's local agent, John Downie (who died in an explosives accident five years later)—the Home Office in 1870 eased the restrictions on storage and transport. By April 1871, Nobel had procured a license to manufacture dynamite in Scotland. He founded the British Dynamite Company (retaining a 50 percent share) in Ardeer, on Scotland's western coast, later that year.[54]

The 1869 Nitro-Glycerine Act also made it difficult to distribute dynamite within Britain, because manufacturers had to ensure that everyone handling it was licensed, and some railways flatly refused to carry it at all. Sometimes this meant that dynamite made in Scotland was loaded onto horse-drawn carts, then bumped along exposed to the elements on uneven roads, hardly a safer alternative and probably much easier to steal.[55] Some dynamite was sent by rail, but it was labeled "slate" or something else.[56]

Responding to continuing accidents, and also to the explosive industry's desire for more judicious regulation, the British Parliament passed a much more sophisticated and thoughtful piece of legislation, the 1875 Explosives Act.[57] The Act established a thorough system of inspection and control of all explosives in Britain, not just high explosives such as nitroglycerine, dynamite, and guncotton, but also including gunpowder.[58] The well-paid and well-trained government inspectors (who were often former Royal Artillery officers, backed up by a scientific adviser, who was an academic) had considerable discretion in how they interpreted the act—for example, during factory and warehouse inspections—and they worked with local authorities, who were responsible for issuing licenses.[59] At first some railroads continued to refuse to carry explosives, but they slowly gained confidence in the safety inspections regime and permitted it.[60] Ultimately the act made it much easier for Nobel, and other producers, to operate in England, and

Nobel became a proponent of the high-quality inspection and transport system there.

Inexorable Downward Pressure on Price

As his explosives empire grew, another problem Nobel faced was that his plants began to compete against one another, driving the price of dynamite in their sales territories down.[61] Nobel had some success in driving the price artificially high for periods, but it required some strenuous business machinations. His French partner, Barbe, suggested in November 1878 that he merge his British and German companies, an amalgamation and structural change that Nobel initially resisted.[62] However, only a few years later a new wave of competitors popped up throughout Europe, again producing knockoff types of dynamite blending nitroglycerine oil with things like charcoal and sawdust, driving prices down, including three new companies in Germany alone: Dresdner Dynamit-Fabrik, Kolner Dynamit-Fabrik, and Deutsche Sprengstoff. So in 1886 Nobel accepted Barbe's suggestion, uniting those three competitors, along with his German and British conglomerate, into a cartel that he named the Nobel Dynamite Trust Company Ltd.[63] Member companies divided their territories among themselves, and collaborated to set the price of dynamite above what would have been the market rate.

In building his cartel, Nobel was following the example of the major US manufacturers of gunpowder. Powder firms had pocketed more than $1 million selling high-quality gunpowder to the North throughout the Civil War.[64] In 1872, the three largest manufacturers—DuPont, the Hazard Powder Company, and the Laflin & Rand Company—aligned with four smaller US firms to form a powerful monopoly that became known as the Powder Trust. The express purpose of the trust was to fix the price of gunpowder.[65] When upstart competitors took advantage of the opportunity to undercut the trust's pricing, selling powder at lower prices, the trust companies would temporarily lower their price in that territory, often selling below cost, forcing the competitor into financial straits. Trust members would then buy a controlling share of the competitor firm's stock, forcing it to do the trust's bidding. An early example of this strategy was carried out with the California Powder Works, which in 1875 was forced to sell 43 percent of its stock to trust member E. I. du Pont de Nemours and Company, and was then pressured to raise its prices and also to refrain from shipping to any competing territory.[66] The same strategy was deployed against the Chattanooga Powder Company in 1890, the Phoenix Powder Manufacturing Company in 1891, and the Southern Powder Company in 1894.

The US Powder Trust, especially the DuPont company, flexed its muscles against Nobel's efforts, as well. In the 1860s, members of the trust fought nitroglycerine-based explosives altogether. Then, realizing that gunpowder could not hold the newer technology off, they wanted the right to control and profit from its sale in the United States. In 1888 they bought all the interests in Nobel's US factories. In a market-sharing and price-fixing arrangement, the Powder Trust divided the US market between the six principal members in California and the Eastern seaboard. The deal stipulated that the American companies would stay out of Europe, Africa, Australia, and Asia, in return for Nobel ceding the rights to sell dynamite in the United States.[67] Nobel continued to receive royalties from his patents, but grumbled in 1889 correspondence that the US patent fees were so small that they could be "seen only under a microscope."[68]

Although the Sherman Antitrust Act, which was the first US law prohibiting monopolistic practices and empowering the Justice Department to break trusts up, was passed in 1890, the Powder Trust held on until 1912. The government finally took decisive action against the cartel that year in light of growing political and economic tensions between the major European powers, portending that vast quantities of gunpowder might shortly be needed for large-scale war. The government was prescient, as war broke out in Europe two years later.[69]

The European Nobel Dynamite Trust differed from the US Powder Trust in that it did not operate strictly within one country. According to Swedish economic historian Ragnhild Lundström, it was the first international holding company in the world, meaning it was the first company whose sole purpose was to own shares in other companies throughout the world and control their joint business. A year after the Nobel Trust was born, in 1887, Nobel and Barbe also founded La Société Centrale de Dynamite (Society for the Centralized Control of Dynamite), which incorporated the French, Spanish, and Italian-Swiss dynamite companies. The headquarters of these two new multinationals, in London and Paris respectively, coordinated prices, investments, sales, purchases of raw materials, and production levels between the factories and each other.[70]

The goal was not to drive up prices but to both reduce competition between trust companies and prevent outsiders from gaining market share, thus profiting from high-volume sales of dynamite at low prices.

The remarkable speed at which the dynamite business grew and the low cost and relative safety of the explosive explain the readily accessible supply of dynamite for use in attacks. But what explains the intensifying demand for this particular explosive versus other, superior, options?

The Stoking of Discontent

Not unlike today, the late nineteenth century was a period of great contradictions: an era of unprecedented technological progress, which brought an abundance of new conveniences and greatly improved the quality of life for many, but which also led to great hardship for many in the lower classes.

Maturing industrialization provoked growing social discontent. Monarchies, some with elected Parliaments, some without, still predominated in most regions of the globe; but even where republics had replaced them—as in France and the United States—universal suffrage in practice meant voting restricted to non-Muslim males in France, or adult white males in the American south, for example. The rest of the public had no viable avenue for political change. Meanwhile, even as overall global standards of living were rising, wealth inequality between low-wage factory or railroad or oil workers, on the one hand, and industrialists such as Andrew Carnegie, J. P. Morgan, and John D. Rockefeller, on the other, provoked anger. Appalling working conditions particularly fed workers' anger and rallied them to violent protest.

Globalization also contributed. The disruptions to familiar ways of life due to late nineteenth-century globalization were in many respects even more dramatic than those of today. People, commodities, goods, and capital all moved more easily between states than they do now. Steamships, railroads, and the Suez and Panama canals all abruptly shrank the distances between people, and the cost of travel plummeted over the century, from about $150 in 1826 for an average first-class advertised fare from London to New York, to $60 in 1900.[71] The development of the transportation infrastructure led not only to economic globalization but also to the mass resettlement of people looking for a better life. Waves of migration produced an angry backlash to an influx of low-skilled laborers, many of them fleeing famine, religious discrimination, or unemployment at home, going from places like Ireland, Italy, Russia, China, Germany, and Poland to the United States, Great Britain, and Canada. There they accepted lower pay and worse conditions than indigenous workers did, leading to downward pressure on wages and deep local resentment.

Between 1880 and 1920, approximately 23,500,000 immigrants arrived in the United States, for example, more than 12,000,000 from eastern and southern European states. This totaled more than twice as many as had arrived between 1840 and 1880 (9,438,480), most of them northern Europeans.[72] American periodicals like *Harper's Weekly* in 1886 referred to them all as "Hungarians" and decried their supposed tendency toward violence.[73] Many people whose lives were convulsed by these changes were humiliated, disoriented, discriminated against, and felt left behind.

Civility broke down, as attacks that had appalled the public in the 1880s were increasingly condoned or even championed by labor activists who saw no other way to bring about change, protect their livelihoods, or engage the fight between capital and labor, especially in the years before and after the First World War in the United States.

Dynamite was the perfect weapon with which to express the growing rage, and it assumed an almost mystical place of honor as a populist leveler between workingman and boss, monarch and peasant, wealthy and indigent.

A number of particular developments rallied people to action.

The International Anarchist Convention of 1881, and "Propaganda of the Deed"

Anarchist leaders came up with a potent rallying cry that attracted a new wave of devotees. Four months after Tsar Alexander's assassination, the leaders of the movement held an international conference in London, from 14–19 July 1881, attended by all of the major figures, including Peter Kropotkin, Marie Le Compte, and Errico Malatesta. There they formally blessed the concept of "propaganda of the deed" (from the French *propagande par le fait*). It was in some ways an old activist principle—that talk is cheap. Italian revolutionary Carlo Pisacane had written in 1857, "The propaganda of the idea is a chimera. Ideas result from deeds, not the latter from the former."[74] With dynamite now providing the perfect means with which to act, propaganda of the deed came to refer specifically to acts of killing, especially dynamite bombings. The conference passed a formal resolution calling for the study of the technical sciences, such as chemistry, so that people could make bombs "for offensive and defensive purposes."[75] Although the original concept was to use the weapon for a mass insurrection, rather than random acts of terrorism, this quickly changed in implementation.[76] Terrorism was seen as the ultimate anti-organization, anti-authoritarian act, and the leaders were philosophically dedicated to the complete autonomy of local groups and individuals to decide what form their violent activities should take.[77]

Whereas earlier would-be revolutionaries had failed to capture the popular imagination and were easily crushed by the state, dynamite bombings riveted the public's attention, not least because common people were often both the perpetrators and among the victims. Dynamite also enabled the anarchists to distinguish themselves from common criminals, who generally used knives and handguns. Explosives were an impressive new technology. Additionally, dynamite attacks required little planning, and they appealed to lone wolves and small groups of conspirators, who carried out most attacks. The leaders of the movement did not have to become directly involved in organizing the violence; they could limit their role to writing philosophical works, bemoaning the unequal ownership of assets, and laying out a vision for a better society.

From the time of the conference forward, the movement was bifurcated into an "elite" leadership and a far-flung, mostly unorganized mélange of malcontents, many of whom were inspired by the anarchist dream but were not deeply familiar with anarchist thought. Some of them were hangers-on to the movement rather than true believers—thugs, exhibitionists, and mentally unstable individuals who opportunistically jumped on the violence bandwagon.[78]

Bombings spiked dramatically in the aftermath of the London conference. Over the thirteen years following Nobel's invention of dynamite (1867–1880), 17 bombings had been carried out. In 1881 alone there were 16, followed by 13 in 1882, 25 in 1883, and 42 in 1884. (See Figure 4.2, Worldwide Bombings per Year, 1867–1891.) The popularization of the concept of propaganda of the deed had opened the floodgates, ultimately dismaying many of its original proponents.[79] Instead of a mass revolution against the centralized state, envisioned by proponents of propaganda of the deed like Prince Kropotkin, what unfolded was violence that, above all, killed innocent civilians in streets, hotels, houses, factories, theaters, opera houses, and restaurants.

Dynamite Schools and Pamphlets

Publications extolling dynamite as the ultimate populist weapon proliferated. The anarchists were a leading force in this, but Irish nationalists extolled its virtues, too.[80] Writers argued that the benefits of science should not be confined to the upper class and advocated scientific education of the lower classes, including about the use of explosives. In 1885, for example, Gerhard

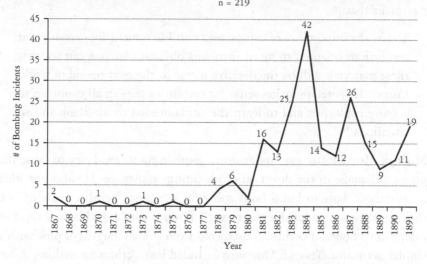

FIGURE 4.2 Worldwide Bombing Incidents per Year (1867–1891).

Lizius, who was an anarchist thought leader and ultimately one of the eight men convicted and executed in the Haymarket massacre, wrote an editorial, published in the anarchist paper *The Alarm*, describing dynamite as the "sublime stuff" and declaring, "In giving dynamite to the downtrodden millions of the globe, science has done its best work."[81] Dynamite was also lauded in songs and poems, such as in the poem "Nihilisten," published in the German socialist journal *Vorbote {Herald}*, the monthly central organ of the German section of the First Socialist International, which proclaimed, "Hurrah for science! Hurrah for dynamite!—the power which in our hands shall make an end of tyranny."[82]

The anarchists not only advocated education about dynamite, they delivered it, opening bomb-making schools and giving public lectures and workshops, as well as distributing instructional pamphlets. The first of these guides was a seventy-four-page pamphlet titled "The Science of Revolutionary Warfare: A Little Handbook of Instruction in the Use and Preparation of Nitroglycerine, Dynamite, Gun-Cotton, Fulminating Mercury, Bombs, Arsons, Poisons, etc," published in the United States in 1884 by Johann Most.[83] He had been exiled from his home country of Germany, ending up first in the United Kingdom, where he attended the 1881 London conference. Moving to the United States, Most got a job in a New Jersey explosives factory. His hands-on daily experience there enabled him to write simple step-by-step instructions for bomb-making. He opened the guide with the statement that:

> Today, the importance of explosives as an instrument for carrying out revolutions oriented to social justice is obvious. Anyone can see that these materials will be the decisive factor in the next period of world history. It therefore makes sense for revolutionaries in all countries to acquire explosives and to learn the skills needed to use them in real situations.[84]

Much of Most's focus was on dynamite, explaining its tendency to cause the greatest damage in the direction of maximum resistance. He also provided details about how to build both spherical dynamite bombs, made of two thin metal halves soldered together with a hole drilled for a blasting cap and fuse, and rectangular bombs, constructed from 6-inch lengths of pipe with a similar detonator. Though Most also included instructions for making dynamite, he recommended buying or stealing it, because "dynamite is used for many purposes, so that it is nonsense to believe that it cannot be obtained from conventional suppliers."[85] The author did not mince words or advocate selective targeting in killing people. He wrote openly of creating "maximum destruction in all directions, for example, inside a house—and particularly in the middle of a large group of human targets."[86]

Most's pamphlet had great influence. It was serialized by the Chicago anarchist newspaper *Arbeiter-Zeitung*, which sold copies of Most's pamphlet at ten cents apiece at meetings, picnics, and in newspaper offices. A charismatic speaker, Most also traveled throughout the United States, hawking the pamphlet and exhorting audiences to arm themselves with explosives for the revolution to come. In 1889, the budding anarchist Emma Goldman attended one of his lectures and thereafter began giving her own fiery talks throughout the United States, exhorting her audience members to take action. Polish immigrant Leon Czolgosz reportedly attended one of her lectures, and two weeks later, on 6 September 1901, assassinated President William McKinley.[87] Upon his arrest, Czolgosz, who was enraged about harsh working conditions and claimed to be an anarchist, said that Goldman had inspired him.[88]

Anarchist Newspapers and Periodicals Worldwide

Joseph Most's pamphlet was but the first of a number of such publications, which proliferated across the globe. In all, during the Anarchist Period

(1878–1934) nearly 250 anarchist publications sprang up. Most of them flourished for a short time and then folded (see Appendix B),[89] and often their brief appearance was accompanied by a bevy of local bombings, which were then publicized by these rags.

In 1892, following the theft of a large amount of dynamite from a Paris suburb, an anarchist newspaper published a formula for making mercury fulminate and simple illustrations for how to construct a bomb. Between March and November 1892, France had an unprecedented twenty-six bombings, twelve of them in Paris. Ten of these were part of a concentrated one-month spree (29 February–30 March), including at least three bombings by the French anarchist known as Ravachol (François Claudius Koenigstein). The bomber was captured by police on 30 March, tried, and executed on 11 July 1892, and then a wave of revenge attacks (or attempts) followed. Over this difficult year, 300 bombs were discovered by Paris police and gingerly analyzed in the reinforced basement laboratory of the municipal headquarters.[90] The city's police force was besieged by reports of imminent threats, with officers sent digging through the garbage for devices and reacting to anonymous letters threatening to bomb nasty landlords. "Long Live Anarchy" was found scribbled on walls and sidewalks. Many police were so exhausted and fearful for their lives that they asked for transfers or resigned. Some coerced "anarchist" confessions from ordinary crooks.

In Barcelona, the "City of Bombs," over twenty local anarchist publications appeared, many during the height of the bombing period there from 1902 to 1909. Likewise, forty anarchist journals and newspapers were published in France, and their periods of publication coincided with the spike of anarchist bombings in Paris, Lyon, and other French cities. Across Europe as a whole, 73 percent of all bombings occurred within 200 miles (322 km) of an existing anarchist publication (Figure 4.3).

Probably the most striking illustration of this pattern comes from Buenos Aires and the nearby city of Rosario in Argentina. Seventeen of the 27 identified South American anarchist bombings (63 percent) occurred in these two localities, which were home to more than 50 anarchist publications between them, including 24 in Buenos Aires and 30 in Argentina, all circulated at the peak of the anarchist South American bombing campaign in the 1920s.

Mass-Market Sensationalism

The emergence of the first wave of terrorism coincided with the appearance of the mass-market newspaper. Lurid coverage of attacks ran in both

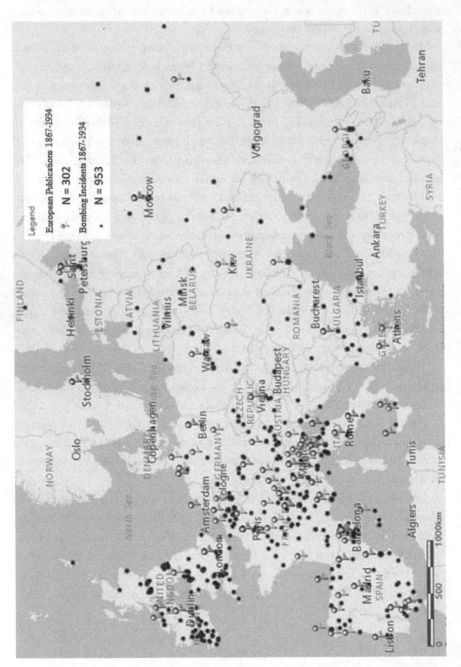

FIGURE 4.3 European Anarchist Publications and Proximity to Bombing Incidents, (1867–1934).

mainstream newspapers and tabloids, which boomed in circulation at the end of the nineteenth century.

During this era of "yellow journalism," papers competed furiously to attract readers, and they often did so through shameless means. In 1896, William Jennings Bryan, running as a populist candidate for the US presidency, warned of an "epidemic of fake news."[91] After he lost the election to William McKinley, Bryan started his own newspaper, *The Commoner,* which he used as a platform for his own "fake news" (the phrase that he used at the time), including that he would have won the election had the Republicans not engaged in fraud, supported by the corrupt media.[92] Also in 1896 a columnist complained in *The Davenport Daily Republican,* "It could not be foreseen that a time would come when whole columns of fake news would be published, that whole columns of sensational stuff would be printed and read."[93] There was so much malarkey passing for facts, often attributed to "friends," that to protect itself from deception, *The Polk County Republican* of Tennessee carried a notice near its masthead, "All Communications, to receive attention, must have the writer's name to it. This is our only protection against 'fake' news and the rule will not be broken under any circumstances."[94]

Fear and violence sell, and the papers also made the most of the news of "dynamitings." In addition to stories of homegrown bombings, news crossed the Atlantic via newly laid transcontinental underwater telegraph cables, bringing the global scale of the bombings vividly to life. Stories with graphic details of the bloodshed lured readers with sensationalist headlines, such as this sampling of headlines from the *New York Times*: "Anarchist Blown to Pieces: Killed in London by an Explosion of Some Sort of Bomb," 16 February 1894; "Reign of Terror in Paris: Ancaer Bomb Explosion in a Small Hotel," 21 February 1894; and the formidable "The Guillotine's Sure Work; Details of the Execution of Vaillant, The Anarchist. 'Long Live Anarchy! the Last Words of the Condemned Man—He Refuses the Ministrations of a Chaplain and Walks to His Death Without a Tremor—The Body, After a Pretended Burial, Turned Over to the Doctors for Dissection," 6 February 1894, describing the public execution of French anarchist August Vaillant following his attack on the Chamber of Deputies.

Newsboys loudly hawked papers on street corners, broadcasting the most salacious stories. The coverage attracted new readers in droves, and papers cut their prices to optimize sales. In England, the cost of many newspapers was cut to a halfpenny an issue and their circulation soared. *Lloyd's Weekly Newspaper* of London, the best-selling weekly, reached a million in circulation during 1896, up from 200,000 in 1861.[95] In Paris the circulation of newspapers doubled between 1871 and 1910, and the cost dropped by half.[96]

In the United States, the Pulitzer and Hearst empires flourished. Between 1883 and 1895, the circulation of Pulitzer's *New York World* grew from 15,000 to 600,000,[97] and though the paper cost only five cents an issue, it was the most profitable newspaper ever published. Hearst's *New York Journal* was not far behind.

Popular fear and fascination with dynamite spread rapidly, resulting in what became known as dynamite psychosis, especially in major cities like New York, Paris, and London. Widespread sensationalist reports and anxiety about anarchists in the wake of dynamite attacks triggered repressive police powers and sometime executions of the perpetrators, particularly in France. Fanatics like twenty-two-year-old Emile Henry, shocked by the widely reported public execution of August Vaillant for throwing a bomb into the French Chamber of Deputies in 1893, then carried out a series of copycat bombings in Paris in 1894. In his room, Henry had copies of five anarchist newspapers and instructions for making a bomb.[98] Mario Buda, who was reportedly responsible for the 1920 Wall Street bombing, had read about the 1916 Preparedness Day bombing in San Francisco, as well as the spate of letter bombings sent to government officials in June 1919.[99] Stories and pictures of Buda's Wall Street bombing then led to widespread hysteria and political support of the US anti-immigrant Palmer Raids in its aftermath. It was a cycle of escalating fear and reaction. Over and over, the sensationalist publicity after dynamite bombings resulted in popular fear, then government repression, then outrage or inspiration, and then additional bombings.

Patterns in Numbers of Attacks

A close analysis of nearly 1,300 bombing incidents throughout the world during this first wave of violence shows that the number of attacks was directly related to a key set of economic and legal factors: fluctuations in the price of dynamite, regulations on transporting nitroglycerine products by rail, local trust activities, and whether or not there were strict factory export controls. (See Appendix B for further information on bombing incidents and explosive control legislation at the time.[100]) The number of bombings in Europe was on the rise between 1881 and 1885 (a time when the price of dynamite was falling), and then dropped off precipitously in 1886 (coinciding with the formation of the Nobel Dynamite Trust). (See Figure 4.4.) Bombings in France, Spain, and Italy peaked in 1892, following the theft of a large amount of dynamite and the execution of Ravachol. Likewise, there was a significant increase in the number of bombings in the United States beginning in 1905,

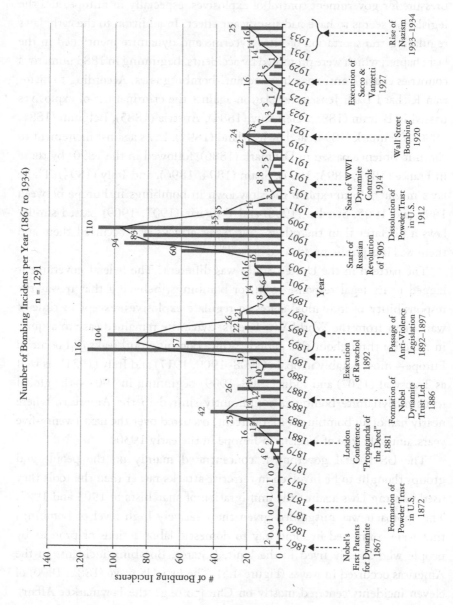

FIGURE 4.4 Number of Worldwide Bombing Incidents by Year with Significant Events (1867–1934).

probably related to declining economic conditions, the rising influence of activist labor unions, and an influx of immigrants spreading anarchist and socialist philosophies from countries including Russia, Poland, and Italy.

Large bombings that harmed civilians usually resulted in greater political pressure for government control of explosives, especially in Europe, and the legislation seems to have had significant effect. In addition to the early laws regulating transportation of nitroglycerine and dynamite mentioned in the last chapter, which were prompted by accidents, beginning in 1883 numerous countries enacted anti-anarchist and anti-bombing laws. According to historian Richard Bach Jensen, legislation against the criminal use of explosives passed in Britain (1883), Germany (1884), Austria (1885), Belgium (1881, 1886), Denmark (1894), and Switzerland (1894). Laws against incitement to commit violence passed in Denmark (1886), followed in the 1890s by those in France (1892, 1893, 1894), Spain (1894, 1896), and Italy (1894). Those laws might help to explain the slowdown in bombings in Europe between 1897 and 1905. Sweden (1906) and Bulgaria (1907, 1909) passed similar laws a bit later than the others, but fewer attacks had occurred there and there was less public pressure to do so.[101]

The pattern in the United States was different. The federal government lagged in its legal response to major bombings, believing that it was the responsibility of individual states to regulate explosives, except in time of war. Apart from the World War I years, dynamite remained easy to acquire in America throughout this period. While bombing outbreaks did occur in Europe—most notably in Spain (1908–1909, 1917) and Italy (1921), as well as Transvaal (1907) and India (1908–1909) beginning in 1907—the global focus of dynamite bombings significantly shifted to the Americas, where nearly half of all bombings (46.4 percent) occurred over the next twenty-five years, until the rise of Nazism in Europe in the early 1930s.

The US federal government concentrated mainly on the people and groups thought to be involved in terrorist attacks rather than the tools they used, passing laws against the immigration of anarchists in 1903 and 1907. The approach was misguided given the relatively high level of bombings that were connected in some way to domestic labor actions or protests by people who already lived in the United States. Bombing incidents in the Americas occurred in waves (Figure 4.5). The first US wave (1884–1890) of eleven incidents centered mostly on Chicago (e.g., the Haymarket Affair). The second wave (1892–1896) of twenty-one incidents was centered mostly in the Northeast (New York, New Jersey, Pennsylvania), and it involved steelworkers (e.g., the Homestead Affair). There was a quiet period between 1897 and 1903 (only six incidents, mostly involving miners), and then a very

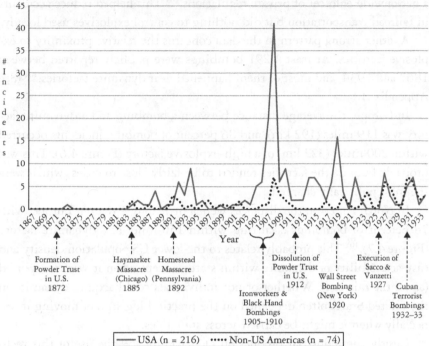

Number of Bombing Incidents per Year in the Americas (1867 to 1934)
n = 290

Formation of Powder Trust in U.S. 1872

Haymarket Massacre (Chicago) 1885

Homestead Massacre (Pennsylvania) 1892

Dissolution of Powder Trust in U.S. 1912

Execution of Sacco & Vanzetti 1927

Wall Street Bombing (New York) 1920

Ironworkers & Black Hand Bombings 1905–1910

Cuban Terrorist Bombings 1932–33

——— USA (n = 216) ••••• Non-US Americas (n = 74)

FIGURE 4.5 Number of Bombing Incidents in the Americas by Year with Significant Events (1867–1934).

major outbreak (eighty-six incidents) from 1905 to 1910, which involved the ironworkers and, reportedly, the Italian Black Hand. This period of violence (usually on the part of both security forces and workers) culminated with the bombing of the *Los Angeles Times* building in 1910, and the prosecution of Ironworker Union heads in 1911.

Unlike in Europe by this point, in the United States no permit was required to buy explosives, and any regulation fell under state, not federal, law.[102] But many states had no restrictions at all, and the laws that did exist were rarely enforced. As accidents continued to occur, some self-regulation by industry emerged. For example, the US railroad industry put in place its own system to control explosives. Following a major dynamite accident on 1 November 1903, in which a large load exploded in Crestline, Ohio, injuring scores of people and setting 500 freight cars on fire, the Pennsylvania Railroad developed some of the first regulations against the transportation of dynamite in the United States. The company then led the American Railway Association (ARA) to establish the ARA Bureau of Explosives in 1907, which, backed by the Interstate Commerce Commission a year later, became a nationwide enforcer of private regulations.[103] This helped reduce accidents in railroad transportation but did nothing to control explosives used locally.

Another strong pattern in the data concerns the relative proximity to explosive factories. At least 1,291 bombings were publicly reported between 1867 and 1934, and most of them happened near dynamite factories.[104] (See Appendix B.)

In Europe, the average distance between a bombing and a dynamite factory was 119 miles (192 km), and 86 percent of bombing incidents occurred within 200 miles (322 km) of a high-explosive factory (Figure 4.6). This was probably because the factories tended to be fairly close to cities, which were the most politically charged.

In North America, the distances were a little longer than in Europe, with 78 percent of bombings occurring within 150 miles of a dynamite factory (Figure 4.7).[105] This probably relates to the lower US population density and easy accessibility of dynamite within states, except when it was transported (as on the railroad). Whether or not individuals could acquire dynamite in the United States often depended on the practical logistics of moving it, especially when it might be carried across state lines.

Clearly the industrial production of dynamite and the use of this technology in violent attacks were closely connected. In other words, most attacks were not carried out using homemade dynamite (or dynamite transported over long distances) but rather dynamite that was produced, sold, and bought relatively nearby.

How the Wave of Global Dynamitings Ended

Important to tamping down this first wave of terrorism was the establishment of international police and intelligence cooperation, especially in Europe. This had begun back in 1898, following the assassination of Empress Elisabeth of Austria. Fifty-four delegates from twenty-one countries attended a highly secret international anti-anarchist conference in Rome, followed by an agreement hammered out between the European foreign ministries. Anti-anarchist legislation and anti-explosive laws in the United Kingdom, Germany, Switzerland, Belgium, and to a certain extent France helped to reduce the frequency of bombings, especially after 1900. These efforts were important precursors to the 1923 establishment of Interpol, the network of international law enforcement agencies whose headquarters is in Lyon,

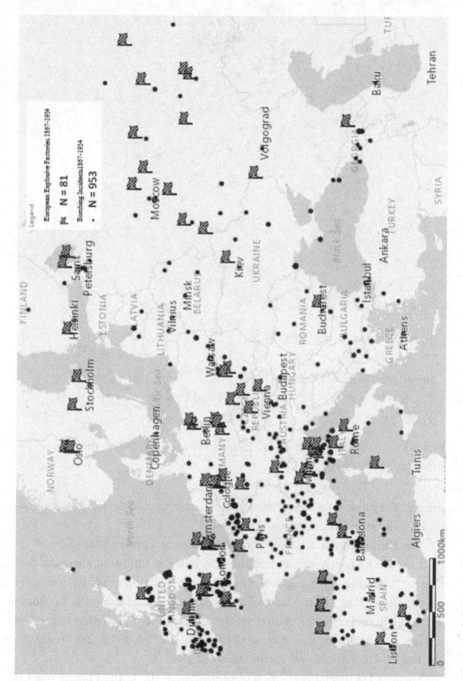

FIGURE 4.6 Proximity of Bombing Incidents to High Explosives Factories in Europe (1867–1934).

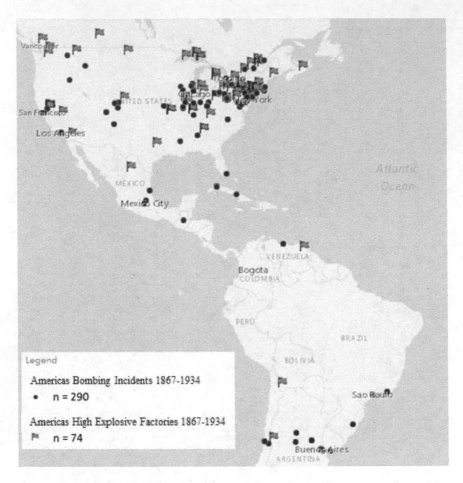

FIGURE 4.7 Proximity of Bombing Incidents to High Explosives Factories in the Americas (1867–1934).

France. They also foreshadow the more extensive global network of international counterterrorism cooperation today.

The United States, as always, took a different approach. It did not attend the Rome conference and declined to join the European anti-anarchist league.[106] Among other reasons, the American policing tradition was very decentralized. The United States had no federal police force until President Theodore Roosevelt in 1908 directed the establishment of a federal detective service, which became the "Bureau of Investigation" in 1909 and ultimately the Federal Bureau of Investigation (FBI) in 1935.[107]

The first US federal regulation of explosives in the United States was the Explosives Act of 1917, which restricted them only for the duration of the First World War, and was enforced by inspectors with the Bureau of Mines.[108]

In 1920, William Chenery of the *New York Times* commented on the folly of lifting federal regulation: "[Y]ou could buy a stick of high explosives quite as legally as you could purchase a paper of pins, for all the Government hindered you."[109] The Explosives Act was reactivated in 1941, during the Second World War, and then deactivated again by President Harry Truman in 1947. It was not until 1970 that there was a consistent set of explosive regulations nationwide in the United States.

Once the genie of a potentially lethal, profitable new technology is out of the bottle, it can be difficult to control its spread, especially if the new technology is commercially produced, has positive uses, and stands to make many people lots of money. Intelligent regulations do have a track record of limited success, however. The British example of highly experienced inspectors backed up by expert chemists, in a carefully crafted program regulating the explosive's handling and transportation, helped Nobel sell much more of it, with the Scottish plant becoming one of his most profitable ventures. Highly publicized accidents had reflected poorly on his business, and the British government strongly supported his patent. In other countries, federal regulations were either too stringent, as in France and Russia, or missing altogether, as in the United States and Australia. Individuals could still make "bootleg" dynamite, but due to its ready availability, reliability, and low cost, most of the dynamite used in violent attacks was factory-made. In short, intelligent regulations may have made terrorist attacks harder to orchestrate.

Nobel's Remorse

Alfred Nobel had not anticipated the civic violence his invention inspired, or the staying power of dynamite. He invented some 355 patented products during his lifetime, including for explosives that were superior to dynamite. These included blasting gelatin, or gelignite, which was more resistant to impact and less likely to sweat, created in 1875, and ballistite, a smokeless powder made from nitroglycerin and guncotton, that was better suited as a propellant, invented in 1887.[110] Any objective expert assessment of his subsequent explosive creations would have to conclude that dynamite was not the most powerful, effective, or reliable of them. Yet dynamite—in both its construction of modern infrastructure and its disruptive role on the world stage—would go down in history as his primary legacy.

Blasting gelatin not only solved the problem of seeping nitroglycerine oil, its higher density and greater plasticity gave it more force, making it more efficient in cutting through rock.[111] It could even be used under water. Yet its great power prohibited its use in munitions and for many other applications of explosives, and it was more expensive than dynamite. Demand proved disappointing. Ballistite was more successful than blasting gelatin, quickly adopted by the munitions industry because it left little residue in gun barrels. In fact, some experts consider ballistite Nobel's most *disturbing* invention because it combined devastating power with controlled burning. It spurred the development of repeating rifles, rapid-fire artillery, and eventually machine guns.[112] A version of it continues to be used today as solid rocket fuel propellant. But even with the success of ballistite, an invention that had a smaller initial market but kicked off an entire class of high explosives used by the military, and even though Nobel desperately wanted to move on from the production of dynamite, he felt that to keep his business strong he had to continue manufacturing and selling it at high volume throughout the world.[113]

Nobel was a brilliant scientist and entrepreneur, but as with so many inventors, he had not been able to predict how his most famous invention would be used. As the wave of terrorism unleashed increasing mayhem, he was appalled that dynamite was killing thousands of people, and he reportedly became obsessed with his legacy. A story often repeated in accounts of his life asserts that his decision to endow the Nobel prizes was triggered by a mistaken obituary. Supposedly, upon the death of Alfred's brother Ludvig, in 1888 at the age of fifty-seven, a Paris newspaper confused the two and published an obituary for Alfred instead, describing him as the merchant of death who became rich by finding ways to kill more people faster than ever before.[114] Alfred read it and was shocked. This obituary story has become a part of the Nobel legend and was even repeated by Al Gore when he accepted his Nobel Peace Prize in 2007. The problem is that historians have been unable to find any such newspaper account. No one knows exactly where (or if) it was published. It is just as likely that his lifelong devotion to science, political activities late in his life, and a general sense of regret explain the bequest.

Filled with remorse about how his achievements would be recorded in history, Nobel became deeply interested in the peace movements of the day and even considered himself a pacifist. Perhaps indulging in wishful thinking, in 1892 he conjectured to his friend Countess Bertha von Suttner, a prominent Austrian peace activist and author of *Die Waffen Nieder* (Lay Down Your Arms), "Perhaps my factories will put an end to war sooner than your congresses: on the day that two army corps can mutually annihilate each other in a second, all civilized nations will surely recoil with horror and disband their troops."[115] He thus laid out an early version of deterrence theory.[116]

Alas, he was sadly mistaken, as the outbreak of World War One made all too painfully clear.

In 1895, a year before he died, Nobel hand-drafted a four-page will, leaving virtually his entire fortune, amounting to about $215 million in today's terms, to endow the famous Nobel prizes, with the Peace Prize meant to honor "those who, during the preceding year, shall have conferred the greatest benefit on mankind." Well-intentioned as the endowment may have been, he was criticized in the press for trying to reframe his legacy.

Technology evolves in the hands of human beings. As Alfred Nobel found out toward the end of his life, once an invention captures the popular imagination, it changes form or purpose in ways that surprise its creator. This was also the case for Mikhail Kalashnikov, whose brilliant assault rifle was meant to protect the Soviet Union from invasion and ended up killing more civilians than any other firearm.

CHAPTER 5 | The Kalashnikov and the Global Wave of Insurgencies

I'm proud of my invention, but I'm sad that it is used by terrorists. I would
prefer to have invented a machine that people could use and that would help
farmers with their work—for example, a lawnmower.[1]

Mikhail Kalashnikov, 1919–2013

THE KALASHNIKOV ASSAULT rifle, officially Kalashnikov's Automatic, or Автомат Калашникова in Russian, commonly referred to as the AK-47, is the disruptive technology that facilitated the second global wave of destabilizing political violence, becoming the weapon of choice of insurgents, terrorists, "freedom fighters," organized crime groups, and lone mass shooters.[2] The AK-47 is a mass-appeal weapon of mass destruction.[3] It and its derivatives have killed millions, facilitated the overthrow of governments, turned the tide in wars, leveled the field for insurgents fighting large-scale conventional armies, and inspired its own distinctive breed of attack on innocents. Experts estimate that this one type of weapon today slaughters a quarter of a million people each year—the culprit in devastation from the battlefields of Syria to a schoolhouse in Beslan, from a hotel lobby in Mumbai to a concert theater in Paris.[4]

Though handguns were often used alongside explosives in the first global wave of terrorism, as in the assassination of Archduke Franz Ferdinand, guns lacked the kind of cult-like following reserved for dynamite. They were expensive, difficult to maintain, and most effective in the hands of well-trained, professional soldiers. Anarchists and other social revolutionaries scoffed at firearms, extolling the vastly superior killing efficiency of dynamite. "One man armed with a dynamite bomb," wrote anarchist paper *The Alarm* in 1884, "is equal to one regiment of militia. . . . Anarchists are of the opinion that the bayonet and the Gatling gun will cut but a sorry part in the social revolution."[5]

The AK-47 and derivative rifles changed the equation. Relatively cheap, requiring virtually no training to use, able to unleash a devastating barrage of fire, and virtually indestructible, the rifle rapidly diffused around the globe, becoming the most widely distributed and available firearm in history, outstripping the next family of rifles, the American-made M-16, by five or ten times.[6] Tens of millions have been manufactured, though initially, unlike dynamite, the AK-47 was not commercially sold; it was created, promoted, and disseminated by the Soviet government, which eventually lost its tight control over the technology.

The original Type 1 rifle, produced by the Soviet Union from 1947 to about 1950, became the basis for a plethora of successor models, including Soviet improvements, such as the AKM and AK-74, but also the Chinese Type 56, the Egyptian Misr, and the Iraqi Tabuk. In total, some 200 variants have been fashioned, with the whole class of these weapons commonly referred to with the AK-47 moniker.[7] Such appeal does the name Kalashnikov have that even some rifle models that are not truly automatic (meaning continuing to fire when the trigger is depressed), but rather semi-automatic (producing one shot with each pull of the trigger), are also generally referred to as AK-47s. All of these rifles use one of three types of bullet cartridges, all commonly produced in factories throughout the world.[8] Except when discussing the original 1947 rifle (AK-47) or other specific models (e.g., AKM, AK-74, etc.), they will all be referred to here as Kalashnikovs.

While exact numbers are unknown, in part because the Soviet Union for years kept the numbers it manufactured classified, arms experts estimate there are between 70 and 100 million operational Kalashnikovs worldwide today.[9] Key to the high number is the weapon's longevity; once born, the rifles never seem to die. Fifty-year-old models are still firing today. Rebels in the Lord's Resistance Army in central Africa have wrapped the rifles in plastic and buried them, only to dig them up years later to fire them again without a hitch. Fathers reverentially pass them down to sons, and a black market of used guns thrives.[10]

This chapter will not discuss the technical advantages and disadvantages or distinguishing features of AK-47s and related Kalashnikovs in detail;

much has been written in that vein elsewhere.[11] The focus here will be on why the rifle was disruptive, changing the nature of the market for guns, making firearms with mass lethality available to non-state actors, and instigating an outbreak of revolutionary violence globally. The AK-47 became a political weapon, which, like dynamite, attracted violent individuals and groups not just with its effectiveness and accessibility, but also with its cross-cultural symbolism. Many other guns have spread internationally in a robust global trade in small arms, but the AK-47 and its derivatives have been by far the most popular.

Why? As with dynamite, numerous rifles of superior design have been produced but have not been nearly so widely adopted. Why did those better guns *not* diffuse? A look back at the checkered history of firearms will prove illuminating.

The Evolution of Firearms and Introduction of the Machine Gun

Based on a sculpture from the 1100s of a Sichuan man apparently holding a gun, most historians accept that firearms were invented by the Chinese in the twelfth century, but Europeans were the first to perfect them for individual use, in the fourteenth century.[12] From the arquebus (a smoothbore muzzle-loaded matchlock gun, often perched on a single-leg rest) to muskets to rifles and pistols, firearms evolved gradually from the fifteenth to the eighteenth centuries, following the pattern of sustaining technological innovation. For centuries, guns were individually handmade, cumbersome, hard to maintain, and costly for ordinary people to supply with gunpowder and shot. Advances in design did not generally tip the balance in warfare; during this early period, superior training or better supply chains more often decided the outcome of European wars than differences in armaments did.[13] Beyond professional soldiers, due to the expense of firearms, only members of the aristocracy were likely to own them, except in the American colonies.

Guns on the American frontier were a matter of survival, so European settlers of all classes had them. In 1607, the Virginia colony on the James River required all men to be armed, and those who could not afford their own weapons were given them.[14] Beginning in the 1620s, the English settlers formed citizen militias to protect against Native Americans and potential Spanish competitors.[15] Settlers were encouraged to develop their skills as marksmen, not only for security but for hunting, a vital source of food. The Revolutionary War (1775–1781) was initially fought largely by volunteer part-time militiamen, the famous Minutemen, while over time the Continental Army was professionalized.

Even in a country of democratized gun ownership, though, supplies of gunpowder, shot, and guns ran severely short. George Washington, who transformed the Continental Army into a well-trained fighting force, and who described militia members as ill-disciplined, greedy, and disorderly, had to repeatedly beg the Continental Congress for more weapons.[16] In a letter dated 10 February 1776 to the legislature of the Massachusetts colony he wrote, "I have taken every method my Judgement could Suggest to procure a sufficient number of Firelocks for the Soldiers of this Army. . . . I am constrained by necessity to Inform you, that the deficiency is amazingly great, and that there are not nigh enough to Arm the Troops already here."[17] As the war dragged on, the shortage was partly remedied by stealing firearms from the British and importing them from the French, who, anxious to undermine the power of their hated English rivals, treated the American Revolution as a proxy war. The French supplied the Charleville Musket, which made a major contribution to the army's effectiveness and became the model for the first US manufactured musket, produced by the Springfield armory in Massachusetts in 1795.

The demand of militaries for better firearms technology drove the progression from individually handcrafted pieces to standardized machine production, and in this the United States was a world leader.[18] Samuel Colt manufactured the first large quantity multi-shot revolvers (handguns with revolving multi-chamber cylinders) beginning in 1836, employing industrial-age machining tools that allowed for mass production.[19] That brought the cost of the weapons down enough that they could be readily purchased by individuals for private use. Rifle manufacturers Remington and Winchester soon joined Colt in large-scale production, and their popular revolvers and rifles became the stuff of legend due to their popularity on the Western frontier. The next major innovation was interchangeable parts, which allowed for easy repair by simply replacing a broken component, often cannibalizing it from another weapon. This was spurred by the US War Department, which sought better ways to repair and maintain small arms, with the first mass-produced small arms of this type manufactured in New England during the 1840s.[20]

The four years of the US Civil War (1861–1865) saw many crucial transitions in firearms, especially rifles. Muzzle-loaded smoothbore muskets (loaded at the front end of the weapon, with smooth barrels) and rifle muskets (with spiral-shaped grooves inside the barrel, or "rifling") were used together in the war. Rifling made the guns much more accurate: smoothbore muskets were accurate only to 75 yards; muzzle-loaded rifles were accurate to 200 or 300 yards or more.[21] They were joined by breech-loaded rifles (loaded at the back of the barrel, near the firing mechanism and trigger), which were three times faster to load and greatly preferred by the cavalry. The Spencer repeating carbine rifle, the first repeating rifle suitable for use in combat, was produced at large scale during the war.

Most important to our story was the invention of the machine gun. One early model, developed in 1861 for the Union Army, was the Ager gun, also called the Coffee Mill gun because it looked rather like a big coffee mill on wheels. It fired 120 rounds per minute but was prone to jamming and both heavy and expensive. The Union Army ordered sixty Ager guns but the barrels kept overheating and jamming. At more than $700 each, the gun was deemed a waste of precious funds.[22] A better version was needed, and in 1862, Richard Jordan Gatling invented the first reliably workable machine gun, which was introduced onto the American battlefield that year.[23]

Gatling's purpose, as he described it in an 1877 letter, was to give one man the killing power of a hundred, which he portrayed as a humanitarian goal. Like Alfred Nobel (and many others since) Gatling argued, or at least convinced himself to believe, that the high lethality of the weapon would deter the outbreak of war, as well as making wars shorter and more humane. Not only would the side with the Gatling gun quickly win, he argued, but the power of the weapon would lower the number of troops needed for battle. Since more soldiers died of disease than combat-related injury anyway, that would greatly reduce the death toll.[24]

The Gatling gun was a hand-cranked, multi-barreled revolving cylindrical contraption situated somewhere between an individual firearm and military artillery. Its first prototype fired 200 shots per minute out of six revolving barrels that an operator cranked by hand.[25] The US Army initially hated the Gatling gun: it was cumbersome, often misfired, and had a ferocious appetite for ammunition. The key concern was that keeping the gun supplied would pull funds away from providing individual soldiers with adequate ammunition, which was seen as the higher priority. The experience with the troublesome Ager gun had also soured them on expensive, risky, and unreliable gadgets introduced hastily in the midst of a war. Money would be better spent on standard weapons and ammunition for the soldiers doing the fighting and dying *now*. The gun also violated the military's sense of ethics. One soldier, operating the earlier Ager gun, complained, "It does not seem like soldiers' work."[26]

When Gatling hit a brick wall with Lt. Col. James Ripley, the Union Army's chief of ordnance, he lobbied individual officers, with some success.

Gen. Benjamin F. Butler personally bought twelve Gatlings, at $1,000 apiece, and Admiral David Dixon Porter bought one as well; but the guns did not make a big impression. Though the gun did see some action during the war, the US Army did not officially adopt the Gatling into its arsenal until 1866, and even then was reluctant to use it.[27]

The gun was more enthusiastically embraced in the United States as a tool of political intimidation, particularly for putting down labor strikes and other protests. Between the 1860s and 1920s, many American police departments, state militias, and factory owners purchased Gatling guns and regularly used them for these purposes.[28] When a mob threatened to set fire to the *New York Times* building during Civil War draft riots of 1863, editor Henry Jarvis Raymond ordered three Gatling guns to be set up—two in windows and one on the roof—leading the crowd to back down.[29] Ordered into action by state governors, the National Guard often used Gatlings for crowd control. Responding to an 1877 railroad strike in Pittsburgh, the Philadelphia National Guard brought at least one Gatling gun, although the guardsmen refused to fire it. The Tennessee National Guard deployed a Gatling gun in 1891–1892 against Briceville coal miners, who were protesting the use of convict labor that undercut their wages.

The most notorious use of the guns against striking workers was in response to the Homestead Steel Strike, in the town of Homestead on the outskirts of Pittsburgh, from 30 June to 6 July 1892, at a mill owned by Andrew Carnegie. The chairman of the company, Henry Clay Frick, hired 300 Pinkerton private security guards to put the strike down; but they were met by some 10,000 strikers and quickly overwhelmed. Notably, some of the workers threw dynamite. Eventually the governor called in 8,000 Pennsylvania national guardsmen, who fired on the strikers with Gatling guns.[30] About 21 people were killed and 39 wounded, both workers and private security guards, over the course of the strike.[31] The Gatling gun became a symbol of the suppression of the growing labor movement, and the alignment of state institutions with commercial interests in opposition to populist demands for political and economic reforms.

Eventually the gun gained wider acceptance in the US Army, used to horrifying effect against Native Americans, including in the Great Sioux War of 1876–1877. The gun was still not a decisive factor in most battlefield combat, however, until the Spanish-American war (April–August 1898).[32] Gatling fire paralyzed the Spanish infantry and famously provided cover for Col. Theodore Roosevelt and his "Rough Riders" in the Battle of San Juan Hill. This conferred iconic status on the gun that greatly exaggerated its military utility. The guns still frequently jammed, which meant they required expertly trained infantry to provide a shield and complement their firepower. Teddy Roosevelt was nonetheless an enthusiastic fan and heartily championed the Gatling.

As for the European militaries, they used the Gatling gun as a powerful tool of colonial suppression. Their colonial forces were significantly outnumbered in combat with home armies, and the gun more than leveled the field. In Africa, the British used Gatling guns in the Ashanti war (1873–1874) and, most famously, in the battle of Ulundi in the Zulu war (1879), during which the gun's superior firepower, combined with effective infantry tactics, enabled the British expeditionary force to defeat in just a matter of hours an army of Zulu warriors that outnumbered it four-to-one.[33] The Germans later used the gun to put down mass rebellions in South West Africa and Tanganyika (1904–1907).[34] A vicious racism was expressed in press editorials of the day applauding the use of the gun to inspire awe and obeisance in the "barbarous or semi-civilized foe" and teach a lesson to "savages."[35]

The Gatling gun opened the way to the development of a slew of more powerful and reliable machine guns, whose provenance we cannot fully detail here. Particularly notable was the gun designed by American-born, British inventor Hiram Maxim,[36] which could fire as many as 600 rounds per minute versus the Gatling's original 200 per minute and was a key instrument of slaughter during the First World War. During that continental cataclysm, norms of "civilized" warfare gave way entirely to machine-powered butchery. The Maxim was fully automatic, meaning it used the energy of its own recoil action after firing to eject the empty cartridge case, chamber a new round, and fire again. It worked with ruthless efficiency, and Europeans on both sides soon learned that when their infantry charged over open ground in the face of machine gun fire, they could die just as efficiently as the "savages" had.[37]

Though machine guns became an integral component of every advanced military arsenal, they were still quite cumbersome. The Gatling gun had started out at about 224 pounds, which increased to 500 pounds when loaded with ammunition.[38] That made it essentially an artillery weapon. Designers tried various ways to make it lighter and more usable to infantry. The 135-pound "Camel gun" of 1874 was a miniature Gatling gun with only an 18-inch barrel, mounted on a portable tripod and designed to be transported on a pack animal (such as a camel). By 1892 design improvements had further reduced the weight to about 74 pounds. The Maxim gun weighed only 40 pounds by 1895: 25 pounds for the gun and 15 pounds for the tripod.[39] That was still too heavy for one man to carry, however, especially along with the heavy large-caliber ammunition the gun used, which was now the bigger encumbrance. So machine guns were still closer to artillery weapons than personal firearms appropriate for infantrymen.

During the First World War, finding a way to give individual soldiers greater firepower and agility became the pressing priority. American John Browning developed the first of the hand-held Browning Automatic Rifle (BAR) series in 1917 for use by the American Expeditionary Force, to replace the French-made Chauchat (which easily overheated or clogged) and M1909 Benét–Mercié machine guns (which regularly jammed).[40] The BAR could be slung from the shoulder and fired from the hip to provide "walking fire," but it was never very effective in this role. It was still too heavy and was difficult to maneuver, especially in the confines of a trench.[41]

The Germans, who were the first to deploy a portable automatic firearm in combat, developed the somewhat more effective Bergmann Maschinenpistole 18, or "MP-18." In 1918, a small number of this gun were used by the Stosstruppen (shock troops), who were sent to breach weak points in enemy lines, but since battlefield communications were virtually nonexistent, tactical direction of the troops was sorely limited. Communications with headquarters in the rear was the key limiting factor, as higher commanders struggled to know where to exploit successful attacks or even where the advancing troops were located. Casualties were high and the shock troops struggled to hold ground long enough for follow-on units to capitalize upon their gains.

Despite these problems, the final German offensive, launched in spring 1918, initially made great progress using shock troop tactics, primarily because the violence and precision of the attacks disrupted Allied command and control. Some historians argue that the MP-18 was important to that success, although given the limited numbers available this assertion is debatable.[42] In any case, the Germans understood the weapon's value and continued the development of lightweight automatic firearms, ultimately leading to production of the first true assault rifle.

An assault rifle is a rapid-fire intermediate-range short rifle designed for infantry use, capable of semi-automatic (one trigger pull per round) or automatic fire (one trigger hold for a continuous stream), fed by a high-capacity magazine that shoots medium- or small-caliber cartridges. The first true assault rifle to be mass-produced was designed for German troops by Hugo Schmeisser twenty-five years after he created the MP-18. This Sturmgewehr, or storm rifle, was midway between a machine gun and what came to be called the submachine gun. It was fully automatic, with ammunition that was bigger than for a pistol but smaller than for a standard rifle. It had more range and power than a submachine gun, without the tripod, weight, and ammunition-feeding device of a full machine gun.

The Germans first used the Sturmgewehr very effectively in combat to cover infantry assaults and engage in deadly close combat toward the end of the Second World War (1943–1945), although by this point the war's momentum had already turned against them.[43] These rifles were retrieved on the Eastern front by the Red Army, and in 1943, one was demonstrated before the Soviet People's Commissariat of Arms. The Commissariat launched an intensive innovation program, involving at least a dozen well-established professional Soviet arms designers, as well as one lowly tank driver who was a novice to firearm design, Mikhail Kalashnikov.[44]

Kalashnikov's Invention of the AK-47

Determining how much of the official history of the invention of the AK-47 is true and how much a product of Soviet propaganda is difficult.[45] The Second World War, called the Great Patriotic War in the USSR, killed 26 million Soviet citizens, both military and civilians, roughly 14 percent of the population, ravaging hundreds of cities and towns.[46] Joseph Stalin well knew that wars can provide an opening for domestic unrest: the First World War had catalyzed the 1917 revolution that brought the Bolsheviks to power. He was not about to risk a similar scenario. To shore up domestic morale, the USSR needed inspirational role models, and Stalin particularly sought to glorify the peasantry. So it was that Mikhail Kalashnikov, a Red Army sergeant with intense blue eyes, a pleasing countenance, and a fabricated peasant's pedigree, became a national symbol of proletarian grit and peasant ingenuity, and the subject of fierce Russian pride.

Senior Sergeant Mikhail Timofeyevich Kalashnikov was born in the village of Kurya, on 10 November 1919. In the southern central Altai region of Siberia, Kurya is just across the border from what is now Kazakhstan. Disfiguring pockmarks on Kalashnikov's face testified to a brutal childhood, during which he barely survived smallpox and endured the deaths of eleven of his eighteen siblings. What was not widely known about his early life was that his father was a kulak, meaning a privileged peasant who owned his own farm. Though life could be quite difficult for these landowners, they at least could grow their own food and keep a small number of farm animals.[47] They were prime targets of the Soviet state and were stripped of their land holdings in Stalin's collectivization of agriculture (1929–1933). The Kalashnikov family lost all its possessions in 1930, their cows and sheep senselessly slaughtered in front of them, and they were exiled to a large collective farm (*kolkhoz*) in western Siberia.[48] Shortly thereafter, during a bitter winter, Kalashnikov's father, Timofey Aleksandrovich, died. Mikhail was eleven.

This part of his biography was left out of the official government story of his life, and Kalashnikov wrote in his memoirs that for decades, even after he'd become highly celebrated, he feared that his family's deportation and less than ideal Communist lineage would come to light and that he would be denounced and drummed out of his sensitive work.[49]

When he was nineteen years old, in 1938, Kalashnikov was drafted into the Red Army. Soviet Army boot camp was never a picnic, but Kalashnikov was brutally bullied because of his short stature (5'3"). Yet his height also made him perfectly suited for the cramped interior of a tank. He had not finished high school but he was bright and showed a keen aptitude for machinery.[50] So the Red Army sent him to the Tank Mechanical School in Kiev for training as a tank driver and mechanic.

Three years later, in June 1941, the Germans launched Operation Barbarossa, a massive invasion of the Soviet Union. Over 3 million Nazi soldiers were supported by some 650,000 troops from Finland and Romania, who were later supplemented by troops from Italy, Croatia, Slovakia, and Hungary. Stalin had ignored warnings of Hitler's imminent attack and was in the midst of a paranoid purge of the Red Army's senior officer corps, executing some 15,000 to 30,000 officers, including many of the most skilled personnel, at just the time they were most badly needed. Partly as a result, the Germans swept across the western Soviet Union in a matter of weeks, attacking Leningrad (St. Petersburg) in August and encircling Kiev, in Ukraine, by the end of September.

In October 1941, now twenty-one-year-old Kalashnikov took part in the Battle of Bryansk, a town on the Desna River about 200 miles southwest of Moscow. The engagement was part of the Red Army's desperate attempt to repel the Nazi invasion and protect the capital city. In Bryansk, the Nazis killed some 80,000 Red Army defenders, as well as 80 percent of the townspeople.[51] When a Nazi shell hit Kalashnikov's T34 tank, it drove pieces of the tank's armor into his chest and arm, causing serious injury.[52]

As the city was overrun, Kalashnikov joined a group of wounded Red Army soldiers being trucked to a hospital. Reaching a deserted village, Kalashnikov, his lieutenant, and the driver left the vehicle to reconnoiter the local area, and in their absence, the Germans discovered the truck and executed all the wounded Soviet soldiers being transported, slaughtering them at close range with automatic fire. The image of this carnage seared into Kalashnikov's memory, and in later years he had recurrent nightmares about it. For days Kalashnikov walked, bleeding from his wounds, until finally reaching the Red Army line. He was sent to a hospital in Yelets, which is about 250 miles due south of Moscow.

According to the official story, Kalashnikov was inspired to design a better weapon by the horror of watching his comrades so viciously slaughtered. He made his first sketches for a new firearm while he was convalescing from his wounds in 1941 and 1942.[53] All around him lay wounded and traumatized Russian soldiers recounting stories of Nazi rapid-assault tactics and the disadvantages of Soviet firearms in meeting them. Their long rifles were no match for the Germans' hail of bullets. Some Soviet troops were equipped by this point with a submachine gun, the PPsh41, which was highly prized, by German as well as Soviet troops, for its high rate of fire. But it used pistol ammunition and was noisy, very heavy, hard to steady, hard to load, and short-ranged.[54] A better gun was needed.

Here is where the story becomes difficult to disentangle from the spin of the great Soviet propaganda machine. The AK-47 legend claims that, despite the fact that Kalashnikov had never finished high school nor studied physics, engineering, or metallurgy, his natural mechanical talent led him to the initial innovative yet simple design of the weapon, conceived on his own from his hospital bed. Over subsequent years, while he worked in a railroad depot and at various other armament sites, Kalashnikov purportedly refined the concept and shepherded it into production.

In his account, Kalashnikov credited others with helping with refinements but claimed to be the one most responsible for the gun's development and various improvements over the course of the Cold War (Figure 5.1). Others dispute that, characterizing the design as the collective work of many engineers, and the weapon as a hybrid of available technologies.[55] In the absence of archival records, the exact role played by Kalashnikov versus others cannot be determined. But there is no doubt that Kalashnikov was talented and played a crucial role as part of a team that designed and modified the gun in a trial-and-error process involving more than 100 modifications between 1947 and 1949.[56]

A key factor in the rifle's design was that the Soviet government had already determined what the AK-47's ammunition would be, and that later played a vital role in the gun's diffusion. To understand why, we have to backtrack a bit.

For decades, a heated debate had raged among military leadership about whether to place priority in rifle development on lethality or agility. During the First World War, soldiers in the trenches were equipped with high-velocity, heavy bullets that could strike and kill a man a mile away.[57] Both the rifles and the cartridges weighed soldiers down and tied them to supply

FIGURE 5.1 Mikhail Timofeyevich Kalashnikov, State History Museum, Moscow.

lines, since they couldn't carry much ammunition.[58] The powerful recoil (or kickback) of these rifles also meant soldiers had to be strong and well-trained. But a shift to smaller cartridges would require expensive retooling of ammunition factories, and would also make rifles shorter-ranged and less accurate, which many military leaders thought was a move in the wrong direction.[59] Smaller-gauge rifles would actually be *less* lethal, and why would anyone want their soldiers' weapons to be less deadly?

The Germans disagreed. During the interwar years, they developed their mobile assault tactics, which placed a premium on troop agility. In defiance of the Versailles treaty, they secretly began to develop smaller, lighter cartridges.[60] They were able to devise a cartridge that was the same diameter as their current standard ammunition, but cheaper, shorter, and lighter, so very little retooling was necessary.[61] German factories were ready to produce it in 1938, although the rifles that used them took longer to develop and were only fielded in late 1943.

The United States also developed a smaller cartridge, shot by the M1 Carbine semiautomatic rifle, which was distributed to troops beginning in

1942.[62] The M1 Carbine was a lightweight weapon, more powerful than a submachine gun but less powerful than the German Sturmgewehr; in fact, some American soldiers derisively called it a "pea shooter." But others felt the speed and ease of handling it outweighed its disadvantages.[63] The M1 Carbine went into large-scale manufacture in 1942 and more than 6.2 million variants were produced over the course of the war, exceeding the manufacture of any other American small arm.[64]

In the Soviet Union's desperate fight for survival, Soviet intelligence officers took note of these designs and shared them with arms teams, who had been experimenting with lighter rifle ammunition but had faced stiff opposition from their leadership.[65] That opposition dissolved. In 1943, they designed the M43, which was quite similar to the German cartridges.[66] They also crafted a family of rifles to fire the cartridges, including the Degtyarov RPD light machine gun[67] and the Simonov SKS semi-automatic carbine.[68]

As the Second World War ended, and in the wake of the atomic bombings of Hiroshima and Nagasaki, Stalin feared a Western invasion of the shattered Russian homeland, a replay of Operation Barbarossa that this time would likely succeed. In 1945, he intensified the pressure on both the Soviet nuclear development teams and conventional arms designers to conceive a new weapon that would ensure successful civil defense. A design contest challenged engineers to create a new automatic rifle that could fire as rapidly as a Maxim gun (600 rounds per minute) but be handled by a single person. It had to be lightweight, reliable, simple to operate, and easy to manufacture, with a small number of parts.[69] It should also use the intermediate 7.62 × 39 mm ammunition, by this point being produced at high volume in Soviet arms factories in Barnaul (in western Siberia), and Tula (south of Moscow).[70]

The Soviet arms bureaucracy bore down to meet the challenge. Kalashnikov was assigned to the Schurovsky weapons-testing center near the town of Kolomna, 70 miles southeast of Moscow, one of many armaments factories that had been hastily relocated away from the vicinity of Moscow and Leningrad as Nazi troops closed in. There he joined esteemed Soviet arms designers such as Sergei Simonov, Aleksei Sudayev, Nikolai Rukavishnikov, and Vasili Degtyarev, who contributed to the assault rifle project.[71] While many brilliant minds therefore clearly played a role in creating the AK-47, Kalashnikov's common touch may help explain the simple genius of the ultimate model.

One of three final options considered, culled from a competitive field of some fifteen entries, Kalashnikov's assault rifle prototype was evaluated in comparison to those by famous weapon designers A. D. Bulkin and Vasily Dementyev.[72] Bulkin was the first Soviet designer to incorporate a rotary bolt (the rotating metal piece that ejects a spent cartridge and then picks up a new one) in a rifle design, which was later incorporated into the AK-47 design, one clear case of the contributions of others.

The contest judge, Vasily Lyuty, favored Sergeant Kalashnikov as the perfect, charming proletariat hero, awarding him the win,[73] and in January 1948, Kalashnikov was sent to Izhevsk Motor Plant No. 524, a manufacturing center in the Western Ural Mountains where the first 80,000 AK-47s were mass-produced.[74] Hugo Schmeisser, the German designer of the Nazi Sturmgewehr rifle, had been taken prisoner by the Soviets after the war and forced to work there. Some reports claim that Schmeisser played an important role in modifying the rifle during its transition from design to production, but that is contested and impossible to confirm.[75] In 1949, the Red Army adopted the Kalashnikov as its standard-issue rifle, and that same year, Sergeant Kalashnikov earned the Stalin Prize, one of the highest honors awarded to Soviet citizens. [76]

Kalashnikov was demobilized in 1949, returning to civilian life, but he became a member of the reserves and wore a uniform for the rest of his life. A public hero, he worked as a senior official in the arms-manufacturing bureaucracy for the rest of his career; served six terms on the Supreme Soviet, the Soviet Union's top legislative body; and was hailed as a model Soviet patriot despite his kulak history. He received a number of honorary military promotions in the reserves, the timing of which is interesting. According to Kalashnikov, he started as a staff sergeant and remained at that level for many years, but then an American newspaper article highlighted that a "mere sergeant" was arming the Warsaw Pact. Suddenly a slew of promotions began, raising him to the rank of colonel, the highest allowed under a Council of Ministers decree that forbade honorary promotion to general during peacetime. After the USSR dissolved, President Boris Yeltsin appointed him lieutenant general.[77] Kalashnikov died in 2013 at the age of ninety-four.[78]

Why the AK-47 Was So Widely Adopted

It was not the sophistication of the AK-47 that made it special: all of the component parts were adapted or derived from earlier concepts and models, some of them far more advanced. But the final product reflected a series of compromises that made the gun perfect for mass production by a socialist state with a largely unprofessional conscript army.

The genius of the gun was its simplicity. It was designed for peasants with no prior military training. It was easy to carry, shorter than the infantry rifles that it replaced but longer than the submachines guns that preceded it, and lighter than comparable rifles, weighing only about 10 pounds, less in later models. Its parts (Figure 5.2) were machine-produced and interchangeable from gun to gun, and conscripts could learn how to assemble and disassemble it in a few hours. In short, the gun was ideal for lightly armed, highly mobile troops, traveling long distances.

When the Kalashnikov appeared, it was widely seen as technologically un-impressive.[79] Firearms experts throughout the world ignored or even derided it. Many better firearms were available, such as the British Lee-Enfield rifle[80] or the US M1 Garand, which General George Patton called "the greatest battle implement ever devised."[81]

By comparison, the lowly Soviet gun was inaccurate, shorter-ranged, and had loose-fitting pieces that rattled. But the AK-47 was designed for toughness and durability rather than long-range accuracy. For a fighting force of novice marksman, target discrimination was not a top priority.[82] The intermediate-sized cartridges were lighter than long-rifle cartridges, and soldiers could carry twice as many rounds, even if they were not physically

FIGURE 5.2 The AK-47. © Johanna Parkin/DigitalVision/Getty images.

strong. The smaller cartridge also meant there was less recoil (or kickback), so just about anyone could fire it without being knocked down.

Reliability was a priority, and the Kalashnikov was truly remarkable in that respect. During the initial development of the gun, Soviet testers tried everything they could think of to jam it. They dragged it through ash, broken bricks, and sand; it fired. They soaked it in bog water; it fired. They tested it with heavy, sustained use to clog it with carbon; the rattle-trap parts shook off the residue and fired. According to Kalashnikov, during a test that fired the weapon choked with sand, his co-worker Sasha Zaitsev exclaimed: "Look, Look! The sand is flying in all directions like a dog shaking off water—look."[83]

Subsequent models were also rust-resistant, because the bore and chamber were chromed and coated with a protective finish during manufacturing. The rifle worked in snow, rain, or desert conditions. Even jungle humidity didn't tax it, as the Viet Cong soon learned. The folding stock Kalashnikov, designed for paratroopers and soldiers operating in cramped spaces (such as a tank), could be hidden in a coat.[84] Kalashnikovs seemed to last forever, tolerating weak springs, rust, dirt, heavy use, and years of neglect, returning to lethal effectiveness with almost no maintenance. Although its durability was sometimes exaggerated, the gun soon became legendary.

The Soviets kept refining the rifle. The updated AKM model quickly followed the original AK-47 and became the standard infantry model for Soviet soldiers beginning in 1959.[85] Stamped steel production was by then viable, making the AKM cheaper to mass-produce as well as about a third lighter than the original AK-47.[86] The simplest Kalashnikovs had no more than eight to ten moving parts, all big enough to be difficult to lose.[87]

The AKM was followed in 1974 by the AK-74, first seen in combat during the 1979 Soviet invasion of Afghanistan.[88] But the AKM remains the most popular Kalashnikov-type assault rifle worldwide. Today, models from the 1960s routinely show up in the arsenals of former Soviet allies and the Russian army—not to mention civil war factions, terrorist organizations, criminal syndicates, insurgent movements, and other non-state groups.

A Humble, yet Disruptive Innovation

From the time Niccolò Machiavelli wrote *The Art of War* in the sixteenth century, European military advancement focused on professionalizing fighting forces. Nation-states developed elaborate bureaucracies to support and train their armies, drilling soldiers to march in concert, stand against fire, quickly reload, and clean and maintain a rifle. Increasingly complicated weapons systems required technical training, and scores of military academies and war colleges were opened during the nineteenth century to train large staffs of highly educated officers. Only the industrialized states could afford this schooling, which included instruction not just in strategy and leadership, but also in logistics, systems analysis and, increasingly, the employment of high technology. As time went by, navies and air forces were especially reliant on advanced technology and weapons systems. The necessity of building large bureaucracies to support such highly professionalized and technically advanced forces was a key driver in the evolution of the modern European industrialized nation-state.

The unpretentious Kalashnikov rifle reversed the equation. By minimizing the expertise needed for ordinary infantryman to care for and shoot their weapons, it obviated professional training. It was a more potent weapon than the American M-1 Garand and British Lee-Enfield because anyone could use it. As the volume of AK-47s and successor Kalashnikovs rapidly multiplied, almost anyone could buy or steal one and then learn to use and care for it in a matter of hours. That is why it would soon change the world.

CHAPTER 6 | How the Kalashnikov Diffused

B Y WIDENING THE range of people who could acquire and use an assault rifle, the Kalashnikov altered the nature of international conflict. The gun became an emblem of the Cold War, one of the most recognizable objects in the world, and its diffusion was driven not only by its technological features, but also the global political, economic, historical, and cultural context into which it emerged. As C. J. Chivers put it, if the AK-47 had been invented in Luxembourg, it would probably never have been heard of again.[1] But it was invented in Russia, and right from its first public appearance, the AK-47 gained worldwide notoriety as a weapon that could disrupt the Cold War balance of power.

The Kalashnikov's Debut and Public Demonstration

At first the Soviet Union took care to keep the rifle, and its light and inexpensive ammunition, a secret. In the early 1950s, Soviet soldiers carried their Kalashnikovs in special pouches that disguised their shape, and they carefully picked up spent cartridges.[2] Although adopted by the Soviet Army in 1949, the gun was not produced in mass quantities for years, and it does not seem to have been used in the Korean War, despite the fact the Soviets supplied armaments to the North Koreans and their Chinese allies. The first time the firearm was used in combat was apparently in the suppression of a small uprising in Berlin in 1953.[3] It may also have been used to crush the

June 1956 Poznan uprising in Poland, when Soviet-Polish generals directed security forces to fire into protesting crowds. We know that Poland got the rifle that year, but we can find no reliable evidence of which firearms Polish military and security forces used against protestors.[4]

The first indisputable use of the gun in a broad conflict was during the Hungarian uprising, from 23 October to 4 November 1956. A photograph featuring the gun hit the pages of *Life* magazine and made news throughout the world. In the 1950s, picture-filled weekly news magazines were the main visual medium in the Western world, as television did not appear in nearly every home until the 1960s.[5] The AK-47 appeared in a strongly pro-American mass-market weekly just a year after the Warsaw Pact between the USSR and seven East European Communist satellites had been signed.[6] For the Soviets, the gun's star turn was embarrassing: it was brandished not by a Soviet soldier but by a Hungarian rebel.

The Hungarian revolt was spurred by a student demonstration in Budapest expressing support for anti-Kremlin, anti-Communist protests in nearby Poland. The Hungarian protestors began to demand "national independence and democracy" and tore down a huge statue of Stalin that stood at the center of Budapest, the capital city.[7] Other cities joined in with their own rallies. When students rushed the state radio station, Hungarian state security forces (AVH) opened fire on the unarmed demonstrators, killing and wounding a number of them.[8]

Anger and unrest quickly spread throughout the country, resulting in a complex political situation involving an ambivalent Hungarian Army, a police force that sided with the demonstrators, and organized civilian militias throughout the country led by local volunteers. Since Hungary was a member of the Warsaw Pact, Soviet forces were stationed there, and they were soon joined by thousands more Soviet troops from Romania and Ukraine in a massive show of firepower.[9] Ultimately around 31,500 Soviet troops and a full retinue of tanks, armored personnel carriers, fighter planes, and bombers converged on Budapest.

On 24 October, they entered the city, and the population reacted in anger. According to declassified Soviet records, even the Presidium of the Central Committee (CPSU), the governing body of the USSR, understood that the show of force had further escalated tensions.[10] On top of that, Soviet tanks became bogged down in the narrow streets and, as they lacked two-way radios, were unable to call for backup.[11] In some areas, crowds of people armed with kitchen implements and gasoline-fueled Molotov cocktails fought them.[12] Members of the Hungarian Army began defecting to the rebel side.[13] By midafternoon, at least 25 protesters had been killed and more than 200 wounded.[14]

The rebel holding the AK-47 in another now-famous photograph taken a few days later was twenty-two-year-old József Tibor Fejes (Figure 6.1), who had probably obtained the gun from a dead Soviet soldier. He is wearing a scarf and a jaunty bowler hat, with the gun's strap slung across his shoulders and the rifle held across his chest, standing defiantly before a captured Soviet tank.[15] Instead of a glorious image of a Red Army soldier, the gun's Western debut was in the hands of a Hungarian insurgent, fighting Soviet soldiers and killing their Hungarian Communist allies. The Kalashnikov was out of the bag.

Unfortunately, the notoriety of the photograph did no favor to young Fejes. According to CPSU documents, the Soviet leadership still thought the crisis could be resolved through negotiation, and it withdrew most troops from Budapest. But on 30 October, a mob attacked the Budapest Communist Party headquarters, seizing the building and sparking alarm in the Kremlin.[16] By the next day, newsreels were publicizing scenes of bloody reprisals by protestors against the Hungarian AVH. Fearful of a domino

FIGURE 6.1 Hungarian Revolutionary Joseph Tibor Fejes holding an AK-47, 1956. Credit: Hungarian National Museum.

effect throughout Eastern Europe, on 1 November the Kremlin sent the withdrawing Soviet troops back to the city for a second violent intervention.[17] This time they crushed the rebels.[18]

Still wielding his Kalashnikov rifle, Fejes was part of a group of rebels roaming the streets on 30 October, looking for members of the AVH. Lieutenant János Balassa was leaving his flat at exactly the wrong time. The rebels asked for his identity card, confirmed that he was AVH, and executed him on the sidewalk. According to the transcript of his trial three years later, Fejes fired the Soviet AK-47, riddling Balassa's body with bullets.

So, shortly after Soviet troops crushed the revolt, the AVH came after Fejes. In April 1959, he was publicly tried, convicted, and hanged for the killing.

Trading in Kalashnikovs

In the 1950s, production of AK-47s was ramped up. Enormous Soviet factories churned them out, and the Kalashnikov became one of the Soviet Union's top exports. At the time, most Soviet non-defense technology and consumer products were widely considered inferior to those of the rest of the developed world. This was not a result of a lack of talent or ingenuity. Soviet scientists were among the best in the world, making brilliant advances in pure physics, biology, chemistry, and mathematics. But innovation was largely limited to the defense sector. In the Soviet command economy, the central government determined the nature, prices, and quantities of goods produced, with the Central Planning Agency (Gosplan) issuing five-year economic plans, designed to build an independent communist state. The domestic economy was deliberately cut off from the rest of the world and protected from the dynamics of free market capitalism.

Without any mechanism for short-term adjustments in supply according to demand, the economy was plagued by both shortages and surpluses, while the lack of free market competition hampered innovation in the production of consumer goods and non-military technologies. As a result, exports amounted to no more than about 4 percent of the total GDP, and they were dominated by raw materials such as oil, minerals, and wood rather than manufactured products.[19] Most technology, other than for defense, was imported.

The lack of earnings from exports was combined with government control of the value of the national currency, the ruble, which was withheld from global currency markets. This left the Soviet economy starved for convertible cash for the purchase of imports because, outside of the USSR, the ruble had no value. Within the USSR, dollars and other foreign currency could be exchanged on a thriving black market, with the ruble garnering a small fraction of its officially set, and grossly inflated, price. Visitors to the Soviet Union found that the official rate for a ruble paid for at a hotel, for example, was easily four or five times its black-market rate discreetly (often riskily) bought on the street or in a park. If on the government exchanges a single ruble cost $1.70, for example, on the black market it would cost just $.25. But while the ruble had no value outside the USSR, Kalashnikovs certainly did, and they became a form of global Soviet currency, especially precious while the USSR was still the primary producer.[20]

Strong international demand for the guns led to Soviet government directives to churn them out by the thousands, at first for barter with trading partners, mostly the communist Eastern European countries of the Warsaw Pact (about 67 percent of trade), but also China and less developed countries. Along with other armaments, and eventually the construction of arms factories and sharing of specifications, AK-47s were traded for manufactured goods, agricultural products, or textiles. The supply of Kalashnikovs became a key form of political influence, strengthening ties with the USSR's allies and helping bring other nations into its sphere of power in the Cold War battle for global supremacy.

The Soviet defense industry also had strong structural incentives to increase the production of armaments regardless of trade in them. Higher production equated to more domestic political clout for the Military Industrial Commission (VPK), which, rather than the armed forces; directed research and development; conducted weapons tests; and decided where, when, and how much of each weapon system would be produced. It also kept defense industry workers gainfully employed at above-average wages in an economy where other sectors struggled. The result was that the Soviet arms industry regularly produced far more weapons than the armed forces wanted.[21] At least 5 million Kalashnikovs were made by the two Soviet ordinance factories in Izhevsk during the first ten years of production (1949–1959), creating a large surplus.[22]

The Diffusion of Kalashnikovs

We cannot follow the diffusion of the AK-47 and the fluctuations in its price with the same precision that we applied to dynamite because the data are not available. The Soviet Union kept no public records of Kalashnikov sales and prices. Also, AK-47s were often bartered rather than sold, and when the

USSR did sell the weapons, it routinely set prices artificially. Good data on the spread of small arms only began to be collected in about 1999, three generations too late in the life of this technology.[23]

Numerous researchers have tried to employ contemporary Kalashnikov price dynamics to predict the outbreak or intensity of conflicts. One very interesting 2006 study by Phillip Killicoat, then a graduate student at Oxford University, used journalist accounts and various small datasets to come to the sensible conclusions that Kalashnikov prices rise when a country regulates the arms trade and fall in countries that have porous borders, and that the rifles destabilize states when large numbers are remaindered from prior conflicts.[24] The well-respected research center Small Arms Survey, established in 1999, headquartered in Geneva, and supported by funding from numerous governments and UN agencies, also tracks the spread of Kalashnikovs around the world. They distribute impartial information about weapons and collect eyewitness sightings and counts from a network of dozens of field researchers, non-governmental organizations, and other informants in major conflict zones including Syria, Iraq, Libya, Bosnia, Nigeria, Afghanistan, Mali, and elsewhere. But even their data are incomplete.

In trying to document the Kalashnikov's diffusion over the decades since its invention and its impact, the best (though imperfect) options are to trace the spread of factories that produce Kalashnikovs and to examine the patterns and outcomes of revolutionary conflicts that have heavily relied on them. That analysis is the basis of the account that follows.

A Proliferation of Factories

By the mid-1950s, the Kremlin began to export the technical schematics and licenses required to construct arms factories that could produce Kalashnikovs and their M1943 ammunition. Stalin had restricted sharing of the AK-47 technology so as to maintain an edge for the Soviet Army. Even when in 1951 Stalin and Mao Zedong signed a secret agreement for the USSR to arm the Chinese People's Liberation Army with Soviet weapons and build a major arms factory in China, Stalin held back the Kalashnikov, yielding only the instructions to make older Soviet firearms such as the Russian Mosin-Nagant rifles. At the time of Stalin's death in March 1953, the Kalashnikov was still produced only in two Russian weapons factories in Izhevsk.

Stalin's successor, Nikita Khrushchev, was responsible for the dramatic proliferation of Kalashnikov factories outside of the USSR. In the immediate post-Stalin succession struggle for power, Khrushchev made it his signature policy to increase the Soviet profile on the global stage, not least so as to strengthen his own political position domestically. A key element in Khrushchev's démarche was shoring up political connections between the Soviet Union and its Communist allies, especially the People's Republic of China (PRC). In 1954, Khrushchev became the first leader of the Soviet Communist Party to travel to the PRC.

Between 1954 and 1959, the USSR gave China a massive influx of aid and direct technology transfer, amounting to about 7 percent of the Soviet Union's national income—an enormous sacrifice when the Soviet economy was still struggling to regain its footing after the Second World War.[25] Even as Soviet citizens suffered, aid to China took precedence over postwar recovery and industrial development in Russia. In 1955, under pressure from Mao, Khrushchev finally shared the technical schematics for the AK-47, along with the SKS-45 carbine rifle (Samozaryadny Karabin sistemy Simonova) and the M1943 ammunition for both. By 1956, the Chinese were churning out their own version of the assault rifle, the Type 56, produced in Factory 626 in Bei'an, a small city in Northeast China.[26]

The USSR had formed the Warsaw Pact in May of 1955 to strengthen its military ties to its Communist allies in response to the rearmament of West Germany and its entry into NATO that year. One of the first priorities was to standardize the equipment of its socialist Eastern European satellite states, especially small arms and cartridges, to make them interoperable. All of the Eastern European allies except Czechoslovakia (which had its own weapons design bureaus) and Albania (which received arms assistance from China) adopted the Kalashnikov as their standard rifle, and technical specifications and licenses to produce the rifle and its ammunition were also shared with them.[27] With the help of Soviet advisers and financial subsidies, armaments factories producing Kalashnikovs sprang up in Poland (1956), East Germany (1957), Bulgaria (1958), Hungary (1958), and Romania (1963).[28]

That same year, the Kremlin arranged a massive sale of Soviet arms to Egypt, under the cover of Czechoslovakia. The Egyptians had been negotiating for arms with the United States and Great Britain, and in this Cold War chess move, Khrushchev brought Egypt into alliance with the socialist bloc instead.[29] By 1958, the Egyptians had acquired not only a remarkable array of Soviet arms but also the specifications for the Kalashnikov, and by later that year a plant in Cairo was churning them out by the thousands. The Kremlin also shared the specifications with North Korea that year, which built a Kalashnikov factory in Kanggye (near the border with China).[30]

Under the Soviet system, intellectual property rights were treated like any other kind of property in a purely communist system: as state-owned public commodities.[31] To get a sense of how different this approach was from that of Nobel Industries and the Powder Trust, consider this sentence from a 1948 Soviet legal textbook:

> However, it should be pointed out that Soviet inventors, in contrast to inventors in a capitalist society, are not interested in retaining a monopoly for their inventions; being advanced men of production, they are concerned with the utmost utilization of their suggestions by the socialist enterprises.[32]

In short, for the Soviets inventing something was cast as an altruistic act.[33] For designers in the Soviet Union, its allies, and elsewhere, the firearm became like an open platform for anyone to adopt and then tinker with.

Being opposed to the capitalist model, the Soviet Union did not abide by the international business standards and agreements that were in place among Western countries. As a result, it would not have been able to patent the AK-47 rifle abroad when it was invented. An inventor's certificate acquired in the Soviet Union had an ambiguous legal standing overseas, and it was not until 1959 that the USSR began to seek foreign patents on domestic inventions at all.[34] The Soviet Union only fully acceded to the Paris Convention for the Protection of Industrial Property, the basic reciprocal international agreement for the protection of patent and trademark rights, in March 1965.[35] (The original convention dates to 20 March 1883.) This meant that during the early period of global diffusion of Kalashnikovs, infringement of any Soviet patent was a moot point, and by the time the USSR might have tried to assert foreign patent rights, the original rifle was no longer a novelty. And as with dynamite, the simplicity of the gun allowed others to easily replicate it.

The specifications for the gun could be derived through reverse-engineering. That is how the government of Josef Tito in Yugoslavia, which was not aligned with the Soviet Union during the Cold War and so was not given either plans or guns, nonetheless designed its own version. Conflicting reports assert that Tito obtained some rifles himself on his 1955 visit to Egypt or that the government acquired some guns from defecting Albanian border guards in 1959.[36] Either way, Yugoslavia began producing its own variant at the Zastava arms factory in Kragujevac (now in Serbia) in 1959.[37] That brought the total number of factories to eleven: two in Russia, six in Eastern Europe, one in the Middle East, and two in East Asia (Figure 6.2).

These secondary factories in turn ratcheted up the diffusion of Kalashnikovs. Mao followed the Soviet lead and provided Kalashnikovs to China's allies, including the Viet Cong and the North Vietnamese Army.[38]

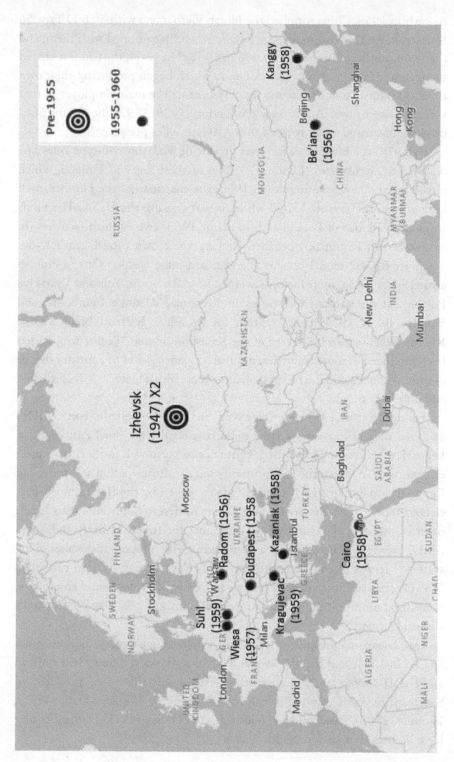

Figure 6.2 Early Kalashnikov Factories (1947–1960).

As early as 1958, China supplied the North Vietnamese with 50,000 Type 56 Kalashnikovs, even before their own army was fully equipped with them, and then sent another shipment of 90,000 in 1962.[39] In 1964, the Chinese shared the technical specifications with Albania, which began producing their own rifles in Gramsh, and the factory quickly became the main employer in that small, isolated town. At its peak production (1978–1992) the Gramsh factory manufactured as many as 24,000 AK-type rifles per year.[40]

By 1970, the number of factories producing Kalashnikov-type rifles had doubled to twenty-two (Figure 6.3). Two of these were in Finland, which now wanted a new assault rifle. In 1956, the chairman of the Finnish small arms committee saw an AK-47 while he was visiting the USSR. The Finns smuggled one out of Poland and, in about 1963, two Finnish factories (in Jyväskylä and Riihimäki) began producing their own model, the RK-60. The later RK-62 model became the standard-issue weapon for the Finnish armed forces.[41] Two more factories were built in Russia, in Tula and Vyatskiye Polyany, along with new factories in Romania, Albania, Cambodia, and Bangladesh. A 1970 report written for President Richard Nixon's Blue Ribbon Defense Panel observed of the Kalashnikov that "[T]his weapon has been produced in more countries, in greater quantity, and to a greater degree of international standardization than any other rifle in history."[42] Its diffusion had only begun.

Following the 1967 Six Day War, the Israeli Defense Forces (IDF) analyzed the performance of their main battle rifle, the Fusil Automatique Léger (FAL) and found it wanting. It did not perform well in the dusty desert environment in which the IDF primarily fought. Examination and testing of Egyptian and Syrian AK-47s and AKMs recovered by the IDF from the battlefield showed that they performed better in desert conditions than Israeli firearms did, and the chief weapons designer for Israeli Military Industries (IMI), Yisrael Galili, designed a new assault rifle around the Kalashnikov system.[43] It was re-engineered to use the NATO 5.56 × 45 mm cartridge, NATO's standard intermediate caliber, so that Israeli rifles would employ the same-sized ammunition as the countries that might give them aid or fight alongside them.[44]

IMI began manufacturing the Galil ARM at its factory in Ramat ha-Sharon in 1971, and the IDF began deploying the new rifle to its troops in 1972. IMI subsequently licensed the Galil assault rifle to be produced in South Africa (1982), Columbia and Italy (later in the 1980s), Croatia (1995), Ukraine (2008), and Chile and Vietnam (2014). Several US companies also have licenses to manufacture the Galil.

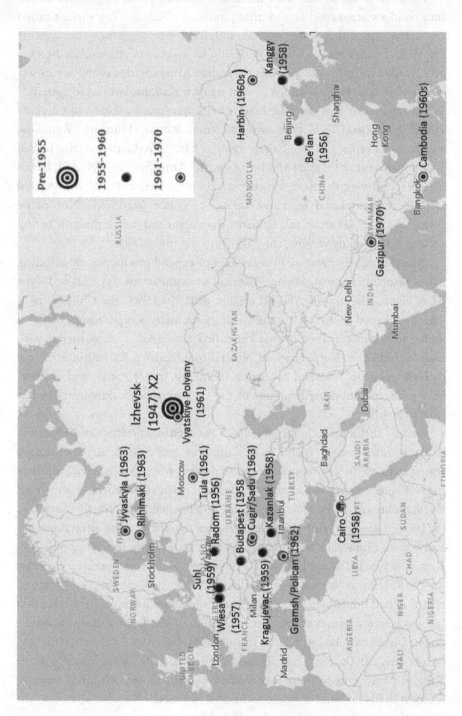

FIGURE 6.3 Kalashnikov Factories (1947–1970).

Diffusion in the Middle East continued when in 1978, Yugoslavia helped Iraq build a state-owned factory that produced a Kalashnikov variant called the Tabuk, rifles that some Afghan and Iraqi insurgents now use against the United States and its allies. Diffusion to another of the world's hotpots proceeded in the mid-1980s, when the Indian Army decided to replace its service assault rifle, the Indian 2A1, with its own Kalashnikov-based system.[45] When production was delayed, the Indian government decided to purchase 100,000 assorted AKM-type weapons from Russia, Hungary, Romania, and Israel to replace some of their aging 2A1s.[46] Production of the Indian Kalashnikov variant, called the Indian Small Arms System (INSAS), began in 1997, and over the next fifteen years an estimated 700,000 to 900,000 INSAS were produced by three Indian ordnance factories.[47] Not to be "outgunned" by its archrival, Pakistan developed and began manufacturing its own Kalashnikov variant, the PK-10, in the mid-2000s.

Also during the 1980s, China decided to expand production by building a new plant in Longyan, Fujian Province, to manufacture Type 56 and Type 81 (Chinese adaption of AK-74) assault rifles. In 1993, the Chinese manufacturer issued a license to the Military Industry Corporation (MIC) of Sudan to produce a Type 56 variant called the MAZ, and partnered with Defense Industries Corporation of Nigeria to establish a Kalashnikov factory in Kano in 2006.[48] Several countries, including Iran, Croatia, and Kosovo, have produced unlicensed variants of the Chinese Kalashnikov over the last thirty years.[49]

Further diffusion followed the 1997 decision of the Russian Federation to try to protect and expand its Kalashnikov brand by finally awarding a state patent to the Izhmash Factory in Izhevsk, where Mikhail Kalashnikov himself continued to work periodically.[50] Izhmash then officially registered the Kalashnikov rifles with the World Intellectual Property Organization (WIPO) in Geneva. But they were closing the stable door after the horse had bolted. Numerous lawsuits ensued against factories in China, Bulgaria, Romania, Israel, Turkey, Egypt, and elsewhere, with Russia seeking royalties for the Kalashnikovs they were producing.[51] Many producers either insisted that they had fundamentally altered the original design or that the AK itself had changed fundamentally after its invention in 1947, particularly when the Soviet Union switched to the 5.45 mm AKM design in 1974.[52] Meanwhile, by the early 2000s, there was an enormous stock of old Kalashnikovs in the Russian Army, even as there was no money available for new arms development. It made no sense to order additional old model Kalashnikovs, and the former Soviet factories in Izhevsk suffered.[53]

From here, the fate of the Kalashnikov enterprise becomes murky, tied to Putin-connected Russian oligarchs and a web of shell companies. With President Putin's blessing, the two original factories in Izhevsk merged in 2012.[54] They then became a Russian manufacturing firm and joint stock company called Kalashnikov Concern, with Rostec, a Russian state-owned holding company, controlling 51 percent of the shares, with 49 percent belonging to a small number of private investors.[55] Kalashnikov Concern issued licenses for production of the AK-103 model to Venezuela (2008) and later Saudi Arabia (2017), with both countries building their own Kalashnikov plants.[56] Kalashnikov Concern also issued a license to manufacture and distribute a range of Kalashnikov rifle variants, importing parts from Russia, to a firm called "Kalashnikov USA," owned by the Russian Weapons Company (RWC Group LLC).[57]

By the end of 2014, nearly sixty plants worldwide were manufacturing licensed or unlicensed Kalashnikovs or variants, including over ten producers of parts, receivers, or variant rifles like the Galil in the United States. (See Figure 6.4.)

The Revolutionary's Weapon of Choice

While the trade in Kalashnikovs had begun chiefly as a means by which the Soviet government could procure goods and support nation states and revolutionary forces friendly to the USSR as a counterbalance to Western power, the funneling of the gun to insurgents and terrorist groups soon outstripped any Kremlin control and took on a life of its own. The gun would later come back to haunt the Soviets in Afghanistan, but before that it delivered a beating to American forces in the Vietnam War.

As mentioned earlier, the Chinese gave the North Vietnamese forces Type 56 Kalashnikovs, which performed well against the US M-16 assault rifle, adopted in 1963 and first used in large-scale combat in Vietnam.[58] The M-16 replaced the M-14, a heavier, large-caliber rifle with a wooden stock and long barrel that from 1959 was standard issue in the US armed forces.[59] Intended as the American response to the AK-47, the lighter, small-caliber, short-barreled M-16 assault rifle was rushed into production, but it proved notoriously prone to jamming, especially between 1966 and 1969, usually because of failure to eject a spent cartridge.[60] The barrels also became fouled by the use of a high-residue powder, which made it difficult to keep the gun clean.[61] This was exacerbated by delays in getting lubricants and cleaning

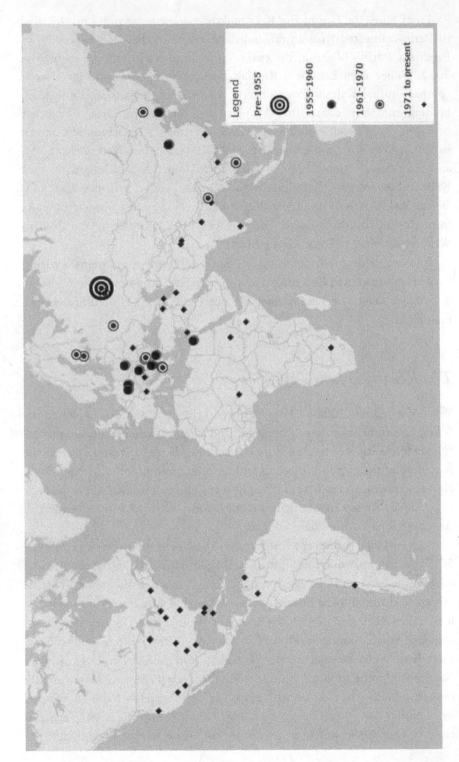

FIGURE 6.4 Kalashnikov Factories (1947–present).

materials to the troops, especially the skinnier rods needed to fit the M-16's narrower barrel.[62] Sometimes in the heat of battle, soldiers were forced to pass a cleaning rod between them to jam down the barrel and remove a stuck cartridge. One soldier later remembered, "When I was with 1/3 in 1967, Marines were writing home to get small paint brushes to clean out the chamber, especially after hitting a sandy landing zone, and lube oil, because neither was available in a timely manner through the supply system. And they were begging to keep their M14s."[63]

After photographs of dead American Marines lying next to their inoperative M-16s appeared in the global press,[64] and US Marines wrote home to their Congressmen, the rifle became the subject of a scathing congressional inquiry.[65] A subcommittee of the House Armed Services Committee in October 1967 concluded that, in addition to the high-residue powder, the rifle's problems were caused by insufficient troop training in how to care for the weapon, lack of proper maintenance kits, and a failure to give the rifles a chromium-plated barrel and chamber.[66] Some US Marines claimed that they preferred to carry the Kalashnikovs that they retrieved when enemy soldiers died or surrendered, because even though the rifles were less accurate than their M-16s, the Kalashnikovs worked more reliably, especially in short-range firefights under jungle conditions with high humidity.

The M-16 was also more expensive and scarcer than the lowly Kalashnikov.[67] In 1967, Colt Industries in West Hartford, Connecticut (descendent of the same company that produced the Gatling gun) was the only firm manufacturing the M-16. The monopoly caused keen shortages during the Vietnam War, including when the plant shut down for a strike in July 1967.[68] By this point in the war, more than 11,000 Americans had died.[69]

Eventually the problems with the M-16 were resolved. The chamber and barrel were chrome-plated and an improved rustproof coating applied. Thorough new training in how to care for the finicky weapon was given to troops before they deployed, with a wide range of informational pamphlets also distributed (including one in a comic book format). Troops were given appropriate cleaning materials, with barrel rods of the correct diameter, and drilled in how to use them. And a buffer was added to the M-16 inside the return spring, to slow the rifle down and reduce the buildup of the dirty ball powder. But the improvements were not fully implemented in the field until late 1968.[70]

The M-16 is now the most highly engineered military assault rifle in the world, technologically superior to the Kalashnikov.[71] But despite the commercial savvy of the US arms industry, it did not diffuse globally to the extent the Kalashnikov did. Estimates put the number of M16-related rifles (including the civilian version, the AR-15) somewhere around 13 million worldwide, a number that pales in comparison to the 70–100 million estimated Kalashnikovs.[72] The M-16 remained more expensive; less available; and much harder to use, repair, and care for.

But there was another reason, linked not to engineering but to human beings. As the Soviet Union and China gave Kalashnikovs to their proxies, who passed them on (and on and on), the gun gained resonance as the emblem of a defiant revolutionary. The revolutionary narrative served Soviet political purposes, and they further disseminated it. Soon the AK-47's symbolism independently went viral and increased demand: win or lose, a Kalashnikov tied its bearer to a powerful story.

A gruesome watershed in building the Kalashnikov's global notoriety was when Palestinian terrorists used AKM models in the televised 1972 Munich Olympics massacre of Israeli athletes also mentioned in chapter 2. Shortly before the operation, members of the Black September group smuggled the AKMs and ammunition, along with grenades and other supplies, into Germany and pre-positioned them.[73] Easily scaling the wall of the Olympic Village, eight Palestinian group members took eleven Israeli athletes and coaches hostage, leading to a 20-hour siege that played out in real time before 900 million television viewers. The operatives had taken a horrified global audience captive as well. One Israeli athlete and a coach were killed resisting capture and then, following negotiations, a helicopter transfer to a local airport, and a botched German rescue operation, all nine remaining Israeli hostages were slaughtered, along with five terrorists and a German police officer. Terrorism had made its global television debut, and the Kalashnikov had played a starring role.

The Kalashnikov spread throughout the Middle East to both state forces and other terrorist and insurgent groups largely due to the Kremlin's relationships with the governments of Syria, Algeria, Tunisia, and Libya, which in turn began to give or sell guns to causes they supported.[74] Libya's leader Muammar Gadhafi was well known for his training and support of terrorist groups, including the Provisional Irish Republican Army. Making use of its factory in Cairo, the Egyptian government also supplied Palestinian militant groups for their fight against Israel.[75]

Christopher Carr in his book *Kalashnikov Culture: Small Arms Proliferation and Irregular Warfare* describes how AK-47s and their variants made their way from revolution to revolution in the twentieth century, migrating from Liberia to Sierra Leone to Ivory Coast and flowing from Cambodia through Thailand and into Bangladesh and India.[76] They often transited through key global arms marketplaces including Cox's Bazaar (Bangladesh); Bakaaraha

(Mogadishu, Somalia); Darra (Peshawar, Pakistan); and the Triborder region where Paraguay, Argentina, and Brazil form a confluence near the Iguazú, Falls.[77]

The fact that the Soviet Union had designed the weapon added to its appeal for revolutionaries aligned with Marxist causes. But the Kalashnikov waged battle on both sides of the Cold War. The US government distributed a large number to proxies it supported, especially in areas of the world where revolutionary factions were already familiar with the gun, already had available ammunition or spare parts for it, or where the government wanted to camouflage its role. Most notably, in the 1980s, the CIA purchased thousands of tons of Kalashnikovs for the Mujahideen in Afghanistan to use against Soviet troops.[78] The rifle thus ironically played an important role in the breakup of the Soviet Union, which began to crumble under domestic economic and political pressure following the failure of the military occupation and its forced withdrawal.

In the anti-Soviet fight, the guns were channeled through the Pakistani intelligence service (ISI), which then diverted them to factions that the ISI sponsored, including anti-India groups in Kashmir. The Mujahideen also sold surplus weapons for cash, flooding the region with them. During this period, the term "Kalashnikov culture" was coined in Pakistan to describe the instability of weak central governments challenged by subversive groups armed with the rifle.[79] In a further bitter irony, al-Qaeda took up Kalashnikovs that had been supplied to the Mujahideen, and Osama bin Laden used an AKS-74U model as a menacing symbol when he leaned one against the wall behind him in numerous news photos and public appearances, including the videotaping of his first post-9/11 speech.[80]

In Africa, Kalashnikovs empowered liberation movements fighting colonial powers and were then used as the primary weapon by many insurgent and rebel groups in civil wars. In Mozambique, the Soviet- and Chinese-supported Communist Mozambican Liberation Front (FRELIMO) fought for independence from Portugal (1962–1975), and then fought the anti-Communist Mozambican National Resistance (RENAMO) in a civil war (1977–1992). Kalashnikovs collected in the aftermath of the Mozambican civil war included Hungarian, Romanian, Bulgarian, East German, Chinese, and Soviet variants.[81]

Following Angola's independence from Portugal (1975), South Africa secretly purchased 35,000 Type 56 Kalashnikovs from China to supply the National Union for the Total Independence of Angola (UNITA) faction, which was fighting the Communist People's Movement for the Liberation of Angola (MPLA).[82] The Rwandans used Russian- and Bulgarian-supplied

AK-47s during their civil war and genocide between 1990 and 1994.[83] The Kalashnikov is so ubiquitous in Africa that it has allowed groups of young men, including child soldiers, to challenge traditional social hierarchies, upending long-standing clan and tribal governance.

The gun continues to be the main weapon in the arsenal of revolutionary and criminal groups in the Middle East, Africa, Asia, and Latin America, as well as the weapon of choice in terrorist attacks. It's the main weapon used by the Islamic State, particularly older Russian and Chinese types from the 1960s and 1970s that were used by the Iraqi Army, as well as some more modern AK-74Ms, probably stolen from Syrian weapons caches.[84] Notably, in his surprise April 2019 video reappearance, ISIS leader Abu Bakr al-Baghdadi sat with what appeared to be a shortened AKS-74U model Kalashnikov propped up beside him, just as Obama bin Laden always did.

Whether in the hands of warlords, terrorists, insurgents, drug lords, or other criminals, the Kalashnikov has facilitated the disruption of the power of the very governments that have distributed it for short-term tactical advantage, not least the United States, which after 2001 became one of the largest buyers and proliferators of the rifles worldwide, handing them out to allies such as the Afghan National Police.[85]

Back to the United States

Meanwhile, in the United States, Kalashnikov USA (KUSA), licensed in 2012 by Kalashnikov Concern (still headquartered in Russia), became the sole US distributor of Russian Kalashnikov variant rifles, which were highly prized by its buyers because Americans considered Russian-built weapons to be more authentic.[86] By 2013, the United States accounted for 40 percent of Kalashnikov Concern's total gun sales—roughly equivalent to the number bought over the same period by the entire Russian military, which equips each of its regular soldiers with a Kalashnikov. In January 2014, Kalashnikov Concern projected exports of 200,000 firearms a year to the United States and announced that 90 percent of its civilian weapons exports overall were going to the US market.[87] The company CEO claimed that orders from Americans were pouring in at three times the rate at which they could be fulfilled.[88]

After US sanctions were imposed on Russia in 2014, in the aftermath of its annexation of Crimea and involvement in the Ukrainian conflict, importing Kalashnikovs from Russia was prohibited. The solution was to stop being Russian. According to its CEO and majority shareholder, Alexey Krivoruchko, Kalashnikov USA completely severed its ties with Kalashnikov

Concern, the Russian mother company, and became an American company, building rifles domestically.[89] KUSA launched a new line of sporting, hunting, and home security firearms aimed at the American market, and in 2015 operations were moved from Tullytown, Pennsylvania, to a manufacturing facility in Pompano Beach, Florida, where the company reportedly received tax incentives and the gun laws were more lenient.[90]

Not everyone believed the overseas ties to Russia had been severed, seeing instead an elaborate effort to bypass sanctions and keep funneling money to Russian oligarchs through shell companies. In April 2018, Congressman Ted Deutch, in whose South Florida district the Kalashnikov USA factory is based, pressed for an investigation by the US Treasury Department, and the US attorney's office in Miami opened a criminal investigation.[91] As of January 2019, neither organization would comment on whether they were investigating, and Kalashnikov USA was briskly selling rifles at prices thirty percent higher than their Russian counterparts, with plans to begin exporting them at the end of 2019.[92] Kalashnikov hunting and sporting rifles continue to be extremely popular among civilians in the United States.[93] Kalashnikov USA's slogan is "Russian Heritage, American innovation."[94]

The Impact on the Power of States

Before the Second World War, nation-states routinely crushed insurgencies, but afterwards, quantitative and qualitative analysis shows, insurgencies were successful significantly more often. Using data from the Correlates of War Project (created in 1963 by political scientist J. David Singer and historian Melvin Small),[95] Boston University professor Ivan Arreguin-Toft in 2001 analyzed the frequency of victory in conflicts between established states and non-state actors, such as guerrilla factions, insurgencies, and civil war forces, in the twentieth century. He found a sizeable uptick in 1950 in defeats of state forces: between 1900 and 1949, rebels won 34.9 percent of the time, while after 1950, that rose to 55 percent.[96] Military historian Max Boot analyzed 443 insurgencies stretching back to 1775 and up to 2012 and also found an increase in their success, beginning after 1945. He found that overall, in contests that had been resolved, incumbents prevailed 63.8 percent of the time, insurgents won in 25.2 percent of campaigns, and the rest were draws (meaning a negotiated ending with no clear-cut winner). But looking only at resolved cases since 1945, the state win record decreased to 51.1 percent, and insurgent success increased to 39.6 percent, with the rest ending in a draw.[97]

Scholars have pointed to a range of reasons for the increasing success of insurgent contests since 1945. These include the role of state sponsorship, such as through weapons provision, training, financial support, and providing sanctuary, as well as more fervent devotion of followers to revolutionary doctrines, such as those of Mao Zedong and Che Guevara.[98] Others have argued that the increasing professionalization of state armies, which has involved their increasing isolation from local populations onto military bases, has meant a lack of understanding of local communities and a deeper degree of support for insurgents, bringing them enhanced power.[99] Insurgencies depend upon a close relationship with the population—to paraphrase Mao, "The guerrilla must move amongst the people as a fish swims in the sea."[100] Their popularity is crucial to their viability. The soundness of the strategies of both sides is, of course, also a key factor.

With so many variables at play in the outcomes of these conflicts, it is impossible to claim that the Kalashnikov alone *caused* the shift. But the growing worldwide availability of Kalashnikovs seems to be an important element, one that is commonly overlooked in both academic and policy studies of civil war, insurgency, and terrorism. This conclusion is reinforced by the fact that in the post–World War II years, state military arsenals were dramatically beefed up, which would suggest the success rate of non-state challengers should have *decreased*. What there is no doubt about is that the Kalashnikov enabled small non-state groups to take on professional armies to a degree that was impossible before the war. While armies have become highly mechanized and operate in what political scientist Stephen Biddle calls "systems," wielding extraordinary precision and reach of power, non-state factions have used the humble Kalashnikov to devastating effect, not least in defeating the Soviet government that commissioned it.[101]

Why the Kalashnikov Spread

The remarkable history of this rifle adds significantly to our understanding of why certain technologies and not others go viral. The key element in the beginning of the AK-47 story was the Soviet Union's desire to dominate the market by displacing other technologies and locking allies into their preferred system. The goal was not profit but politics, to tie allies to the same interoperable technology, including its production, ammunition, spare parts, replacement models, and so forth. That approach worked extremely well. Indeed, it succeeded beyond the government's wildest expectations, and soon slipped completely beyond their control. By the 1960s, they had locked numerous states into the Kalashnikov system. From that point, the dozens of arms factories proliferating worldwide, producing Kalashnikovs and their ubiquitous, cheap ammunition, track with the increasing use of the weapon in revolutionary insurgencies as the decades passed.

Another key factor, the characteristics of the rifle itself, has been explored in depth here, and is well known and routinely discussed among firearms enthusiasts. Built for a peasant conscript army, the AK-47 was inexpensive, lightweight, simple to understand, and easy to operate for a wide range of users. Its brilliance was its accessibility and reliability. But it was inaccurate and could easily have been overtaken by other, better firearms, had it not been for the first factor: its global ubiquity and self-reinforcing technological lock-in, among clients of the Soviet Union, then proxies of their clients, including illicit users and revolutionaries.

Just as important as all of these traits, the rifle was durable. Long after the Cold War and its proxy wars were over, fresh civil wars, insurgencies, and terrorist movements emerged, at least in part thanks to the technology itself. New technologies do not necessarily have to be exquisite to have a broad and long-lasting global impact.

In addition, the Kalashnikov became essentially an open platform for firearms inventors to exploit and build on. The lack of a patent meant that those who produced the rifle, or gained access to it in other ways, could reverse-engineer it, tinker with its simple elements, and either reproduce the original Kalashnikov or build their own variants. This caused Mikhail Kalashnikov a great deal of angst. In dozens of interviews and several personal memoirs he makes it clear that the question of others exploiting his handiwork or degrading his reputation and honor as an inventor loomed far greater in his mind than the lack of earnings from royalties did.

Lastly, the symbolism and culture of the gun built momentum in an historical context that was crucial to its worldwide diffusion. Most important here was the change in global communications: the arrival of global television coverage in the 1960s was crucial to the AK-47 mystique. Postwar movements for decolonization and self-determination in Africa, Asia, Latin America, and beyond provided the perfect global political setting. Whatever the odds, the narrative of a fight makes a difference, because human beings want to attach themselves to a broader legacy. Groups from the Viet Cong to al-Qaeda to the Islamic State have tapped into the Kalashnikov's historical symbolism to project an image of strength, attract sympathizers, and enhance their political legitimacy.

The Floodgates Opened

After the Soviet Union dissolved, massive stockpiles of Kalashnikovs flooded the global arms market. The Soviet arms-manufacturing plants that were the original beneficiaries of the Kalashnikov's popularity ran into tremendous difficulty navigating the harsh economic realities of global supply and demand. Soviet managers had no experience in marketing or salesmanship, and not only were many countries producing them for their own use and for export, but huge numbers of Kalashnikovs and ammunition that had been stored in the former Soviet Union and Warsaw Pact countries were sold off, now more for financial gain than for political influence.[102]

Russian factories could not compete with the glut of weapons, and a number went bankrupt. Those that remain, including the two factories in Izhevsk, have struggled to survive, in part because of American sanctions on Russian imports.[103] A particularly hard blow came in 2013 when, despite Mikhail Kalashnikov's personal advocacy, their top customer, the Russian Army, chose not to buy the newly developed AK-12 rifle because it still had millions of AK-74s in its arsenal.

Mikhail Kalashnikov himself was more enterprising. After the Soviet Union dissolved, he insisted that he had no regrets that the original AK-47 design never earned him any royalties. "At the time, patenting inventions wasn't an issue in our country," he said in a 2006 newspaper interview. "We worked for socialist society, for the good of the people, which I never regret."[104] But he and his team had been rewarded in many other ways, especially during the Soviet period. The Stalin Prize had brought him 150,000 rubles, a veritable fortune in the Soviet Union. In subsequent years, he had what was called a "personalized salary," meaning that it was above the standard rate paid to others who did comparable work. While he did not become fabulously rich, he and his family lived very comfortably.[105]

Ultimately, Kalashnikov further enriched his family by becoming a capitalist. In 2003, he decided to literally cash in on his famous name, selling the rights to use his name commercially to the German marketing firm MM1 in return for one-third of all proceeds from the sale of items with the Kalashnikov trademark.[106] A wide range of Kalashnikov products followed, including vodka, umbrellas, screwdrivers, snowboards, shaving foam, watches, and penknives, sold in airports and high-end shops throughout the world.[107]

As for the sales of Soviet stockpiles of the Kalashnikov, Ukraine was a particularly avid vendor in the 1990s, making illicit deals for sending large quantities throughout the world, but most notably to the Taliban.[108] As the conflict between the pro-Russian Ukrainian faction and the Ukrainian government erupted, the government began holding on to the stores for arming its security forces, while the more recent AK-74M model has flowed in to the pro-Russian groups in eastern Ukraine from outside sources.[109]

Other former Warsaw Pact countries have sold or distributed large numbers of the weapon. In September 2014, for example, the Albanian Ministry of Defense donated 10,000 Kalashnikov-pattern rifles and 22 million 7.62 × 39 mm cartridges to the Iraqi Kurds.[110] And Kalashnikov variants predominated on all sides of the conflicts in Iraq and Syria.

The rifle variants are so plentiful today that they are not just used to target people; they're widely used in hunting.[111] In Central Africa, where poaching of large game, especially elephants and rhinos, is a thriving black market business, the rifles have facilitated increased killing. A group of poachers will form a circle around an animal, caging it with rifle fire, while a single person armed with a more accurate hunting rifle actually kills the animal.[112] Kalashnikovs are both more accessible than large-bore hunting rifles and cheaper to buy and use. Hunting rifles run about $1,365–$2,200 and their cartridges cost between $18 and $34 apiece, while AK-47s can cost as little as $80–$150 each and their cartridges cost only $0.17 to $0.70 each.[113] Not only has the use of assault rifles increased the killing of endangered species considerably, it has intensified the cruelty of the process: a hunting rifle can kill an elephant with one shot (or at most two to five), while a military assault rifle requires several dozen wounds before the animal dies in agony.

Kalashnikovs are sure to continue to be used as a destabilizing political weapon, as well as in hunting, for the foreseeable future. But in recent years, the lethality of the Kalashnikov in conflict against militaries has been waning. As with dynamite, disruptive innovation and diffusion prompted countermeasures. Body armor, better medical care, clotting factors, and cryogenics (exposure to extreme cold to help in recovery from injuries or fatigue) have all led to fewer troop deaths. And, as we will see later in the book, violent actors are likely to embrace other, newer technologies in the Kalashnikov's place. Somewhat ironically, the Izhevsk-based Russian Kalashnikov Concern, which has now shifted away from the business of Kalashnikov rifle manufacture, may play a role in this coming wave of higher-tech political violence. The firm has begun to diversify its holdings by buying other companies, including producers of motorboats, hovercycles, fully autonomous gun turrets, and small electronics. Indeed, going forward, the leadership of the Russian

Kalashnikov company projects that a growing portion of its future business will be in the sale of drones.[114]

Kalashnikov's Regret

Although baptized as a child, Kalashnikov was a loyal Communist and atheist for most of his life. But he worried about the role his rifle had played in the deaths of millions of innocent people. In June 2006, when the United Nations held a conference on the thriving illicit trade in small arms, Kalashnikov wrote a letter to the delegates that was included in their official conference packet. In it he expressed sorrow that his AK-47, designed to protect the motherland from the Nazis during a war in which he said the Soviet Union lost 28 million people, had fallen into the hands of terrorists and thugs worldwide. He put the blame on politicians and urged the delegates to establish a system of accounting and national marking for each unit produced, with end-user certificates preventing transfer to third parties. In addition to the global control measures under discussion, he urged the strengthening of national laws and compliance with intellectual property rights for his weapons. "May weapons become a way to protect peace, rights and democracy, rather than a tool of terror," Kalashnikov's letter concluded.[115]

Kalashnikov remained a hero to Russians. President Vladimir Putin led the mourning at his funeral in 2013. But Kalashnikov was himself troubled by the consequences of his invention. As the end of his life neared, he began to have even deeper anxiety about the role he had played in history and the fate of his soul in whatever comes next. At the age of ninety-one, he became a Christian. In 2012, Kalashnikov wrote a long, plaintive letter to the head of the Russian Orthodox Church, Patriarch Kirill. "My spiritual pain is unbearable. I keep having the same unresolvable question: if my rifle claimed people's lives, then can it be that I . . . a Christian and Orthodox believer, was to blame for their deaths?" The letter, which a local priest had apparently helped him write, was signed, "Slave of God, designer Mikhail Kalashnikov."[116]

Patriarch Kirill wrote back, offering spiritual reassurance. His spokesperson then issued a public statement that said, "The church has a very definite position: when a weapon serves to protect the homeland, the Church supports both its creator and the soldiers who use it. He [Kalashnikov] designed this automatic rifle to defend his country, not so terrorists could use it in Saudi Arabia." If only inventors of lethal technologies could envision the many unexpected ways in which they will be used.

The Power of Unintended Consequences

Neither Alfred Nobel nor Mikhail Kalashnikov anticipated how their inventions would be used for disruptive political violence. In the same way, the creators of the Internet, social media, and the many new and emerging technologies today have not foreseen the nefarious uses to which they're being put.

Today's period of innovation is comparable to the explosive era of open technological innovation at the turn of the nineteenth century, and arguably even more potentially disruptive. Current and emerging technologies were not consciously designed to kill, as the AK-47 was; they are just as subject to popular whim, more tailored for attention-getting, and more wide-ranging in their potential applications, both good and bad. Because the current technological revolution is an open one, and because there is again so much money to be made by the diffusion of the new technologies, they have again spread rapidly and will continue to do so. Due to their accessibility and ease of use, clusters of new technologies will be combined in novel ways that are unanticipated.

In the next part of the book, we will use the insights introduced in the first two sections to examine how today's new and emerging technologies are already being used to raise the stakes of political violence, by enhancing the mobilization and reach of surprise attacks. We'll also investigate how emerging breakthroughs in autonomy, robotics, and artificial intelligence may be harnessed for even more potent leverage.

PART III

Convergence
Widespread Lethal Empowerment

CHAPTER 7

Open Innovation
of Mobilization: Social Media and
Conquering Digital Terrain

T HOUGH ANWAR AWLAKI was killed in 2011, he inspired Tamerlan
and Dzokhar Tsarnaev, the Boston Marathon bombers, to carry out their
attack in 2013, and taught them how. Two years after Awlaki's death, his
soft-spoken sermons were still as easy to find and download from YouTube
as cat videos and cooking demos. The substance of Awlaki's sermons was
familiar. His ghost championed the same homicidal ideology that Islamist
fundamentalists from Sayyid Qutb to Ayman Zawahiri, motivators of the 9/
11 hijackers, had offered in Arabic for decades. His "Constants on the Path
of Jihad," a five-hour series released in 2005, argued that violent jihad was
a personal imperative. In his final recorded sermon "Call to Jihad," released
in 2010, he parroted the familiar message that every Muslim has a religious duty to kill Americans.[1] It was his style of presentation that made his
sermons stand out.

Awlaki was American, born in New Mexico to Yemeni immigrant
parents. He spoke in perfectly colloquial American English, which gave
him special appeal to English-speaking Muslims in the United States, the
United Kingdom, and Australia. He also took advantage of the latest technology to grab attention. In addition to recording his sermons, in 2010 he
collaborated with a young fellow American named Samir Khan to produce a
slick, English-language online magazine called *Inspire*. The inaugural issue

featured a piece chirpily titled "How to Build a Bomb in the Kitchen of Your Mom."

The Tsarnaev brothers, ethnic Chechens whose father sought asylum in the United States, followed *Inspire*'s step-by-step instructions, turning two pressure cookers into IEDs using explosive powder from ordinary fireworks and detonators made from Christmas lights.[2] To maximize the bombs' ability to maim, they were packed with ball bearings and nails. Stowed in two black backpacks, they attracted no attention from the crowd cheering for the Boston Marathon runners.[3] The bombing, on 15 April 2013, killed 3 people and injured 264, including 16 victims who lost legs—one of whom was a seven-year-old girl.[4]

At the time, Dzokhar Tsarnaev was in his sophomore year at the University of Massachusetts at Dartmouth. One of the classes he was enrolled in was a philosophy course titled "Introduction to Ethics." He had become a US citizen on the eleventh anniversary of the 11 September attacks and seemed to be chasing the American Dream. But his fortunes had taken a bad turn. His parents had divorced and then both left for Dagestan, near Chechnya, after which his grades plummeted. He earned a D, a D+, and two Fs during the fall semester and, with an overall GPA of 1.094 (a D average), he lost his financial aid and was about to be suspended. In January 2013 he begged college administrators for another chance, on the grounds that his relatives lived in Chechnya, "[a] Republic that is occupied by Russian soldiers that falsely accuse and abduct innocent men under false pretenses [*sic*] and terrorist accusations."[5] While the college reversed the suspension, he lost his financial aid and was $20,000 in debt at the time of the bombing.[6]

Dzokhar's older brother Tamerlan was an unemployed champion boxer who had been accused of domestic violence, making him ineligible to become an American citizen, his fervent wish. Tamerlan had come to the United States in 2003 at age seventeen and never really settled. Over time he became entangled with illicit drugs and, by some accounts, was implicated in a triple homicide in 2011.[7] He was also alleged to be an FBI informant, reporting on a multinational drug ring in exchange for a shot at US citizenship, though the FBI denies this.[8] What is beyond doubt is that Tamerlan was frustrated with his life in the United States, had a growing interest in jihadist philosophy, and held powerful sway over Dzokhar, who was seven years younger. In short, the brothers perfectly fit the profile Awlaki was targeting.

Having downloaded a digital copy of *Inspire*, Tamerlan traveled to Phantom fireworks depot in Seabrook, New Hampshire, and bought forty-eight mortars in two big boxes (advertised as "barely legal"), containing a total of about eight pounds of low-explosive black powder. This was the same store that sold Times Square bomber Faisal Shahzad the firecrackers he used in his failed car bomb in 2010. After the Marathon bombing, the clerk who waited on Tamerlan recalled that he asked her for the "biggest and loudest" kit, and that she threw in a second one for free.[9] The pressure cookers apparently came from a Macy's department store, the backpacks from Target.

The Tsarnaevs' bombs were unremarkable: we've discussed similar devices used in the nineteenth century. Manuals to build them have been available since then. The brothers could also have pieced together the information needed from textbooks and a wealth of online sources. If Dzhokhar hadn't failed three chemistry classes in his first three semesters, he might even have learned what was needed in college.[10] What was new was the degree of accessibility, convenience, simplicity, global reach, and sophisticated marketing of Awlaki's propaganda. Whether selling Coca-Cola, cigarettes, or smartphones, marketers know that if you wrap a product in an appealing package and make it affordable, ubiquitous, and easy to use, people are more likely to buy it. Awlaki was a shrewd marketer.

He was quick to exploit rapidly developing audio and video technology as it took off. His stories had first become available on cassettes in 1998. They were then distributed in boxed CD sets. Though he was no technical expert, and had, in fact, flunked engineering in college, easily accessible and newly simplified technology enabled him to steadily upgrade to global "broadcast" platforms and eventually produce near professional-quality videos. He progressed from CDs to Paltalk, an online video chat room, then to an interactive website and blog, and finally to Facebook and YouTube.[11] (Figure 7.1) His video *Constants*, more explicitly violent than his earlier work, hit the market in 2005, the year after Facebook was launched and the year that YouTube was founded; it was the perfect timing for viral dissemination.[12] Awlaki understood the potential of the new platforms.

He appreciated that people are more likely to buy a product if you can somehow tie it to their sense of identity, and he deftly tailored the presentation of his message. Speaking in a charismatic professorial style, he explained the tenets of Islam (his own twisted interpretation) to an audience with a shallow understanding of the religion. Unlike Osama bin Laden, Ayman Zawahiri, or even Abu Bakr al-Baghdadi, Awlaki didn't scream or pontificate. "Listen to Anwar al-Awlaki's . . . here after series, you will gain an unbelievable amount of knowledge," Dzohkhar tweeted a few weeks before the attacks.[13] Awlaki's use of American colloquialisms made him more relatable to those in English-speaking countries. His Internet persona was selling jihadi cool. Along with his simple bomb-making instructions, he offered not only a logic for slaughter but a persona to identify with.

FIGURE 7.1 Anwar al-Awlaki speaking in a video released on 13 February 2011. Source: HO/Site Intelligence Group/AFP.

The Tsarnaevs were among several others he inspired to plan attacks in the United States. The Fort Hood shooter Nidal Hasan (2009), Times Square bomber Faisal Shahzad (2010), and underwear bomber Umar Farouk Abdulmutallab (2009) all took inspiration and direction from Awlaki.[14] Because of this he became the first American citizen since the Civil War deliberately targeted and killed by the US government—in a drone strike that was personally authorized by President Barack Obama.[15]

But his killing only added to his notoriety. For those with whom he already had an online relationship, it enhanced his mystique, and dying as a "martyr," at the hands of the American government no less, drew many more followers to him. His audience ballooned after his death. In the years since, his continued online presence has inspired followers plotting major attacks from Singapore to Bangladesh to the United Kingdom. As of November 2016, eighty-nine Western extremists (fifty-five in the United States and thirty-four Europeans) who either carried out terrorist attacks at home or were arrested for trying to join the Islamic State in Syria had proven ties to Awlaki.[16] According to Scott Shane, national security reporter for the *New York Times*, more than half of US-based terrorism cases after 2011 showed evidence of Awlaki's influence upon the perpetrators.[17]

Some plotters were stopped, like Terry Lee Loewen, an aircraft worker who read Awlaki's *Inspire* magazine, bought a gun, and tried to put a truck bomb between Gates 6 and 7 at Mid-Continental Airport in Wichita,

Kansas.[18] Many others were not. The ghost of Awlaki also influenced the Kouachi brothers, who killed twelve people at the Paris headquarters of the satirical magazine *Charlie Hebdo* in January 2015; Syed Farook, who killed fourteen people in the December 2015 San Bernardino, California, shooting; and Omar Mateen, who killed forty-nine people in the 2016 Pulse nightclub shooting in Orlando, Florida.

The Internet has been a double-edged sword in the fight against terrorism. It's a periscope for terrorist hunters, allowing them to peer at the movements and activities of both known and suspected rogue actors. That surveillance facilitated the killing of Awlaki, along with his *Inspire* accomplice Samir Khan, in the same September 2011 US drone attack.[19] Since 2002, the United States has conducted hundreds more drone strikes against terrorists, including over 250 in Yemen, with many more targeting al-Qaeda; the Taliban; al-Shabaab; and other groups in Pakistan, Somalia, and beyond.[20] The Tsarnaevs probably didn't realize that the US National Security Agency (NSA) had been on to Awlaki long before them. Intelligence analysts working the Yemen account at the NSA and other agencies accessed each issue of *Inspire* two or three weeks before it appeared on militant web forums—even sometimes reading successive drafts shared online.[21]

Yet it's impossible to carefully monitor all of the individuals in such a vast audience as the one Awlaki built. Federal authorities knew about Tamerlan Tsarnaev, and even put his name on a terrorist watch list, but still he slipped past them.[22] It's likely also impossible to ever entirely eradicate Awlaki's presence from the web. Though many of his sermons have been removed from YouTube, persistent searchers can still find them. What's more, Awlaki's audio statements continue to be re-edited to sound as if they are referring to current groups. A 2013 English-language video produced by the Islamic State took an audio clip from a May 2008 lecture he gave over the telephone to a conference in South Africa, paired it with his portrait, and invented a celebrity endorsement. In his Internet immortality, he will haunt us long after ISIS and al-Qaeda have disappeared.

New Triggers for Old Passions

The confluence of the Internet and connected online digital-based technologies, especially smartphones, has created a virtual world in which it is possible to engage electronically, and globally, in the full span of human activities, from spreading hope, love, and enlightenment, to inciting hate, abuse, and murder. We have created a virtual reality that is changing actual reality. Digital technologies themselves are not the enemy. As said, online channels can be of great assistance in tracking down malicious actors. But because they are being used to mobilize and manipulate people to support and perpetrate violence, they are making the prediction of future clashes more difficult, and we'd better think hard about exactly how and why they are being harnessed.

This cluster of Internet-connected global communication technologies is more widely diffused today than dynamite was and the AK-47 was and still is. They are also potentially globally destabilizing. Basic human motivations for political violence, such as the fear, honor, and interest Greek historian Thucydides described during the Peloponnesian War of the fifth century B.C., have endured for millennia. But the triggers provoking such violence evolve over time. Today's technologies are powerful new triggers for human conflict, altering who fights, when, and how.

Their rate of diffusion is the key factor in their destabilizing potential. So much coverage of the current wave of technological innovation emphasizes the speed of change, but when put in historical context, today's digital technologies have actually been a long time coming. Most of them have evolved from discoveries made in the 1960s in computer operating systems, battery life, miniaturization, and networking.

What's unprecedented today, however, is that powerful computers are in everyone's pockets, offering both ready-made platforms for open innovation and a potent means of political manipulation. Most people barely touch the full capabilities of their Androids and iPhones. Indeed, it's often said that the computing power in a single iPhone is hundreds of times more capable than the computer that guided the flight of Apollo 11 in 1969, which was a huge IBM System/360 Model 75 mainframe that had to be complemented by the work of more than 3,500 IBM employees (most armed with slide rules).[23]

A good deal of attention focuses on computer technology itself, especially whether or not the long-running increase in processing power may soon slow down, reaching a natural limit. In 1965, Intel cofounder Gordon Moore famously predicted that the number of transistors contained in a single integrated circuit device would double every eighteen months, thereby doubling computing capability, while the cost of manufacturing held steady.[24] For decades, what became known as Moore's Law proved correct, partly because it was inherently astute, and partly because it became an expectation that the semiconductor industry strove to fulfill. The maxim was also extended to a wide range of other technological improvements, including computer memory capacity and sensors, which got smaller even as they were able to collect a wider range of more sophisticated data.

Moore himself forecast that the doubling every eighteen months (the core of his eponymous "Law") would stop by about 2025.[25] He cited the physicist Stephen Hawking, who said that integrated circuit technology would ultimately run up against the finite velocity of light, the upper limit of the speed with which signals can propagate through circuits, as well as the limitations of atomic scale and the quantum nature of materials.[26] In 2015 the Intel Corporation even publicly concurred that the end was near.[27] Some technology experts counter that ongoing human cleverness may postpone the hitting of physical limitations, such as by exploiting higher-density groupings of integrated circuits, transitioning to molecular-based switching devices, and processing more signals at the same time in parallel.[28] Whoever proves correct, when it comes to analyzing how technology is and will be used for purposes of political violence, processing power is not the important issue. It's the pace and scope of human adoption of digital technologies that is at the heart of the danger.

While today's pace of technological innovation isn't unprecedented, and probably lags behind the sweeping technological changes we explored in the late nineteenth century, the rate of diffusion and ease of adoption of digital technologies are exceptional, which is why the current era seems faster to us. The number of people and organizations globally who are tinkering with digital technologies, innovating the human purposes to which they are applied, is growing dramatically, and as a result the ways they are being harnessed are also multiplying geometrically.

Just as the Apollo moon landing was at least as much a feat of human management and teamwork as an accomplishment of technology, today's digital technologies rely on human initiative or intervention to produce outcomes, and these outcomes can be either good or bad.

Mobilization and the Atomization of Violence

In this chapter, the focus is on how Internet-connected technologies are being used to mobilize individuals for political violence. As used here, the concept of mobilization has three dimensions. *Social mobilization* leads people to support a cause, making them feel connected with a group, story, or belief system, often so much so that it becomes part of their identity. *Physical mobilization* drives people to act, organizing them and teaching them how to do so, either singly or in groups. And social and physical mobilizations are dependent upon the *psychological manipulation* of public perceptions.

In examining how non-state actors are using new technologies for mobilization, we will look at both sustaining and disruptive technologies. Recall that disruptive technologies not only produce unexpected results but actually shift power from dominant players to surprise actors in ways that change the overall shape of the marketplace of violence. Clusters of digital technologies are enabling mobilization in both sustaining ways, helping dominant actors reach traditional constituencies, and disruptive ones, helping surprise underdogs reach unexpected, often larger constituencies.

The accessibility of the Internet on mobile devices, especially social media platforms on smartphones, has bypassed traditional avenues of mobilization, from the traditional media, including books, newspapers, television, and radio, to institutions such as the church and the army, and replaced them with direct appeals to individuals. Media and government control over coverage of events has been disrupted, with mobile phones allowing individuals to record their experiences and live-stream them with a tweet. Social media has become a Wild West, with creative new practices for reaching masses of people, and manipulating and deceiving them, ranging from those fully conducted by humans to the fully automated—often further disseminated by human beings.

We have created the perfect technological ecosystem for violent mobilization. The cluster of Internet-connected technologies is ideally suited for igniting emotions, persuading people, connecting them to one another and developing their sense of belonging to a group or movement. They are also exquisitely crafted for disseminating information, and disinformation, enabling the logistics for attacks, and doing so under cover. The Internet has become a powerful facilitator of terrorist mobilization because it has evolved from a platform primarily for information sharing into one focused on grabbing the attention of as many people as possible and monetizing it.

The economic energy driving digital innovation has inspired many wonderful developments. Spreading life-saving medical advances, enabling faster responses to weather or humanitarian emergencies, facilitating better understanding of climate effects, and enhancing government transparency are just a few. But it has also generated extremely powerful tools for manipulating people's emotions and riveting their attention, as well as for targeting information specifically to them and tailoring it to their interests and views. These tools allow any individual with an Internet connection to spread false information, follow people's movements, obtain voluminous information about them, and communicate with people all around the world clandestinely.

The same tools can, of course, be used to mobilize people behind good causes. They were used to drive the 2010–2011 Arab Uprisings, and they've rallied people to many peaceful goals, such as the Black Lives Matter social movement. They are particularly well suited, however, for destructive purposes. Movements seeking to bring about lasting political and social reforms require real-world human organizing and institutional change in order to achieve the lasting effects they desire. With the Arab Spring, we saw how readily such reforms could be co-opted by strongmen and reversed. By contrast, when the goal is to spread fear through episodic violence, and thereby to undermine structures of power for purposes of disruption, these tools facilitate doing so without the hard work of extensive organizing.

A terrorist's whole point is to make him- or herself heard by using outrageous symbolic violence, to force attention to the cause and provoke public outrage at their governments' inability to protect them. This is why when the means to murder and the methods to advertise it both become more accessible, an increase in terrorism inevitably follows.

The recent evolution of Islamist terrorism is a clear case in point. From the earliest years of al-Qaeda, bin Laden devised a media strategy. As cable news stations proliferated, he courted journalists, held personal press conferences, and sent videos via courier to Al Jazeera. Yet cautious news editors often filtered, clipped, or suppressed his messages. The twelve-page 1996 declaration of "holy war" entitled "Message from Osama bin Laden to His Muslim Brothers in the Whole World and Especially in the Arabian Peninsula: Declaration of Jihad against the Americans" barely registered with the public. Attacks on the Khobar Towers (25 June 1996), the US embassies in Tanzania and Kenya (7 August 1998), and the USS *Cole* (12 October 2000) failed to "inspire." An al-Qaeda courier explained bin Laden's frustration: "Every time I took a new tape, he told me how important my mission was, how this time, the Muslims of the world would finally listen, and how I must absolutely deliver the tape to the right people."[29]

Bin Laden's signature focus on long-planned, simultaneous attacks was designed to be spectacular enough that any editor would be forced to air coverage of them. In targeting media-rich Washington and New York, the 9/11 attacks were not only about killing a large number of people but also drawing worldwide attention. From his hideout in Abbottabad, bin Laden sent grainy videotapes and handwritten letters to trusted news sites and subordinates, including one in which he wrote, "[W]e have to undertake a large intensive media campaign, part of it through American media sources if possible."[30] Other letters convey his determination to shape and disseminate al-Qaeda's global message.[31] Bin Laden understood very well that for terrorists, attention becomes power.

As the digital age began to reshape the media environment, such large-scale, meticulously planned attacks were no longer required for gaining attention, or for spreading fear and building a following. In contrast to bin

Laden, and just as Awlaki had done, Daesh (aka ISIS) developed a model of production that cut out the editors and went directly to the audience. Under withering military attack, the territorially based organization in Iraq and Syria soon disappeared and the so-called Islamic State shifted its focus to decentralized attacks throughout the world, recorded and shared on social media via smartphones.

Just as with the Anarchist movement, the atomization of violence has today spread to include lone-wolf attackers and the mentally deranged. Take the case of how online conspiracy theories have incited such attackers. They triggered Richard Poplawski's killing of three Pittsburgh police officers with a Kalashnikov in 2009, when a white supremacist website said that the Obama administration was about to ban all guns;[32] Oscar Ortega's 2011 attempt to assassinate Barack Obama, shooting a Kalashnikov at the White House apparently inspired by an Internet-streamed Alex Jones film;[33] and Jared Loughner's killing of six people and wounding of Representative Gabby Giffords outside a Tucson, Arizona, grocery store, evidently motivated by on-line right-wing extremist groups.[34] James Thomas Hodgkinson, member of left-wing Facebook groups such as "Terminate the Republican Party" and "The Road to Hell is Paved with Republicans," went from online rants to opening fire on the congressional Republican baseball team, wounding four before the police killed him.[35] Such bizarre incitements and fabrications are nothing new; they have long had outsized effects on certain individuals, fu-eling a perverse form of populism.[36] But social media is greatly enhancing the speed and scope of their spread and thereby elevating their power, with potential effects on governance that we are only beginning to understand.[37]

Digital online technologies are, in short, atomizing mobilization and shifting the balance of power between states, private actors, and violent groups. Following the *levée en masse* of the Napoleonic Wars, the storyline of mobilization over the next two centuries was about the Western nation-state gradually consolidating power over the masses and directing their mobili-zation toward conducting wars. The state's monopoly never became com-plete, but especially with the massive industrial mobilizations of the two world wars, the state's grip was greatly tightened. Since about 1995 that pro-cess has been reversing, with Internet-based technologies driving a gradual undermining of the nation-state's ability to control and channel popular vio-lence. We are likely still in the early stages of this reversal.

I am not arguing that terrorist groups, and other malicious non-state ac-tors, are becoming the masters of the online universe.[38] Private technology firms and state security and military organizations have driven and will con-tinue to drive innovation in digital technology, and for the most part the mass public has determined the course, with developers responding to their needs and desires. Terrorists and other non-state actors have mostly adapted to the changing environment rather than shaping it. Jihadists were neither early adopters of social media nor innovators with it; they picked up on ways to utilize it as others crafted them.

Would-be disruptors don't have to be innovators; they don't even have to be truly technologically literate, or even connected. As was the case with the French Revolution, in which most of those fighting were illiterate, even if people are not online—the situation with the majority of those who participated in the 2011 protests in Egypt, for example—they can still be influenced to act by those who are. The Internet-connected agitators can use the tools to enhance the power of traditional means of mobilizing. Word of mouth is still an extremely powerful force for persuasion, as are television and radio, and especially so in *combination* with social media.[39]

The highly sophisticated threat of cyberwarfare has rightfully grabbed attention, as major powers spar online in an ongoing contest that is indeed deeply problematic and dangerous. But it must not distract us from also paying close attention to the ways in which non-state actors are harnessing widely available and seemingly non-lethal tools, with far-reaching implications. During the Cold War, the United States and the Soviet Union were fixated on their nuclear rivalry. Meanwhile, the accessible AK-47 facilitated dozens of insurgencies that changed the global political map, such as by undermining the Soviet occupation of Afghanistan and, arguably, the stability of the Soviet Union.

It's not just high-end weapons that have outsized results. Computer scientist Dorothy Denning wisely pointed out in May 2001 (pre-9/11) that terrorists "still prefer bombs to bytes," and this is still true.[40] Groups like ISIS, al-Qaeda, Hamas, and Hezbollah have been inept at wielding cyber warfare tools, for example. Criminals exploit them more successfully, but even they are well behind states.[41] And yet, carefully targeted terrorist acts, using digital technologies to stoke public fear or intimidate opponents, hold the potential to prompt a cascade of state actions more rapidly than ever. Governments may lash out in outsized fashion in response to domestic pressures to crack down, especially in democracies. They may also act in inflammatory ways out of fear that other states will exploit apparent weakness or vulnerability, especially in autocracies like Russia or China. Strong reactions to major terrorist attacks can lead to counter-reactions, as we know from the First World War and, to some extent, the 9/11 attacks.

A framework is needed that digs deeper into how the new tools have been changing the nature of mobilization and will likely continue to do so.[42]

While specific apps and other tools will continue to evolve, the broad *effects* of digital technologies on mobilization will persist as the power of the tools keeps growing.

The fundamental elements of mobilization today are not new; Johann Most would recognize the basic nature of the techniques used. They are: information warfare, recruitment, training, logistics, and fundraising. Most would also recognize some of the specific methods being used to achieve these aims, such as publishing propagandistic online magazines. These are evolutionary, sustaining innovations in mobilization, though the web has put them on steroids. Because they are familiar, these sustaining uses of digital technology will each be discussed relatively briefly. Six disruptive innovations will be examined more extensively: mass interactivity, mobile- and live-streaming, first-person filmmaking technology, viral disinformation, accessible end-to-end encryption, and attention-grabbing psychological tactics.

The most transformative element in this mobilization landscape is social media. Facebook's motto is "move fast and break things." In what follows we'll see where they and other major tech companies (along with their customers) are moving fast, that is, accelerating classic methods of mobilizing for political violence with deep historical roots, and where they are actually breaking things, spurring disruptive global conflict that is changing the world.

Moving Fast: New Tools for Old Tactics

Social media and video channels, primarily YouTube, have been used, in part, in the same way pamphlets, books, speeches, and newspapers were by the Anarchists. Groups have created a host of websites and online magazines, like Awlaki's *Inspire*, to draw people to the cause through highly persuasive messaging and by humanizing themselves, as ISIS did with a website devoted to cat photos, featuring shots of kittens curling up around grenades and fighters holding a kitten in one hand and a Kalashnikov in the other. Groups have also, of course, used these channels to post photos and videos of horrific acts of violence, such as beheadings. Both charismatic inspirational communications and the publicizing of violence are evolutionary versions of the ancient art of informational warfare.

In the fifth century B.C., the Chinese military theorist Sun Tzu wrote, "All warfare is based on deception."[43] Psychological operations have long been integral to war, used to soften up the battlefield, confound the enemy, or panic populations. Terrorists have always understood the power of sowing fear. Publicizing luridly violent acts is as old as terrorism itself. As mentioned in chapter 2, the first century A.D. Sicarii deliberately chose prominent pro-Roman priestly aristocracy in Jerusalem as their victims and killed them with daggers (*sicae*) in the public square, in full view of hundreds of onlookers. Symbolic killing is at the heart of terrorism.

With the emergence and progression of computer technology, new methods of information warfare were steadily developed. During the Cold War, for example, both the United States and the Soviet Union crafted sophisticated information operations, which were incorporated into war plans beginning in the 1970s, though the United States did so less coherently due to institutional tensions between psychological operations and public diplomacy.[44] Russia was the first to deploy what are called distributed denial of service cyber-attacks (DDoS), which it used against Estonia in 2007, followed by more powerful DDoS attacks aimed at Lithuania, Georgia, Kyrgyzstan, Kazakhstan, Ukraine, Germany, Finland, the Netherlands, the United States, and other countries.[45] These involved shutdowns of social media, defacement of government websites, hacking election systems, stealing proprietary documents, and disrupting critical infrastructure like nuclear power plants. In a 2013 article, Chief of the Russian General Staff Valery Gerasimov reiterated that the "information space" is a key part of war, as demonstrated by the "use of technologies for influencing state structure and the populations with the help of information networks."[46] Russia's sophisticated use of information operations to interfere with the 2016 US presidential election was another incremental step in this game plan.

One difference now is that some non-state actors are getting as good at information warfare as state militaries are. When ISIS marched on Mosul in June 2014, for example, the group also launched a marketing campaign on Twitter called #AllEyesonISIS.[47] And so they were. Their campaign was devastatingly effective, despite the fact that only about 20 percent of the Iraqi population had access to the Internet. Of course, it was not ISIS information operations and social media alone that resulted in gains: the commanding Iraqi general stated that the government had left the city undermanned and the troops short of weapons and ammunition.[48] The impact of the campaign of terror was clearly substantial, though. Staged photos of atrocities and videos of Iraqi soldiers being executed posted on Instagram and Twitter sowed panic among Iraq security forces. Many shed their uniforms, dropped their weapons, and joined a stream of fleeing civilians, leaving the territory undefended.[49] In six days, some 2,000 ISIS fighters captured a city that 25,000 Iraqi security forces were meant to defend.[50]

The other key differences with information warfare of the past are of scope, speed, and reach. Even television's immediacy has been surpassed.

And photos, footage, and audio recordings that would not be broadcast by most mainstream media companies can be easily uploaded to these online channels, which mostly lack editorial gatekeepers. Conscientious news sites try to vet their sources, but that is increasingly difficult when deadlines are instantaneous, live interviewees spread misinformation, and everyone fights for audiences who prefer consuming news à la carte. Plus direct postings bypass the traditional competition over broadcast time in a news lineup.

The global reach of this propaganda extends the range of recruitment. Groups have increasingly appealed to potential fighters all around the world, inviting them to travel to terrorist hotspots such as Syria (ISIS), Somalia (al-Shabaab), or Afghanistan (al-Qaeda). Boko Haram in Nigeria also routinely uploads coverage of its activities on YouTube and calls for recruits. Adam Gadahn traveled from the United States to Pakistan to become a high-profile spokesman for al-Qaeda and then incited other American Muslims to carry out random attacks in a video entitled "Do Not Rely on Others, Take the Task upon Yourself." Tareq Kamleh, an Australian doctor, traveled to Syria and recruited other professionals online. Aqsa Mahmood (alias Umm Layth), a twenty-year-old Scottish woman from an affluent Glasgow neighborhood, drew other women to Syria by promoting *hijrah* (migration) on her Twitter and Tumblr accounts.[51] British authorities believe that Mahmood helped to radicalize British teenagers Shamima Begum (age fifteen), Khadiza Sultana (age sixteen), and Amira Abase (age fifteen), three East London classmates who fled to Syria in 2015.[52] Loss of physical territory is offset by a shift in virtual messaging, as when groups step up their recruitment of foreigners to carry out attacks in their own countries.

The training of recruits has also been enhanced. Online college courses have become popular, with many universities claiming that they replace the classroom experience. It is a small leap to online courses in killing people. Groups post technical instructions for attacks and suggestions for targets, and information about where to get explosives and firearms, especially assault rifles. Al-Qaeda, ISIS, al-Qaeda in the Arabian Peninsula (AQAP), and other terrorist organizations also post online video demonstrations for making explosive devices, with training offered in many languages (twenty-eight at last count)[53] and on dozens of platforms (Twitter, YouTube, Facebook), although all of these platforms are becoming more aggressive at removing terrorist content.[54] In December 2015, the Islamic State published its first issue of a German-language magazine specifically devoted to "jihadi Tech" called *Kybernetiq*. As for designating targets for attack, ISIS has used an online chat room to provide a list of suggested tourist spots to attack so as to create "pictures of horror." These have included Times Square, "the State of Texas," and travel infrastructure in the United States and Europe.[55]

Weapons manuals such as *The Science of Revolutionary Warfare* (1885), *The Blaster's Handbook* (1918), and *The Anarchist Cookbook* (1971) did not need the Internet to reach thousands who wanted to learn about "infernal machines" and Molotov cocktails.[56] One difference today is that, with a click of a mouse, the information potentially reaches billions of people, and it will be available in perpetuity. Today's manuals are also provided in carefully tailored narrow-cast marketing packages, with sophisticated real-time demonstrations of how exactly to kill people. Marketers of violence no longer need to go through publishers, or television stations, or even the postal service.

The bomber at the Ariana Grande concert in Manchester, Salman Abedi, learned on YouTube and other websites how to make the bomb he planted, which used a difficult-to-detect peroxide-based explosive, triacetone triperoxide (TATP), and was packed with small metal objects that killed and maimed indiscriminately. The TATP IED, which killed twenty-two people (twelve of them children under age sixteen) at a pop concert was the same type of explosive used in the 7 July 2005, London underground attacks, the 2015 Paris attacks, and the 2016 Brussels attacks. The failed suicide bomb attack on the Brussels Central Station in June 2017 also used a "badly prepared" TATP bomb.[57] The bomb Abedi made in Manchester was more accomplished. Initially authorities thought that a more experienced explosives expert had made the device. In light of the attack's lethality, a short video from a series called "Jihadi Ideas for Lone Lions" was posted shortly afterwards on YouTube and Facebook, explaining how to make a household bomb "just like Abedi's." The video was eventually taken down.[58]

Some digital technologies have also been used to facilitate more effective operational logistics in planning and carrying out attacks. Internet tools such as Google Earth, LinkedIn, or Facebook are invaluable for intelligence gathering, such as casing potential targets or tracking the movements of security forces. And new capabilities are often combined with established ones. Much as combining old-fashioned clocks with dynamite bombs gave the Clan na Gael unprecedented lethality in the 1880s, the combination of old and new means today opens up a larger variety of options. For example, the chief planner of the 2015 Paris attacks distributed a USB stick with an encryption key, and operatives used protected services such as WhatsApp and Skype to communicate with higher-ups in Syria.

Lastly regarding the use of Internet-connected technologies in sustaining ways, terrorists and insurgents are continually innovating with new schemes for acquiring resources. Post-9/11 efforts to crack down on terrorist financing have largely shifted fundraising away from traceable online methods. At this point, the Internet's role has more to do with matching buyers and sellers, legal and illicit, than collecting online contributions through banking channels. The dark web (meaning content that is not indexed by standard search engines) heightens the incentives to use low-cost, high-speed, unregulated virtual currencies such as Bitcoin and Monero; a few groups in Gaza have solicited donations in Bitcoin, for example. But thus far the savviest groups trend the other way, toward old-fashioned measures.[59]

Like smugglers during the time of Alexander the Great, groups like Hezbollah, Hamas, Jamaah Islamiya, Lashkar-e-Toiba, al-Shabaab, and al-Qaeda rely on well-worn trade routes and untraceable cash transfers, informal systems like Hawala, money service businesses, false trade invoicing, and high value commodities (diamonds, gold, ivory).[60] People are drawn to the latest banking apps, but if you want secrecy and a reliable supply chain, you must avoid traditional financial channels altogether. Traceable transfer of funds from states and larger terrorist groups to smaller factions or individuals is limited. A report by the Norwegian Defence Research Establishment studied all of the "jihadi" cells operating in Europe between 1994 and 2013 and found that half were entirely self-financed, with only a small proportion (25 percent) receiving any money from an international terrorist organization.[61]

Holding territory has been the more lucrative recent source of funding. One study estimates that in 2014, ISIS stockpiled $1.9 billion, gained from taxing people in occupied territory, selling oil, and looting antiquities and other items.[62] Though efforts to fight terrorist financing date to the early years after the attacks of 9/11, today there are more terrorist and criminal organizations collecting more resources through tactics such as kidnapping, bartering, smuggling, and human trafficking than ever before.[63]

These sustaining uses of digital technology, while challenging to combat, are at least in the wheelhouse of traditional government institutions. Security forces can adapt their expertise to fighting them. For example, intelligence agencies can monitor online chat rooms, observe patterns of behavior, or track the movements of known operatives, often even when they transition to the dark web.[64] With warrants, police or security forces can engage in legal collection of data on cell phones; the seizure of hard drives; or the monitoring of online activity to gain a window on recruitment, planning, or operations. Military organizations can use cell phone signatures to target individuals or launch online psychological operations. In this cat-and-mouse game, authorities have far more resources than non-state groups. If they're wisely employed, established countermeasures can effectively combat sustaining uses of new technologies by non-state actors.

More problematic to address are uses of the new technologies that are *disruptive* because they are going well beyond prior capabilities to identify potential recruits, gather them (either physically or virtually), and persuade them to support and engage in violence. Existing laws, jurisdictions, and institutions, especially in democracies, are largely unable to cope with them. These disruptive techniques are mobilizing people in unprecedented ways, spreading disinformation and reshaping the marketplace for violence.

Breaking Things: New Tools Used in New Ways

Driven by the powerful business model behind social media, online technologies have transitioned from being mainly information-sharing platforms into digital juggernauts aimed at capturing attention, collecting data, and facilitating like-minded groups. As a result, their ability to politically mobilize and manipulate human beings has become globally disruptive.

Violent actors have been greatly empowered by these technologies because there are few controls on how people's attention can be solicited, and how their data can be used. While most governments and corporations are at least subject to some regulations or guidelines, malicious non-state actors have evaded most constraints. What's more, security systems designed to protect enormous data caches are vulnerable. The upshot is that extraordinary new means of manipulating people, invading their privacy, tracking their movements, and actually launching attacks have been handed to malicious actors, just because collecting data is so good for doing business. Those who study terrorism have long spoken of "radicalization on the Internet," but that concept does not capture the scale, scope or complexity of algorithmically-driven methods of mobilization and manipulation. Six disruptive features of open innovation in digital media are changing political violence.

BOUNDLESS INTERACTIVITY

While newspapers and television are one-way communication technologies, and the telegram and landline telephone were limited to single points of blind contact, the Internet allows for interactive communication that is almost exactly like person-to-person conversation, including visuals, carried on across the full expanse of the globe. Terrorists have seized on this capability to transform their inspiration, recruitment, and training efforts into a form of virtual one-on-one coaching in a highly persuasive personalized grooming process.

For example, in 2014, "Alex," a twenty-three-year-old Sunday school teacher living in a remote part of Washington State, saw images of the beheading of American journalist James Foley. She was shocked, but also strangely fascinated. Alex sought to find out why anyone would do such a thing. Searching the Internet, she stumbled into contact with ISIS operatives, eventually connecting with a fifty-one-year-old recruiter named Faisal Mustafa, living in Manchester, England. In a polite and soothing style, Mustafa interacted with Alex over Skype, beaming into her bedroom, for as many as seven hours a day, even as her grandparents sat watching Fox News in the adjacent living room. Soon Alex had converted to Islam and declared her new faith over Twitter. Mustafa taught her to pray each day, and he primed her by sending gifts such as pretty headscarves, books, prayer rugs, and chocolate. He urged Alex to move to "a Muslim land," which she understood to mean Syria, and succeeded in persuading her to agree to marry a forty-six-year-old friend of his, whom he described as "a good Muslim." She was supposed to meet him in Austria. All the while, Alex continued to teach Sunday school and work as a nanny two days a week.[65]

To a bored, lonely young woman living in rural America, Mustafa's attention was flattering, and he was introducing her to what seemed like an exciting new world. He groomed her online for about six months, and Alex prepared to leave home. At this point, citing a strict interpretation of Islam, Mustafa told her that she could not travel without a Mahram, or male relative. When he suggested that Alex's eleven-year-old brother, then in foster care, come with her, her protective instincts were triggered. She realized the potential danger to her little brother and confided everything to her baffled grandparents. They contacted law enforcement.[66]

The number of people to which groups can gain such easy, instant, and persuasive access is unprecedented, as is their ability to use interactivity to lead recruits through numerous technological platforms, such as various sites and apps, that take training well beyond that offered by classic weapons manuals. Online sources now display curated, polished accounts of the most reliable means to act, along with cleverly edited videos of actual attacks, with operational mistakes, evidence of reluctance or remorse, and unflattering shots of dead mujahideen removed, for example, but with great care taken to show exactly how others have followed the instructions and become heroes.

The interactivity of recruitment has also gone social, with individuals often egging each other on in chat rooms, forums, email, texts, and tweets.[67] For example, US and European counterterrorism officials allege that ISIS used chat rooms and bulletin boards available through the messaging service Telegram to encourage the December 2016 truck attack on Christmas shoppers in a Berlin marketplace that killed twelve people and injured fifty-six others. A 6 December Telegram bulletin board posting stated, "Christmas, Hanukkah, and New Years Day is very soon . . . so let's prepare a gift for the filthy pigs/apes."[68]

MOBILE STREAMING VIDEOS AND LIVE-STREAMING

Mobile streaming technology encompasses any type of visual and audio streaming that can be done with a smartphone. Live-streaming is recorded and broadcast in real time, and viewers can even provide live feedback. The footage can then be retained online and viewed thereafter, and many live-streamed clips garner massive additional viewership after broadcast. Indeed, once an event has been live-streamed it is difficult to erase it entirely from the web.[69] Numerous apps offer these functions, including YouTube Live, Facebook Live, Meerkat, Blab, Periscope, and Instagram Live, and they are proliferating: by the time you read this there will be others. Over 500 million people watch video on Facebook every day, and *Forbes* estimates that by 2021, more than 80 percent of all consumer Internet traffic will be videos.[70] What is unprecedented here is the quality of mobile phone cameras, the simplicity of their use, and the easy access to live-streaming apps.[71]

For terrorists or criminals, live-streaming moves the use of social media beyond "publicity" to an unprecedented and disruptive level of instant engagement of audiences, both potential recruits and potential victims. Instead of relying on onlookers or victims to generate coverage, it allows a group to tailor its own narrative of events and to hold audience attention. Another source of its appeal for attackers is that it is harder to block from viewers. A video that is downloaded to a device and then posted involves at least a few minutes' delay, so is more likely to be detected and taken down before going viral. Live-streaming also cuts the cost and heightens the impact for groups unable to afford their own media operation.[72] It provides instant notoriety, and therefore gratification, to perpetrators, increasing the perverse appeal of carrying out attacks.

A rash of live-streamed or interactive terrorist attacks began in 2013. Al-Shabaab, the al-Qaeda affiliate based in East Africa, live-tweeted its September 2013 attack on the Westgate mall in Nairobi, Kenya, explaining and justifying its actions as it killed sixty-seven people.[73] Four al-Shabaab terrorists carrying Kalashnikovs and hand grenades stormed the mall, live-tweeting their actions and posting audio online.[74] Three years later, on 12 June 2016, Omar Mateen searched for online references to the Pulse nightclub massacre in Orlando while he was actually carrying it out, adjusting his actions to the reactions.[75] The next day, Larossi Abballa killed a policeman and his partner in Magnanville, a small town near Paris, stabbing the officer and slitting his partner's throat. He then began a thirteen-minute Facebook Live broadcast on his mobile phone while he stood in the couple's home. The victims' three-year-old son could be seen, sitting on a sofa in the background, as Abballa told his audience, "I don't know what I'm going to do with him yet."[76] The boy was ultimately saved in a raid by French police.

Dissemination of streamed video is extremely easy. Users of apps such as Periscope can share streams either by posting them immediately on Twitter (by tapping a button) or sending a link through their social media network. Depending upon how many followers an individual has, this can mean the streamed content instantly reaches thousands of viewers, who may then re-tweet it to thousands or hundreds of thousands more.

It happened again in Christchurch, New Zealand, when on 15 March 2019 Brenton Harrison Tarrant murdered fifty-one people and wounded forty-nine others as he streamed a live broadcast of the massacre. By the time Facebook removed the seventeen-minute video, it had more than 4,000 views, then went viral across the Internet. The first user report came in twelve minutes afterwards; not a single viewer reported it during the ramage.[77]

The damage that can be done even in short-term posting of this content cannot be overestimated—to the victims and their families, and also in terms of encouraging others to stream their own horrific activities. In the beginning, social media companies relied mainly upon viewers to flag offensive content, but that was far from adequate. They then hired roughly 150,000 content monitors, mostly in India and the Philippines, although the effort is still well behind the level of threat, plus some of those forced to watch horrors have sued for post-traumatic stress, with one petitioner describing a violent "internal video screen" that he can't flick off.[78] The newer approach is to use automatic, algorithmically driven tools to identify and squelch horrific content, with major implications that we'll discuss in chapter 9.

Even short appearances of violent material on sites such YouTube routinely result in hundreds of thousands of views. The toxic publicity is also often elaborated on afterwards, with organizations "spinning" events for their purposes. For example, Syed Rizwan Farook and Tashfeen Malik killed fourteen people who were attending an office training event and holiday party in San Bernardino on December 2015, and were then killed by police. Ms. Malik had posted a pledge of loyalty to Abu Bakr al-Baghdadi, leader of the Islamic State, on Facebook. After the shooting, the couple's landlord led a gaggle of media on a tour of their apartment. Farook and his wife had been parents of a six-month-old baby daughter. Broadcast media accounts featured the baby's room, with a crib full of stuffed animals and fuzzy blankets, begging the obvious question: How could they so cold-heartedly leave her an orphan? Then the Islamic State, never one to miss a chance to manipulate emotional triggers, featured the same crib in its English-language magazine *Dabiq*, captioned, "Syed and his wife did not hold back from fulfilling their obligation, despite having a daughter to care for."[79] Thus their abandonment of the child was turned into an occasion for hagiography.

Live-streaming has also been used for many admirable purposes, of course. For example, it has empowered activists in their fights against oppression throughout the world. It first became popular during the 2010–2011 Arab Spring and the 2011–2012 Occupy Wall Street protests. The most famous use in the United States is the #BlackLivesMatter movement, which live-streamed protests from Ferguson, Missouri, and gained an enormous following.[80] The 2016 police killing of Philando Castile at a traffic stop was live-streamed by his girlfriend Diamond Reynolds. Live-streamed video from Syria has allowed citizen journalists to document horrendous killings by ISIS and by the Syrian government.

But live-streaming has also put activists at mortal risk. The Syrian live-streamer, Rami Al-Sayed (aka Syria Pioneer), broadcast the heavy bombardment of the besieged Syrian city of Homs by forces loyal to Bashar al-Assad. He posted nearly 800 videos of civilian injuries and deaths, drawing international media attention. But his signal drew Syrian military targeting, and he was killed with mortar fire as he tried to help a family reach medical treatment.[81] Due to this risk, human rights activists have learned to encrypt their videos through Signal and other apps. Organizations like Witness are trying to determine guidelines for when to share live accounts and when to archive them, to protect the activist.[82]

QUALITY FIRST-PERSON FILMMAKING TECHNOLOGY

Sometimes the videographer is less interested in drawing attention to ongoing events and more interested in being the hero in his own twisted kind of action movie. Combining high-quality, cheap, portable cameras and easy-to-use digital streaming services is an alluring new way to do that.

Mohamed Merah was a twenty-three-year-old petty criminal living in Toulouse, France, who claimed a connection to al-Qaeda. In March 2012, he bought a GoPro action camera—the kind that lots of people use to film their mountain climbing or extreme sports in real time. Attaching it to his body armor, Merah filmed himself making preparations, carrying through with the murders of seven people, and escaping on a motorbike. His first three victims were off-duty soldiers. Then he attacked a Jewish school, killing a rabbi and three children. One of the kids, eight-year-old Miriam Monsonego, hesitated for a second, reluctant to flee without her school bag. He chased her down, grabbed her by the hair, changed weapons when the first one jammed, and then shot her in the head, capturing it all on film. The police caught up with him and laid siege to his apartment building. Merah spent his last thirty-six hours alive editing the twenty-four-minute video file and transferring it to a USB stick. He somehow slipped through the security cordon, mailed the film to Al Jazeera in Qatar, and then returned to his apartment, where police killed him in a firefight.[83]

In September 2015, Vester Lee Flanagan used a GoPro-type camera to capture his perspective while he shot Alison Parker and her cameraman Adam Ward during a live morning broadcast of affiliate WDBJ in Roanoke, Virginia. Flanagan, an African American, wrote in a suicide note that he believed he had been racially discriminated against, and that the murder of nine black churchgoers at the Emanuel African Methodist Episcopal Church a few months earlier in Charleston, North Carolina, had "sent me over the top." The network cut live transmission, but a few hours later Flanagan posted the fifty-six-second video on Facebook, tweeting, "I filmed the shooting see Facebook." Twitter and Facebook quickly took it down, but not before millions of people throughout the world saw it and became unwitting witnesses to the killing.[84]

Law enforcement used the GPS locator on his Apple iPhone to track Flanagan's general whereabouts, and then an automatic license plate identifier in a state trooper's car picked out Flanagan's escape car. Flanagan shot himself during a highway car chase later that morning.

Clamping down on these various means of disseminating visual footage of attacks would not be enough to stop its rapid spread. The perpetrators of violence know that smartphone-equipped witnesses will likely publicize their nefarious work for them. During the 9/11 attacks, hijackers encouraged passengers on the doomed flights to use their cell phones to call relatives, to heighten fear and publicity about what was happening. At that time, passengers' phones didn't have cameras and their calls were not instantly accessible to a broad audience. Fast-forward fourteen years. The Paris attackers who stormed the Bataclan theater in November 2015 made no provisions to capture images of their slaughter of eighty-nine concertgoers, probably because they knew they wouldn't need to. As Jason Burke observes, "They, or more likely their commanders, knew that they could rely instead on the unprecedented prevalence of cameras, and our apparently insatiable appetite for sharing the images they produce, to do the job for them."[85]

These various communications technologies can also be used to launch faux stories aimed at mass scale disinformation, or stoking public panic. This was done on 11 September 2014, when thousands of residents of St. Mary Parish in Louisiana received a text message alert via a fake Twitter account that read, "Toxic fume hazard warning in this area until 1:30 PM. Take Shelter. Check Local Media and columbiachemical.com." Hundreds of tweets publicized faux eyewitness accounts of a powerful explosion heard at a chemical plant in Centerville, Louisiana, with the hashtag #ColumbianChemicals. Dozens of journalists throughout the country found their Twitter accounts flooded with messages about the accident. A false YouTube video showed a masked ISIS fighter standing next to looping footage of an explosion. Alarmed tweets from "Anna MClaren (@zpokodon9)" and "Eric Trapp (@Eric TraPPP)" were personally addressed to Karl Rove and *New Orleans Times-Picayune* reporter Heather Nolan, among many others. The perpetrators created a phony website of Louisiana TV stations and newspapers, false video of a horrific chemical explosion, and even a fake Wikipedia page called "Columbian Chemical disaster."[86]

An army of well-paid "trolls," individuals who send out a high volume of posts, often harassing targeted people, was responsible for the hoax. They were working for the "Internet Research Agency," a supposedly private organization that operated out of a building in St. Petersburg, Russia. The chemical plant hoax was followed by many other online ruses designed to exploit real-world anxieties and manipulate public fear.

For example, on 13 December 2014, a couple of months after the discovery of four US-based Ebola cases, a fake news campaign reporting an outbreak of Ebola in Atlanta with the hashtag #EbolaInAtlanta trended briefly around the Atlanta metropolitan community. On the same day, accounts using the hashtag #shockingmurderinatlanta said that an unarmed black woman had been shot by police, accompanied by a narrated video of the alleged shooing. The lie was disseminated shortly after the killing of Michael Brown in Ferguson, Missouri. An organization called Infosurfing, well known for posting pro-Kremlin infographics on Instagram and VKontakte (the Russian equivalent of Facebook), produced the videos. According to Adrian Chen of the *New York Times*, all of these US-directed postings used an automated web tool called Masss Post, associated with the domain Add1.ru, run by a young tech entrepreneur named Mikhail Burchik, whose address matched the Internet Research Agency's St. Petersburg location.[87]

Fake accounts are a major vehicle for such campaigns. Creating fake Twitter accounts is easy and has been big business for years. The ruse first became public during the 2012 US presidential campaign, when the Twitter account of candidate Mitt Romney suddenly gained 116,000 new followers in a single day. Prior to that, the day-to-day growth of his following had averaged 3,000–4,000. Most were reportedly new to Twitter, a telltale indicator of fake accounts, though the Romney campaign denied buying followers.[88] Scores of cheap software programs are available for creating fake Twitter followers by generating names and picking random information from existing profiles. After accounts are created, other software can automatically generate retweets, magnifying the profile of whatever is posted.[89]

This tactic can also be used with true content, of course. ISIS used this method in 2014 to spread videos of the beheading of journalist James Foley. They were disseminated through thousands of fake Twitter accounts.[90] One study estimated that ISIS had 46,000 to 70,000 accounts, with an average of 1,000 followers each, which is a pretty big megaphone.[91] Having followers tells you nothing about the legitimacy of accounts, as fake accounts usually have fake followers, too. A core group of about 2,000 ISIS users tweeted the same gruesome beheading, which was then retweeted.[92] ISIS has also piggybacked off of existing innocuous hashtags to spread their messages and reach a wider audience. For example, during the 2014 World Cup they showed a picture of a decapitated head with the caption "This is our football. It's made of skin," and sent it out with the hashtag #WorldCup, infiltrating the football conversation on Twitter.[93]

Users can also employ "bots," or automated false accounts that are preprogrammed to follow instructions and flood the Internet with information that privileges certain narratives and drowns out other users. Most of the time their instructions are very simple, for example, to retweet or "like" topics from a designated group or tweets from a particular list of people. Some bots are programmed to spam targeted hashtags (like #WorldCup), and others are designed to amplify certain conversations (posting videos or memes) based on specific wordings or hashtags. Related to bots are sockpuppets, accounts created by one individual made to look as if they belong to a second individual whose identity may be stolen or invented.

While sockpuppets are inherently unethical, we should note that there is nothing necessarily wrong with using bots.[94] Many people, including academics, celebrities, and politicians, set up or buy their own bots to automate their online activity and, often, to bring broader attention to it. Some bots are set up to tweet 150 times per hour. How they are used depends on the ethics of the person who uses them.

Organizations and individuals of all complexions are using bots for both good and nefarious purposes. Computational propaganda, for example, is the use of bots to manipulate people's political views via social media,[95] such as the Russian attempt to influence the 2016 presidential election. But bots can also be helpful to groups trying, for example, to flood the Internet with positive messages aimed at quelling public panic in the aftermath of terrorist attacks. Decisions about whether and how to regulate their use, as well as about policing of their use by the technology companies, will be challenging.

More straightforward is the need to think hard about another type of auto-mated online communication. Newspapers are starting to employ technology that generates and publishes short pieces without a human author for data-heavy stories, such as election results and sporting events. For example, the *Washington Post* uses a tool called Heliograf. Meant to save time in the news-room, this technology, if it were hacked by rogue actors, would give them an-other powerful mechanism for spreading malicious news stories. Imagine for instance that non-state (or state) actors want to influence an election. They might hijack this technology to try to manipulate results midway through election day, such as by reporting fake polling data on the *Washington Post* site with the intent to deter people from voting, thinking that the election has already been decided so there is no point.

END-TO-END ENCRYPTION

Another unprecedented, and disruptive, aspect of technologies now on offer is the widespread accessibility of cheap, end-to-end online encryption tools. These are more widely available, easily downloaded, and more user-friendly than ever before.

Encryption is important protection for all of us for online banking and shopping, but when something is merely "encrypted," it means that someone other than the user might have a key to read it. Your internet service pro-vider (ISP) can encrypt and decrypt your email messages as a service using their keys, for example. Under US federal law, government officials with proper warrants can order your ISP to use their key to decrypt your email messages. If the ISP encryption algorithm is weak or network security lacks proper safeguards, hackers, terrorists, or criminals can also access your email messages.[96]

End-to-end encryption, on the other hand, is meant to be impervious. No one can access it except whoever's at either end. Only the people directly communicating can decrypt and read each other's messages and no one in be-tween, not the company whose server you're using, not anyone else.[97]

This kind of end-to-end encryption has become accessible and cheap, so it is also being taken advantage of by terrorist groups in various ways. ISIS used secure messaging services such as Skype, Silent Circle, Telegram, and WhatsApp for regular communications, and encrypted message platforms like Kik for battlefield communications.[98] The site Telegram hosted a pro-ISIS channel called "Lethal Dose," for example, that features step-by-step tutorials in how to make toxins, from cyanide to ricin.[99] Terrorist recruiters still quickly transition potential recruits to end-to-end encrypted sites for ongoing communications.

Like covert actors of all types, terrorists and criminal groups have long used a wide range of clandestine methods to evade monitoring and counter-measures, from coded language to disposable devices. Pretending to be students, the 9/11 hijackers exchanged emails referring to the World Trade Center as "architecture," the Pentagon as "arts," and the White House as "politics."[100] Emails may be sent on public computers, in libraries or cybercafes where it's hard to trace specific users. Criminals can create email accounts on web-based programs like Outlook, share the login information, then leave messages in the drafts folder—an electronic version of the "dead drop." Drug dealers have used disposable burner phones for decades; in authoritarian countries, activists use them if their phones might be confiscated and searched. Easily accessible burner apps anonymize texts and cell phones by routing messages through a new phone number. If things get hot, users delete the app and it goes out of service.[101]

What's new with end-to-end encryption for apps on smartphones is that they up the ante. Intelligence organizations have ways of monitoring computer keystrokes or accessing phones if they can actually watch them being used by a human. They can also hack into physical devices, usually smartphones, and record the sounds, images, and text messages stored or made on that phone, even if they're sent with end-to-end encryption.[102] But they do need to know who the humans are—meaning, they need to know which people they should be watching or whose phones they should be trying to get access to. That's gotten harder.[103] With the increasing use of remote control operatives, "lone wolves," "wolf packs," and "clean skin" volunteers (operatives with no prior record or involvement in illicit activity), it's become more difficult to disrupt terrorist plots.

As with the other new technologies, encryption tools are being used both for bad and good purposes, and coming to an optimal resolution about how they should be used will be complex. End-to-end encryption safeguards individual privacy and security, and is extremely important for everyone. Enthusiasm among the public for end-to-end encryption grew after Edward Snowden's exposure of the National Security Agency's widespread monitoring of US citizens' private communications. Many believed that was an overreach in state power, and that the development of widespread encryption was an appropriate response. In the aftermath of those revelations, all of the major technology companies, including Apple, Google, Microsoft, and Facebook, announced major new encryption initiatives. Encryption protects citizens who might otherwise be subject to the abuses of authoritarian governments.

On the other hand, those who argue that encryption should make it technically impossible for law enforcement (even with probable cause and a search warrant) to access online communications under *any* circumstances reject centuries of legal practice in democratic states. The question then is why and how, from a legal perspective, Internet-connected communication should be treated differently from every other type of communication that historically preceded it, including the telegraph and the telephone. The answer requires new thinking, because historical precedent doesn't help us here.

HIJACKING PSYCHOLOGICAL TACTICS

Technology companies have employed a number of means of exploiting psychological vulnerabilities in order to make their services irresistible, even addictive. They've done this not because they're evil, but because the industry's business model depends upon a platform or app demonstrating increases in the amount of time people give it. According to ethicist and former Google software designer Tristan Harris, "[T]here [are] a thousand people on the other side of the screen whose job it is to break down the self-regulation you have."[104] In the legitimate economy of the web, whether or not we watch, click on, or "like" leads to someone earning cash. The many sophisticated means by which software programmers have learned to grab our attention and motivate us to act—whether that is to make a purchase, to retweet a message, or to binge watch a drama—are all ripe for exploitation by unsavory actors.

One manipulative technique is targeting. By gathering voluminous data about users and their online behavior, social media and other sites, such as retailers, are able to serve up to advertisers groups of users whom they deem most likely to respond to their ads. Facebook, for example, built its business on selling advertising that can be targeted according to an almost infinite list of parameters, such as "big-city moms" and "veterans in home." Ads can also be served up to people through the tactic called "retargeting." First, the advertiser serves an ad to their target audience, such as on a retail site or on social media. When people click on the ad, the programming "cookies" their web browser, meaning it downloads a small bit of text to their browser that tracks their online behavior. That information is used to serve them the same ad, or a new one, when they visit other sites. Showing ads to people repeatedly is thought to increase the chance that they will make a purchase. While it may seem innocuous when Zara targets you with ads for a new blouse or Williams Sonoma reminds you of that espresso machine you like, when those ads are filled with inflammatory suggestions targeted to the most angry or emotionally susceptible people, they can lead them to join a cause, buy into conspiracies, or take violent action.

Targeting can also lead to "filter bubbles," with people being shown content that is tailored to them according to the information collected about them. While this may sound appealing, it can lead to people viewing only content that comports with their views. The tactic can also be used to shape their views by systematically keeping information from them and feeding them disinformation. This is a means of fueling support for extremist causes.

Nefarious actors can also exploit the psychology that fuels people's engagement with social media. The designers of social media sites have built on people's yearning for self-expression and validation. We are driven to know what other people think of us and to give them a good impression of us. Studies show that we get a hit of dopamine, a neurochemical that causes a feeling of well-being or pleasure—the same chemical that leads to nicotine, gambling, and cocaine addictions—from sharing good news about ourselves.[105] This mechanism can be harnessed for mobilization in surprising ways. For example, even victims of trafficking find it irresistible to put a positive spin on the facts of their lives. Young Nigerian and Gambian women who have been trafficked to Europe and end up in refugee camps in Sicily often end up as sex workers, yet they post Facebook pictures of themselves as glamorous and happy, ending up as advertisers for others to follow them.[106] This is an ideal free marketing strategy for traffickers and is also used effectively by terrorist groups such as ISIS, which in 2014–2016 promoted recruits' personal videos showing them purportedly enjoying the good life in the Caliphate, drawing still more people, including women and children, to Iraq and Syria.

Another cause of concern regards the use of "persuasive technology." Professor B. J. Fogg of Stanford University founded the Persuasive Technology Lab, which has been a crucial influence on Silicon Valley's software designers. Fogg most famously introduced a model of how human behavior can be influenced, which argues that three conditions must hold in order for people to take action: they must be motivated, have the ability to act, and be prompted to do so by some trigger.[107] Software designers have experimented with all sorts of triggers, which push people to take actions, like clicking on a link or making a purchase. For example, the color red has been found to create a sense of urgency in users' minds, so new message notifications are in red. Triggers can be quite helpful, such as by encouraging people to develop healthy new habits like exercising more regularly. But in the absence of ethical and strategic guidelines and considerations, they can also be used for undesirable purposes, such as enticing people to view manipulative content.

Another technique for seizing people's attention online is bundling things together so you stumble across something unintentionally. This is like putting candy near the supermarket cash register, or expensive clothes at the front of the store so you pass them to get to the sale rack. The goal is to encourage impulsiveness, and it works the same way online. Services such as news, videos, baseball games, and friend requests are grouped together so that the user will be exposed to more content, become distracted, and click on it. This technique could also lure viewers toward extremist sites.

All of these brain hacks are working: The average adult checks his or her phone 30 times per day; the average millennial does so more than 150 times per day.[108] Facebook understood the potential and made a huge push to move members from desktop use to its mobile app. The same potential to engage and persuade people can be harnessed by malicious actors. While ISIS harnessed some of the mechanisms of the attention economy in crude fashion with its postings of videos of immolations, drownings, and beheadings, others will follow that example with more-refined approaches.[109]

Unintended Consequences Redux

Those who have built the cluster of new technologies that facilitate our networked world and make it so compelling—from Microsoft to Apple to Facebook and Twitter—were no more able to foresee the negative outcomes of their brilliant innovations than Alfred Nobel and Mikhail Kalashnikov were. Governments were also slow to appreciate the dangers, with the result that most democratic states have yet to create adequate laws, regulations, guidelines, and protections to combat the threats. If we want to preserve the promise of connectivity, we must more effectively contend with its dark side. Preventing the mobilization of individuals for terrorism and other violent attacks is one major challenge; keeping a host of new technologies that can be used for committing attacks out of the hands of violent actors is another. That is the subject to which we turn next.

CHAPTER 8 | Open Innovation of Reach: From
AK-47s to Drones, Robots,
Smartphones, and 3D Printing

I N SEPTEMBER 2014, students at the University of Virginia 3D printed
a military-grade drone capable of carrying a 1.5-pound payload. It was
powered by batteries and a motor and they added an Android phone as its
"brain," hijacking the phone's processor, 3G LTE wireless connectivity, and
camera to make the drone "smart." It had basic "autonomous" capability,
meaning they could program where they wanted it to fly, enter the flight
plan into a tablet, and send the commands to the drone's cell phone. With
inexpensive GPS receivers, a smartphone allowed for navigation without
line-of-sight radio contact or an expensive navigation system aboard.[1] The
"Razor," as they named the craft, was built in a four-foot "flying wing" design
emulating a B-2 stealth bomber, comprising nine printed parts that snapped
together like Lego blocks. A slick YouTube video shared the results of their
experiment worldwide—launching Razor first from a car, then with a bungee
cord used like a slingshot, and finally by hand, with repeated crashes and ultimate triumph, as it zoomed off skyward.[2] The simple design accommodated
printing the drone smaller or larger to make it suitable for carrying anything
from mail to an explosive, all in about thirty hours for less than $2,500.

The revolutionary aspect of the self-manufactured Razor smart drone
was easy to overlook. After all, three years earlier, Southampton University
in the United Kingdom had created the world's first 3D-printed plane, the

Southampton University Laser Sintered Aircraft (SULSA).[3] Made in seven days for less than $6,500, SULSA had elliptical wings and looked more like a traditional fixed-wing aircraft than a stealth bomber. The airframe was printed by a laser that fused nylon powder into solid substances ("sintered nylon"), the strength of which allowed the Royal Navy to fly SULSA to pinpoint safe ship navigation routes through Arctic ice flows.[4] Like Razor, it was constructed of simple parts that could be snapped together by hand, in this case just four. A key distinction from Razor, however, was that SULSA was equipped with a miniature autopilot developed specifically for it, rather than an off-the-shelf smartphone that anyone could buy. Razor's smartphone "brain," allowing it to run entirely by Android control,[5] made it a breakthrough that could advance the creation of private armies or empower amateur hobbyists.

The inspiration for Razor came from Michael Balazs and Jonathan Rotner, two researchers at the MITRE Corporation, headquartered in McLean, Virginia. They had for several years been exploring how the military could build small drones more cheaply. Together they launched the Android Control and Sensor System initiative in 2012, and then partnered with the University of Virginia Mechanical Engineering program to experiment with easy-to-build 3D-printed aerial craft. For the onboard guidance, Balazs and Rotner could have taken the standard approach of buying a $10,000 autopilot system that communicated over Wi-Fi with a dedicated ground station that cost another $20,000. They decided instead to purchase a smartphone for less than $500.[6] "We're using a phone because it's an inexpensive, ubiquitous object that can be used anywhere," Michael Balazs explained. "It can be made very secure, and because it's being commercially driven, it's going to get better and cheaper every day."[7] Commercially available products draw upon the tech industry's massive resources and remove the problem of cumbersome upgrades, which are costly, inefficient, made to military specifications, need long logistical trails for replacement parts, and follow detailed requirements tailored for the protection of each individual system.[8]

The Razor and its subsequent versions combined technologies that are newly accessible at low cost to everyday people and simple to use. This humble craft with its autonomous potential, enabled by the fast-evolving smartphone-driven cluster of technologies, has the potential to reshape the marketplace of violence, spreading to a wide range of new users, including non-state actors.

Of course, small, even nano-sized, drones have long been used by state militaries. Drones like the Raven (RQ-11B), a small unmanned aerial vehicle (SUAV), were used extensively by the US Army in Iraq from about 2006, and by Special Operations Forces earlier than that.[9] Each Raven unit costs about $260,000 (depending on specific capabilities).[10] Comparable small military drones are built by the United Kingdom, Canada, China, Iran, Italy, Spain, Israel, Russia, North Korea, and many other countries. They are sold to a wide range of global government customers. Among advanced militaries, Ravens and other comparable craft are commonplace. But at a quarter-of-a-million-dollars a pop, they are out of the reach of most individuals and small groups.[11] Now the same military capability can be manufactured at a fraction of the price by a bunch of students.

Many engineers predicted that the technology for drones would quickly be scaled up for creating 3D-printed manned aircraft, a logical next step. That is still possible. In today's smartphone-driven economy, however, it is more likely the trend will go in the other direction, toward printing large quantities of small, personalized crafts, owned by anyone and able to perform a wide range of tasks, from delivering medicine to filming forest fires, taking wedding pictures, and dropping lightweight explosives or crashing headlong into crowds of civilians.[12]

The convergence of new technologies is not just eroding the government's traditional monopoly on systems associated with the use of force. Due to the constant flow of technological upgrades in commercial products, drones like the Razor that depend on smartphones may soon be higher performing than many commonly used small military drones whose software cannot be as routinely upgraded.

Convergent Technologies and Extended Reach

This chapter examines how a cluster of emerging technologies—specifically UAVs[13]—referred to with the loose colloquial term drone[14]—other robots, 3D printing (additive manufacturing), and nascent autonomy—intersect to offer extended *reach* to a much broader range of actors. They are grouped together because they interact with and build upon each other, enabling unexpected popular innovation, and with that, power. These technologies, with distinct historical origins and differing degrees of maturation, are now pinging off each other, opening up a broad scope for creating new breakthrough systems. In this world of distributed, interactive learning, individuals can achieve breakthroughs with relatively little technological expertise.

The concept of *reach*, here meaning the ability to attack, defend, or influence through the use of violence, is at the heart of this chapter. Developed by late nineteenth-century military thinkers, especially Halford Mackinder and Alfred Thayer Mahan, the term was originally applied to geopolitics, referring to the ability of states to project force and influence around the world. Although it is still related to geography, reach is no longer defined by it, as now it includes the projection of force through global digital connections. Clusters of new technologies are extending the reach not just of states, but also individuals and groups.

New and emerging technologies are heightening both personal lethality and personal vulnerability. In combination with passionate ideologies and grievances, they are delivering power from below that is as ubiquitous and as destabilizing today as dynamite and the AK-47 became.

These technologies hold potential for both good and evil, and we should be judicious in analyzing their implications and minimizing their risks. UAVs, robots, and 3D printing are being used for many valuable purposes by states and some corporations and individuals—delivering packages, bringing medicine to remote places, helping farmers tend their crops, assisting first responders in emergencies, and monitoring the health of the Amazon River, to name a few. Of course, if that were the end of the story, we could all simply invest in technology stocks and call it a day. Unfortunately, digital-based technologies are also expanding the marketplace for violence in unexpected ways.

To understand the full nature and implications of how power is shifting, we will start with a brief overview of fast-developing unmanned systems on the ground, sea, and air.

The Scope of Robotic Systems

The public debate about drones has barely scratched the surface of the variety of unmanned technologies and autonomous capabilities that will affect the future of political violence.[15] The focus has been on the proliferation of large UAV platforms like the US Predator and Reaper to states throughout the world. Aerial systems are the most advanced in development, but they have no monopoly on remotely piloted or autonomous technologies—or on the vast strategic and doctrinal challenges regarding how best to use them.[16]

A wide range of unmanned military systems, from enormous armored vehicles to tiny nano-systems, operates on the ground. They are often used for activities that are dirty, dull, or dangerous for humans, such as explosive ordnance disposal (EOD) or activities in contaminated environments, as in the aftermath of a chemical attack. The Pentagon calls cooperation between manned and unmanned systems "centaur warfighting," named for the half-man and half-horse creature in Greek mythology.[17] Currently two basic types of ground-based unmanned vehicles (UGVs) are common: those driven through remote control, and those that have autonomous systems that drive the vehicle using algorithms, sensors, and predetermined routes.

UGVs include huge armored trucks used to clear roads of improvised explosive devices (IEDs)[18] and so-called Robotic Wingmen, autonomously driven vehicles from Humvees to tanks that are designed to operate alongside manned tanks and other vehicles.[19] It is worth noting, however, that these driverless vehicles are not *cheaper* than human-driven ones; their savings is in potential human casualties. Their development has come at a steep price, borne largely by the private sector, and they are also manufactured by the private sector and sold to the military at a commensurate cost. The software and systems of all of these military vehicles are driven by the faster, cutting-edge research underway on driverless cars in the civilian world, for example Waymo, Google's self-driving car project.[20]

A large range of small, ultra-light reconnaissance robots has also been de-veloped, such as the Dragon Runner 10, which looks like a toy tractor and at only ten pounds can be carried in a backpack and thrown over hills or into buildings for reconnaissance.[21] The National Robotics Engineering Center at Carnegie Mellon University originally developed Dragon Runner under con-tract with the US Marine Corps, beginning in 2002. These devices were used during the invasion and occupation of Iraq, and they've been acquired by the UK Ministry of Defense and the Arizona Police Department, among others.[22] Dozens more ground-based robotic systems are in use or under development, including a range of ultra-light reconnaissance robots that are designed to go through doorways or climb walls, and nano/micro robots that can follow designated targets, swarm, collect intelligence, and even mimic mosquitos, infecting enemy bloodstreams. The United States, Russia, and China are all investing heavily in them.

Also mostly overlooked in public drone debates is the fact that there is almost as much robotic activity on and under the water as there is on land or in the air. Unmanned maritime systems first appeared at least fifty years ago. In the United States, the University of Washington's Applied Physics Lab, supported by the Office of Naval Research, developed the Special Purpose Underwater Research Vehicle in 1957, used until 1979. The *Argo*, an un-derwater vehicle that became famous for finding the wreckage of the *Titanic*, followed in 1985.[23] While original models were tethered to a mother ship, newer ones have been cut loose. They include drone ships and drone subs or torpedoes.[24] All are equipped with a range of remote sensors, and some are also loaded with weapons.

Underwater drones are performing a range of valuable nonmilitary functions. They've been used to locate the flight recorders of downed passenger planes, including Korean Airline 007 (lost near the Kamchatka Peninsula 1983), EgyptAir 990 (lost over the coast of Nantucket in 1999), and Air France 447 (lost over the Atlantic in 2011).[25]

Commercial underwater craft have become so small and cheap, with the price of low-end underwater camera drones starting at about $600, that they're widely used in undersea exploration.[26] Entrepreneur treasure hunters bundle a range of higher-end undersea drones onto commercial airlines and use them in place of deep-sea scuba equipment. Some longtime undersea explorers fret that the age of human-occupied submersibles is over, because robotic subs are cheaper, can dive deeper, and stay underwater longer.[27] More surprising, senior officials in the US Navy's submarine community have argued that the technology of unmanned military systems will shape the future of the manned submarine program, which is a cultural sea change.[28]

Still, it's important not to overstate the capabilities of both commercial and military unmanned systems as of yet, because they have many vulnerabilities and weaknesses. Small aerial UAVs are affected by high winds and other atmospheric conditions like rain, snow, and ice (although improved stabilization technologies mean they're a lot better than they used to be). Military UAVs are built to withstand harsh weather conditions, and the most weatherproof commercial UAVs tend to be the most expensive. Unmanned marine systems are strongly affected by currents and high seas, not to mention vulnerable data links. All reconnaissance drones are constricted by battery life and by the range over which a signal and an image can be transmitted. The capacity for automation of most commercially available UAVs is currently limited either to preset GPS determined programs or by autopilot settings that enable the UAV to follow its operator's prior instruction.

Those who want UAVs with larger payload capacity, extended flight time, and longer range must build them as individual customized systems.[29] That reduces access because few people have the time and skills to do so. And in known conflict situations, traditional military forces can destroy UAV launch sites and target their operators. Driving a drone can be a risky business.

All of this said, the technology is developing rapidly, and vigorous examination of the likely trajectory of innovation must be quickened. Because aerial systems have been the trailblazers, we will zoom in on them.

How Unmanned Aerial Vehicles Extend Private Reach

Understanding the full range of unmanned aerial systems is important to grasping their full significance—and projecting their longer-term intersections with land and sea systems, too.[30] The vast majority of military UAVs do not transport weapons and are primarily used for intelligence, surveillance, and reconnaissance (ISR) missions. They may also be used to draw enemy fire, thereby revealing the location of anti-aircraft defenses, or as practice targets in training. Just because they're unarmed does not mean they're benign, as UAVs routinely guide attacks by manned aircraft and cruise missiles.

For the US military, UAVs fall into five basic categories, divided mainly by size, range, speed, and mission (Figure 8.1).[31]

The US Air Force and CIA operate aerial craft such as the Predator and the Reaper. These large, long-endurance, armed UAVs have become both famous and infamous, featured in popular books, movies, and plays.[32] They are regularly used on the battlefield. Predators were crucial to US and allied combat operations in Afghanistan and Iraq, for example. But their use for tactical counterterrorism has drawn the most attention, especially their targeted killings of members of al-Qaeda and other terrorist groups in Pakistan, Yemen, Somalia, and Libya.[33] A voluminous academic literature discusses the ethical, legal, and strategic implications of remote aerial targeted killings across sovereign borders in countries that are not at war.[34] Those important arguments will not be repeated here, except to observe that the widespread focus on long-range, high-altitude drones has given a false impression that the technological lineage of these unmanned platforms is new.[35]

In fact, armed unmanned aerial platforms predate manned airplanes by more than fifty years. The first version appeared in 1849 during the Italian War of Independence, when the Austrians sent 100 hot air balloons, each trailing a long copper wire to remotely trigger a bomb, floating toward the besieged city of Venice.[36] Although the Austrians carefully calculated the wind velocity and direction before setting the balloons aloft, the wind shifted, and only one bomb fell within the city. The rest apparently exploded in the air.[37] The Austrians were embarrassed (one account says the Venetian defenders laughed about it), but from the Austrian perspective at least none of their soldiers had died. Manned balloons were common by this point (the French began using them at the end of the eighteenth century) and they were much more dangerous to the operator, with pilots perishing when the balloon ignited, landed in water, or was hit by ground fire and dropped like a rock.

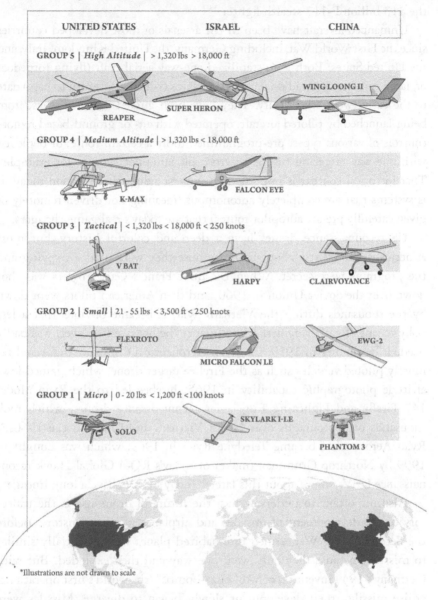

	UNITED STATES	ISRAEL	CHINA
GROUP 5 | High Altitude | > 1,320 lbs > 18,000 ft	REAPER	SUPER HERON	WING LOONG II
GROUP 4 | Medium Altitude | > 1,320 lbs < 18,000 ft	K-MAX	FALCON EYE	
GROUP 3 | Tactical | < 1,320 lbs < 18,000 ft < 250 knots	V BAT	HARPY	CLAIRVOYANCE
GROUP 2 | Small | 21 - 55 lbs < 3,500 ft < 250 knots	FLEXROTO	MICRO FALCON LE	EWG-2
GROUP 1 | Micro | 0 - 20 lbs < 1,200 ft < 100 knots	SOLO	SKYLARK I-LE	PHANTOM 3

*Illustrations are not drawn to scale

FIGURE 8.1

The advantages of uninhabited systems have long been obvious, because keeping a human being breathing, alert, and invulnerable high above the earth's atmosphere is expensive and difficult. The strong preference for pricey inhabited aircraft that emerged during the twentieth century often had more to do with budget politics than technology, such as when the US Air Force canceled all remotely piloted vehicles in the 1980s in order to save cash for the $13 billion F-117 stealth fighter.[38]

Unmanned aircraft have been in the arsenals of most developed countries since the First World War, including Germany, the United Kingdom, Italy, and the United States. Both gyro-stabilized one-way aerial craft (flying torpedoes or, later, missiles) and radio-controlled airships (which can return to base) date to the early twentieth century. They have been guided in many ways, from being launched by piloted aircraft; operated with air- or ground-based remote controls of various types; pre-programmed to fly a certain route; or deployed with one-way targeting (using a gyroscopic autopilot system, for example). That historical context is important to understanding the situation today: it is systems that are completely autonomous (meaning not driven remotely or given carefully pre-set autopilot routes) that are today's real game changers.

US reconnaissance drones have a deep and colorful history, but most Americans know little about them because they were used for spying and the programs were secret. When U2 pilot Francis Gary Powers was shot down over the Soviet Union in 1960, and then American pilots went down by the thousands during the Vietnam War, it became obvious that a less risky means of gathering aerial intelligence (below satellite level) was badly needed. Beginning in 1959, the Ryan Aeronautical Company pioneered remotely piloted vehicles such as the Firebee target drone, which gained low-altitude photographic capability in 1965. Firebee led to the Ryan Model 147 Firefly and Lightning Bug series reconnaissance drones, which took thousands of photographs over North Vietnam and China in the 1960s.[39] Ryan Aeronautical became Teledyne Ryan in 1969, which was bought in 1999 by Northrop Grumman (maker of today's RQ-4 Global Hawk reconnaissance UAV—more about this later). Today's UAVs have a long lineage.

Also important to understanding the future of drone use is the trajectory of the development of missiles and airplanes in recent history. Before the Second World War, armed uninhabited planes were essentially similar to missiles, because they only went one way and then exploded. But with Germany's 1944 invention of V-1 "buzz-bombs" (the world's first operational cruise missile), their development slowly began to diverge. Missiles were designed for one-way, single use and integrated their warheads into the airframe, but armed unmanned airplanes started to be designed for two-way return and reuse, with a separable weapon carried as a payload that could be dropped so that the unmanned aircraft might return for another mission.[40] This divergence is reversing itself again now, because of the cheapness of new technologies. Due to new ways to produce devices, miniaturized components, better and cheaper sensors, and other factors, some vehicles can now be cost-effectively used either way.

A sharp drop in price for small UAVs means that they are becoming expendable. This changes the purposes and conditions under which they will be used, and by whom. They can be deployed, for example, as robotic suicide bombers. This distinguishes them as a disruptive innovation by contrast to large medium- and high-altitude drones, which are sustaining innovations.

PREDATORS, REAPERS, GLOBAL
HAWK: SUSTAINING TECHNOLOGIES

The differences will become clearer with a quick review of the characteristics and recent evolution of high-altitude American drones.[41] US Predators and Reapers date directly to the highly secret CIA Eagle program, tasked with developing armed pilotless drones, which produced five prototypes in the 1980s at a total cost of $7 million.[42] Their approval for use in US counterterrorism came fifteen years later as the result of a 17 September 2001 Memorandum of Notification (or "Finding") signed by President George W. Bush in the aftermath of 9/11, which broadly authorized the CIA's deployment and firing of armed drones.[43]

Both the Predator and Reaper airframes are unimpressive technologically; they might have been built forty years ago. The light and flimsy Predator is slow for use in military operations (a maximum 135 mph) and highly vulnerable to ground-based missiles. The Reaper can go faster (around 300 mph), carry more weapons, and loiter longer over a target, and it has now replaced the Predator in the American arsenal. Both are guided in flight by human pilots remotely, but the landing, which is the trickiest part of the flight operation, is always locally controlled. Forward-deployed bases and some 170 personnel are required to launch, recover, process, and maintain each Reaper combat air patrol, which consists of up to four individual aircraft.[44] These are major, expensive aerial systems not far removed from twentieth-century manned aircraft that were launched from aircraft carriers or air bases throughout the world.

Northrop Grumman's RQ-4 Global Hawk is a much larger unmanned aircraft first flown in 1998 and intended to replace the manned U-2 spy plane. Designed for higher altitude reconnaissance (above 30,000 feet and as high as 70,000 feet), Global Hawk is a fixed-wing aircraft without a human aboard, and it is always unarmed. Unlike a manned aircraft, Global Hawk can go for days without needing to land. Global Hawk is mainly automated (i.e., operated by a computer mouse rather than a joystick) and is also an ev-olutionary rather than revolutionary technology. Like Predator and Reaper, Global Hawk requires a local launch and recovery team that is deployed nearby. Even though they peer through very high-tech lenses and their in-telligence disseminates through global digital links, all of these aircraft need places to land that are relatively close by.[45]

What's new in all of these UAVs (we've only discussed a few of the best known) is their sophisticated sensor systems. Each carries advanced dig-ital communications links, high-quality cameras, laser designators, pow-erful radars, and precision munitions—all of which are expensive and not commonly accessible.[46] This means that they can follow human targets more closely and observe their behaviors for longer periods of time. And because UAVs shoot "smart" guided munitions, a drone operator can even divert a missile after firing if necessary.[47] For the United States, the other major change has been the legal frameworks in which they are being used—especially the post-2001 increased freedom of the CIA to engage in remote targeted assassination.

There is no question that large UAVs are affecting, and will continue to affect, how states use force, offering greater targeting discrimination and in some cases lowering the threshold for intervention. The United States has used aerial drones to target terrorists and insurgents in Afghanistan, Pakistan, Libya, Yemen, Somalia, and elsewhere, and nine other countries, including the United Kingdom, Israel, Pakistan, Turkey, Nigeria, Iran, Iraq, Azerbaijan, and the United Arab Emirates have followed suit, using armed drones in combat.[48] Many more will do so. Because most states lack the global infrastructures to support their long-range deployment, however, they will be forced to use them only in shorter-range situations, such as against their neighbors or their own populations.[49]

To sum up, large UAVs like the Reaper have evolved gradually, be-coming more refined, and more capable of serving an established constitu-ency. Their technology is mature and, apart from some potential incremental improvements, such as adding stealth capability to more platforms or extending loiter time by using solar energy, they will not change dramatically.[50] They duplicate aircraft capabilities that many countries already possess—just un-manned, slower, digitally linked, and vulnerable to ground and air defenses.[51]

What's more, the costs of US large high-altitude armed drones are escalating, mainly due to the increasingly high-tech communications and sensor systems in them. The cost of the Global Hawk went from $57.0 mil-lion in 2002 to $222.7 million in 2013.[52] While the Predator has been re-tired, the price of the Reaper airframe is about $16.9 million (not including the ground stations or the armaments) and the cost of its upgraded laser system recently increased 797 percent.[53] American UAVs are the Daimlers, Ferraris, and Rolls Royces of the global marketplace. Chinese drones are generally cheaper and more practical for small and medium powers like Pakistan and Saudi Arabia. But all of these UAVs require a military and intelligence infrastructure to support them.

For all of these reasons, they are a weapon for governments. And with their adoption helped along by high-profile US demonstrations of their use, large UAVs will be an integral part of virtually every country's military arsenal by 2025, as ubiquitous as aircraft or tanks.[54] Governments with less money and less sophisticated military organizations will buy cheaper and less sophisticated models that lack over-the-horizon systems, just as they buy cheaper and less capable aircraft, ships, or tanks; but any government that can shoot a weapon from an aircraft will soon be able to do the same from a drone.[55]

Indeed, state-to-state proliferation of large UAVs is well along and follows established twentieth-century patterns and logic. Research by scholars Matthew Fuhrmann and Michael Horowitz has shown that there is a strong correlation between GDP per capita and whether or not states have UAVs—which is exactly what we would expect.[56] What is new today is that the United States has engaged in a high-profile global demonstration of how to use them more aggressively, giving other countries a model to follow. In long-term significance, the proliferation of *practice* eclipses the proliferation of these platforms.

Looking ahead, more potentially disruptive than state-to-state proliferation is the power of smaller, simpler UAVs (i.e., those in Groups 1, 2, and 3 of Figure 8.1) to give unprecedented capabilities to groups and individuals who would otherwise not have any aerial attack capability at all. These UAVs will affect the relationship between states, groups, and individuals because they will spread the capacity to use lethal force from the air at very low altitudes that are typically unprotected by military air defenses. Surprise attacks on leaders or innocent civilians are already affecting the course of both interstate and intrastate conflicts. And some states share UAV technology with non-state groups.

STATE-TO-GROUP SHARING OF UAVS: HEZBOLLAH, HAMAS, AND THE HOUTHIS

The Iranians have shared large military drone technology with Hamas, Hezbollah, and reportedly the Houthi rebels in Yemen. Iran, which began using the Ababil surveillance drone in 1986 during the Iran-Iraq War, has one of the oldest active drone programs in the world.[57] The Iranians supplied drones to President Assad during the Syrian civil war, and have also given them to Sudan.[58] Jump-started by Iranian drones, since about 2004 Hamas and Hezbollah have engaged in their own drone development programs. And the Houthis' drone program sharply increased in scope, lethality, and range in 2019, though experts disagree as to whether Iranian assistance or indigenous ingenuity is the primary driver.[59]

Hezbollah learned of the value of aerial reconnaissance drones by monitoring the Israelis. The group intercepted unencrypted (or insufficiently encrypted) Israeli UAV reconnaissance feeds in 1997, gaining intelligence on the movements and locations of Israeli forces. As a result, in September 1997 a secret Israeli assault force sent to assassinate a Hezbollah leader suffered an ambush that killed a dozen Israelis.[60] Since then, the Israelis have had periodic problems with feeds being intercepted by Hezbollah and Hamas, leading to a government investigation whose outcome was an initiative by the Israeli Air Force to encrypt their entire fleet of UAVs beginning in 2010.[61]

While derived from Iranian models, Hezbollah's UAV fleet also reflects independent innovation. Hezbollah first breached Israeli air space in 2004 using a 9.5 foot long, unmanned surveillance aircraft that buzzed noisily at about 1,000 feet for a short time, then crashed into the sea. It had apparently flown under Israel's air defense radar system and escaped notice because of its small size. The incident sparked an emergency Knesset meeting and contributed to the strengthening of Israeli air defenses targeted specifically against drones. During the 2006 Lebanon war, Hezbollah launched a Mirsad-1 surveillance drone (believed to be derived from either the Iranian Mohajer-4 or Ababil-T drone) over Israeli territory. At least half a dozen Hezbollah-driven armed and unarmed drones have been destroyed over Israeli territory since then. In 2011, Israel deployed an elaborate missile defense system it calls Iron Dome, which protects against incoming rocket, artillery, mortars, and some UAVs. In October 2012, for example, Hezbollah shot an Iranian-built drone into Israeli airspace and Israeli Air Force aircraft and ground-based anti-aircraft units followed it, shooting it down over the Western Negev desert.[62] This incident alarmed the Israelis because the drone flew near the town of Dimona, the site of Israel's nuclear weapons complex.[63] More recently, in September 2017 the Israelis used a Patriot surface-to-air missile from their Iron Dome defense system to destroy a Hezbollah UAV in the demilitarized zone on the Syrian-Israeli border.[64]

With both Iran's and Hezbollah's help, Hamas likewise began to build and test its own drones, derived from Iranian prototypes. Hamas claims to have three types of Iranian-derived drones: the A1A (strictly reconnaissance), the Ababil A1B (which is armed), and the A1C (armed and intended for one-way suicide missions).[65] During the 2014 Gaza war, an unmanned Hamas aircraft flew over the Israeli port city of Ashdod until a US-supplied Patriot missile in Israel's Iron Dome defense system destroyed it. Since then, the Israeli Air Force has shot down several Hamas drones that have taken off from the Gaza strip, and the Israelis also reportedly assassinated the lead Hamas drone expert in Tunisia.[66] While no Israelis have yet been killed by one of these UAVs, they've had important political and symbolic effects, intimidating the public, who hear them flying ominously above their heads, and inspiring followers and potential supporters of Hezbollah and Hamas.

Israel has taken the threat very seriously and the Israeli Air Force says that it is able to keep aerial drones under full surveillance from the time they take off until Israeli aircraft or ground-based systems destroy them. The ability of the Israelis to fend them off demonstrates the robust defenses states can build against drones, especially those in Groups 3–5. Fighter pilots practice shooting down enemy drones, and in 2014 Israeli contractor Rafael unveiled a new laser air defense system called Iron Beam, which is designed to intercept low-altitude and low-trajectory threats, such as mortars and, potentially, armed drones.[67]

Houthi rebels in Yemen have steadily increased their use of armed drones, threatening Saudi Arabia, the United Arab Emirates, and the Saudi-backed Yemeni government. From about 2017, Houthi UAVs have evolved from primitive propeller-driven surveillance drones to larger aircraft-shaped craft that the UN says can range more than 900 miles and fly 150 miles per hour.[68] On 10 January 2019 the Houthis successfully used an explosive-laden variant of the Iranian Ababil-T (or what the Houthis call the Qasef 2K), a medium-sized (Group 3) model airplane-shaped armed drone, to assassinate Yemeni senior officers as they sat on a dais watching a military parade. Then, beginning in May 2019, a series of Houthi drone attacks were launched against Saudi and UAE targets, including a Saudi Aramco oil refinery outside Riyadh, the radar arrays of Saudi Arabia's Patriot missile batteries, and airports in both countries.[69]

To this point, Iran is the only state that is known to have purposely given UAVs to non-state groups. Others have involuntarily shared them when they were shot down. For example, Hamas has recovered and studied at least one (probably more) Israeli drones that were lost in the Palestinian territories.[70] In 2017 Houthi rebels shot down a US Air Force Reaper over Yemen, a loss confirmed by US authorities.[71] More concerning is the threat from a new breed of smaller smart drones that non-state groups can now develop on their own, without having to shoot them down or steal them.

Most of the research done thus far on the spread of UAVs concerns the large unmanned variety in Groups 3, 4, and 5—the Harpy, K-Max, or Wing Loong II types shown in Figure 8.1. But it's the unmanned systems in Groups 1 and 2—the small and micro drones—that will be more surprising and possibly destabilizing.

Designers of small UAVs shun expensive high-end purpose-built military software and systems, and turn to faster developing open-source platforms and smartphones. Small commercial UAVs offer a range of capabilities that give users enormous flexibility to adapt them to their needs or purposes. They come in three categories: Ready to Fly (RTF) drones arrive off the shelf and are primarily used by novices; Bind and Fly (BNF) drones allow for customizable transmitters; and Plug and Fly (PNF) drones have customizable transmitters, receivers, batteries, and chargers.[72]

The civilian drone market is exploding. At the Paris Air Show in June 2017, the Teal Group, a private analytical firm that provides analysis of the defense and aerospace industry, issued a prediction, based on a market survey it conducted, that Civil Unmanned Aerial Systems (CUAS) would be the most dynamic growth sector of the world aerospace industry in the coming ten years. The company forecast that nonmilitary UAV production would soar from $2.5 billion worldwide in 2017 to $11.8 billion in 2026, reaching $73.5 billion in total spending.[73]

This technology is well suited to the big technology firms. The types of expertise needed to develop drone hardware are advanced camera technology, computer vision, deep learning, and artificial intelligence, all of which the largest firms are deeply involved with.[74] Major companies such as Intel, Verizon, Facebook, and Google have been pouring investment into this sector, and it is their innovative technologies that are powering the increasing sophistication of robotics, including small UAVs.

But the mass *adoption* of UAVs may be driven by China. The platform commonly described as the "Model T" of small UAV technology—the DJI Phantom or Mavic quadcopter—is produced in Shenzhen, China.[75] Founded in 2006, DJI (whose full name is Dai-Jiang Innovation Technology Company) has become the world's biggest consumer drone maker, selling 2.8-pound quadcopter drones that are simple, small, reliable, and cheap. A quadcopter drone is a small flying craft, usually shaped like an "X" (or sometimes an "I"), with a propeller at the end of each of its four arms that provides lift, like a helicopter. DJI now controls some 85 percent of the drone market, crushing its top US competitors, such as California-based 3D Robotics, with aggressive pricing.[76] They are developing high-quality, lower-priced drones at a fast pace, including the Mavic Pro (sold for $999) and two versions of the Phantom 4 (sold for $1,399), all introduced in 2016, as well as smaller, cheaper DJI drones, including Spark, introduced in 2017 and selling for $499, and Tello, a smartphone-driven light mini-drone tailored to beginners, introduced in 2018 and selling for $99.[77]

The technology DJI's drones are equipped with is impressive. The original DJI Phantom uses advanced stabilizing and automatic flight control technology, sophisticated microprocessors, and powerful commercial software. In 2015, the Phantom was upgraded, adding automatic visual positioning in place of GPS guidance for when it hovers indoors.[78] And it can perform autonomous flight: users of the Phantom 2 Vision and Vision + ground station feature can control the drone and view real-time video from their smartphones, setting the route in advance.[79]

In fact, DJI Phantom drones are so good that the US Army used them for a time in training and combat operations, including for ordnance disposal, small unit decision-making training, joint US-NATO exercises, military police operations, and a range of other missions.[80] US Special Operators have also used DJI products in Syria, Iraq, and Afghanistan. The Israeli Defense Forces are likewise purchasing DJI Mavic drones for border security aerial reconnaissance.[81]

The Army had encouraged using off-the-shelf UAVs as a way to keep abreast of commercial advancements and field them quickly. The problem was that every time a DJI system is launched, it sends a ping back to its manufacturer—which in this case, is located in China. If you're monitoring an established border, this is probably not much of a concern. But it means that every time ISIS or the US Army or a range of other actors uses these off-the-shelf systems, data about their activities are captured and collected in DJI's massive commercial databases. While the data do not necessarily indicate who *owns* the UAV, they do pinpoint exactly where and when a device is being used. Because it is generally known whose forces are where, say in Syria or Afghanistan, intelligence operatives with access to the data can determine with reasonable accuracy which actors are operating a device. Militaries and groups that use off-the-shelf technology such as the DJI Phantom are offering major commercial actors, some of them closely aligned or overlapping with their governments, a perfect window onto their activities.[82]

In August 2017, the US Army ordered troops to stop using consumer drones made by DJI, "the most widely used non-program of record commercial off-the-shelf UAS employed by the Army," they claimed. "Cease all use, uninstall all DJI applications, remove all batteries/storage media from devices, and secure equipment for follow on direction," Lt. Gen. Joseph H. Anderson, the US Army's deputy chief of staff for plans and operations, directed in a memo dated 2 August 2017.[83] This means "any system that employs DJI electrical components or software including, but not limited to, flight computers, cameras, radios, batteries, speed controllers, GPS units, handheld control stations, or devices with DJI software applications installed."[84]

The diffusion of DJI drones, and the many other models being sold around the world, and easily purchased online, has only just begun. Most of the time they are used for positive reasons (and they certainly are a lot of fun to fly!); but their capacity to carry out malicious acts is growing. Easily able to carry a small payload of 2 or 3 pounds (or about 1 kilogram), mid-level DJI drones of various types have shown up everywhere from wedding receptions to farms to one of the most highly sensitive locations in the world. In January 2015, a 2-pound DJI Phantom quadcopter landed on the White House grounds. It was flying too low and was too small to be detected by a system designed to detect planes, missiles, and large flying objects. Indeed, security officials claimed at the time that the drone breached the defenses because on radar it looked like a large bird.[85] Leery of the bad publicity, DJI pushed a "mandatory firmware update" for its newer Phantom drone that prevents it from flying within 15.5 miles of the White House.[86]

Many other scares have already happened, often accidental but foreshadowing what we could face in the future. For example, on 22 April 2015, an unmanned quadcopter flown from Fukui Prefecture, Japan, and carrying a radioactive payload, landed on the roof of Prime Minister Abe's office. The owner of the hobbyist drone said he was protesting the Japanese government's nuclear policy. The same month, Broadcaster Tokyo MX accidentally flew a drone onto the grounds of the British Embassy, and another small drone flown by a fifteen-year-old boy fell at the Zekoji Temple in Nagano, disturbing the monks.[87] Thus far most of these small UAV incidents have been more about publicity than payload, but the trends are toward more serious uses.

The capabilities of some currently available UAVs are worrying. According to an exhaustive analysis of hundreds of UAVs by the Remote Control Project, the commercial drones currently capable of carrying the heaviest payloads are in the agricultural and film sectors. Small UAVs especially designed to hoist heavy photography equipment, such as the DJI Agras MG-1, for example, can already handle 22 pounds (or 10 kilograms), and they are getting better.[88] They are broadly available and could be altered to serve other purposes, such as dropping explosives or delivering chemical or biological weapons.[89] Even though most people would never dream of such malicious purposes, the transformative future threat is the potential weaponization of these off-the-shelf devices, which anyone can access and customize.

We'll next consider how UAVs have already been adopted by terrorists and insurgent groups, and then study how a cluster of different types of innovations well underway might turn them into much more lethal tools.

TERRORIST AND INSURGENT GROUPS' LETHAL UAV PROGRAMS

Since at least the 1970s, terrorist groups have shown interest in remote-controlled vehicles, or at least in the off-the-shelf technology behind them. The Provisional IRA (PIRA) had a research and development unit that used model plane radio receivers to wirelessly detonate bombs, for example.[90] When the British jammed the radio frequency, PIRA switched to attaching amateur radar detectors to bombs, triggering them remotely with police radar guns.

Interest in unmanned technology inevitably spread to other groups. In 1993, the wealthy Japanese group Aum Shinrikyo had a sophisticated chemical weapons program whose greatest technical challenge was determining how to effectively distribute the toxins, such as sarin gas. They considered using a remote-controlled helicopter for the dispersal. Instead they ended up spraying the gas from a truck in 1994 (8 died, 200 were injured) and placing sarin-filled plastic bags in the subway in 1995 (13 deaths, 6,252 injured).[91] As the gas is more lethal when released in an enclosed place, whether or not the attack venue is enclosed has proven more important to its effects on human beings than the means of delivering the attack have been.

Since then, the most well-developed non-state group deployment of remote-controlled weapons has been by organizations that have evolved beyond small terrorist groups into insurgencies, thus able to hold territory in which to experiment with, refine, and store them. In 2002, the Colombian FARC acquired nine unmanned aerial systems suitable for carrying explosives, although they did not actually use them. The same year, the Pakistani Lashkar-e-Taiba sought UAVs from an American operative for use in Kashmir. In 2005, a Pakistani military raid on the Haqqani terrorist network in North Waziristan uncovered a UAV that the group had been using for reconnaissance. Since 2006, numerous other groups, as well as individuals, from right-wing extremists to nationalists and Islamists, have sought to use model airplanes or UAVs either for reconnaissance purposes or to carry bombs.[92]

Fortunately, thus far most of these efforts by non-state actors to use small, armed UAVs have been primitive. Drones launched by Hamas have been more like flimsy model airplanes, for example, and the Israelis have easily dealt with them.[93] More troubling is aerial reconnaissance and the weaponization of commercially sold small drones by the Islamic State. In summer 2013 five men associated with Islamic State of Iraq began to experiment with using remote-controlled helicopters to distribute sarin and mustard gas. When Iraqi intelligence services closed in, they found a workshop full of small copters and chemical precursors. A year later the Islamic State released a YouTube video of aerial footage of a captured Syrian airbase in Raqqa, shot from a DJI Phantom commercial drone.

The Islamic State also used increasingly sophisticated drone footage for the intimidation of targets and recruitment of followers. An hour-long video documentary released in summer 2014 opened with an aerial shot from a drone over Falluja, Iraq, and then shots of ISIS fighters heading off to battle.[94] Shooting aerial pictures of places Islamic State were fighting to control became common, with the footage sometimes publicizing the locations of suicide attacks, and battle film being edited to resemble a video game for recruitment purposes.[95] As their operations became more sophisticated, ISIS drones were also used to pinpoint targets of mortar and rocket fire, and to lead vehicle-born suicide attackers to their objectives.[96]

The first lethal use of a drone by the Islamic State was in October 2016, when Kurdish fighters in northern Iraq shot down what they thought was an IS reconnaissance drone, and it exploded when they began to disassemble it, killing two of them.[97] Two French paratroopers were reportedly also injured.[98] According to Don Rassler, a US Army drone researcher at West Point's Counterterrorism Center, this attack was the outcome of at least a year's prior planning and experimentation with off-the-shelf drones, and it was quickly followed by a series of other attacks with commercial quadcopters that dropped small explosives on both military forces and civilians.[99] In November 2016, when Iraqi forces took major parts of Mosul, they found an ISIS workshop dedicated to weaponizing commercial drones, apparently one of several active ISIS drone workshops.[100] The number of ISIS drone attacks culminated in the Spring 2017, when there were between 60 and 100 aerial ISIS drone bombing attacks each month across Iraq and Syria, intimidating the population, killing at least a dozen people and, according to a surgeon in Mosul, sending more than ten injured people to his hospital every day.[101]

ISIS used low-cost, commercially available quadcopter drones that were jury-rigged with plastic tubing, 9V batteries, and a range of other plastic, metal, and wood materials to enable them to drop projectiles.[102] But if the drone went down with the explosive, no worries: it was relatively cheap and expendable. The most common type of drone used by the Islamic State was the DJI Phantom, purchasable on Amazon.com for as little as $450. Jabhat al-Shamiyya (the Levant Front, based near Aleppo, fighting in the Syrian civil war) claimed to shoot down an even cheaper X-UAV camera-carrying Talon drone flown by ISIS—a winged platform made of plywood that can be bought in kit form for about $100.[103]

This is not to argue that commercial drones made ISIS ten feet tall. Iraqi and American forces were able to carry out successful attacks on ISIS drones, and some of their operators. At least nine US airstrikes targeted people or sites that were associated with flying drones. Iraqi police or Kurdish fighters shot others down, or followed them back to their controllers and targeted the human beings. And the Iraqi coalition eventually put countermeasures in place, such as a "no-fly zone" for commercial drones covering all of Mosul (although, notably, ISIS got around that too, sometimes using software tweaks, at other times sticking bits of tinfoil over their drones' GPS devices).[104] ISIS was trying to hold territory: their drones could not tip the balance in a conventional fight against US and Iraqi forces. Still they demonstrated how low-cost, accessible, cheap technologies can have disproportionate effects, especially when combined with surprise, as in a terrorist attack.

CROWD-FUNDED, "GRAY ZONE," AND PRIVATE UAV INTELLIGENCE

Other non-state actors using small commercial drones in the service of conflict are private actors who have been fully trained in their use while in the military. A small number of former intelligence officers, having left service, buy their own drones and shoot reconnaissance footage, which they sell to actors who struggle to get military intelligence in other ways. Services provided include information about enemy troop movements or even damage assessments in the aftermath of military attacks or natural disasters.

For example, when the Kurdish Peshmerga fought ISIS in Sinjar, northern Iraq, in 2014 they had nothing but a handful of radios and personal mobile phones. They were frustrated by the slowness of US-provided intelligence on Islamic State movements, which had to go through the central Iraqi government and usually arrived too late to be relevant to ongoing battles. Jason Rexilius, a former US Air Force intelligence officer, also knew that the US government classifies imagery intelligence as Top Secret, meaning that American forces are reluctant to share it with local counterparts. So in

2013 he decided to form a private company, the Third Block Group, and sell open-source intelligence, meaning intelligence drawn from unclassified sources as well as his own footage, to governments and NGOs. Using the slogan "solutions for an open planet," he has operated in Somalia, Colombia, the Philippines, and Iraq. To evade export restrictions on US technology, he buys his drones and all his equipment in Europe.[105]

Proxy forces affiliated with state militaries have also employed drones. The Donetsk People's Republic militias in eastern Ukraine use Russian-made Eleron-3SV drones for reconnaissance.[106] Russian-linked forces have regularly dropped bomblets from UAVs, hitting military forces or barracks. The most-destructive operations have involved small drones armed with thermite grenades that act as detonators when they land on ammunition depots or arms warehouses.[107]

On the other side of the Ukraine conflict, the Ukrainian armed forces have also formed an aerial UAV reconnaissance unit called Aerorozvidka, relying initially on commercial drones because they lacked the cash to buy more expensive military UAVs.[108] To assist the Ukrainian armed forces, pro-government nonstate groups have teamed up to tap the support of Ukrainians abroad for the purchase of drones. There's a long tradition of diasporas sending money and weapons to the homeland for a cause, but what's innovative here is that it's being done through crowdfunding.[109] A volunteer-led organization called the People's Project has a website that collects contributions with a "donate" button, for example, soliciting $36,750 for the PD-1 (People's Drone-1), which includes a drone, a support van, and control equipment, all designed and built by UkrSpecSystems, a Ukrainian UAV producer.[110]

Such "gray zone" conflicts are increasingly characterized by fractionalization of the warring sides, which the new technologies are facilitating. Warring groups can construct a powerful story about their mission, reach out directly to individuals to join them, crowd-source their military equipment needs, collect highly sophisticated intelligence, and project force employing combinations of cheap commercial off-the-shelf technologies that lend themselves to innovation.

ADVANCES POWERED BY THE SMARTPHONE

The smartphone is the technological driver of small UAV innovation. Going forward, the increase in capability for individuals and small groups building drones with smartphones and open-source software will accelerate, and the results may be destabilizing. Miniaturized, automated technologies will enable small groups and individuals to leverage their use of force in unprecedented ways, just as dynamite and the Kalashnikov did.

Twentieth-century models of proliferation, originally built to track missiles and nuclear weapons, are a poor fit for small UAVs. With a range of less than 300 kilometers, high-technology small UAVs represent a different lineage of technological development that benefits from advances in miniaturization and automation.[111] Traditional air defenses and military capabilities have not been designed to detect and deter such small targets flying at lower heights, and they may surprise, evade, or overwhelm military and law enforcement defenses through sheer numbers. There is as of yet a dearth of analysis, planning, and doctrine about fighting this threat—especially in places not officially "at war."

The cluster of fast-developing new technologies will also be driving rapid evolution of their capabilities, and prices of small UAVs will continue to drop. Microsoft founder Bill Gates argues that robots are experiencing the same cycle of price drops and widespread adoption that characterized personal computers at the turn of the twenty-first century, and this observation applies to most of the new technologies we are considering.[112]

The private sector is driving this innovation, with strong incentives to offer an array of powerful new features. The hardware is fairly mature, though one aspect that is likely to improve significantly is the development of safer, cleaner, cheaper, and more concentrated batteries, largely due to advances in driverless cars.[113] Lithium ion batteries have been the key to progress there so far, and their ability to store more energy has been steadily increased, enabling some electric cars (such as the Tesla Model S, for example) to go more than 300 miles without recharging. But, as owners of the exploding Samsung Galaxy 7 smartphone found out in 2017, lithium batteries can be unsafe if they are heated, damaged, or punctured. Professor Michael Zimmerman at Tufts University has developed a safer, more powerful lithium battery containing solid plastic electrolytes that don't ignite the way current lithium batteries have done, which could be the answer for powering tomorrow's driverless cars, as well as smartphones, robots, and armed drones.[114]

Evolving advances in software will be even more dramatic. Programming innovations driving smartphone advances, driverless vehicles, and artificial intelligence will transfer to drones and other unmanned weapons, allowing them to carry heavier or more powerful payloads, increasing their range from the controller, making their digital links more secure, equipping them with the ability to identify specific targets and to loiter over them longer.[115]

A good example of an advance that can be weaponized is fast-emerging facial recognition software. The pioneers in biometric facial recognition were the gambling industry, who have for at least a decade used biometric facial recognition systems to identify and eject people who cheated, were addicted to gambling, or routinely won too much money.[116] Basic facial recognition capability was built into Android phones starting in 2012, with an owner's face recognition needed for login. Samsung's somewhat more sophisticated, though still quite rudimentary, security suite, used on its Galaxy S8, and Note 8, combines iris scanning with broader facial recognition capability. That software can only recognize the broad contours of a user's face captured on its camera. As of this writing, Apple's iPhone X is the most sophisticated, using an infrared flood light along with an array of sensors that capture finer details of the user's face in three dimensions.

In smartphones, it's not clear that any of these systems is better yet than an old-fashioned password for the purpose of security, but there are many other uses facial recognition can be put to.[117] Law enforcement in developed countries widely employs the software in CCTV cameras to catch criminals, monitor secure areas, and locate missing or vulnerable people. Some police in China use glasses that are enabled with facial recognition in surveillance of the public.[118] Researchers from IBM have been able to use facial-recognition software to set off highly sophisticated malware attacks, triggered only when a certain person sits down at the computer.[119] Facebook is exploring using facial recognition software in stores to scan the crowd, tie shoppers to their Facebook profiles, and relay information to clerks.[120]

The same kind of advanced intelligence software, using facial recognition, can identify targets for killing. In tests during field exercises, the Pentagon has used autonomously flown drones to find and fix sights on mock insurgents, hiding in the shadows and carrying replica Kalashnikovs.[121] The US Marines are already testing facial recognition software in remote weapons systems. If armies and police can use facial recognition software to identify targets, then mercenaries, insurgents, terrorists, and criminals will be able to do the same. They will be able to tap the enormous amount of photographic data on the web to teach the software which faces to look for, and it will likely not be long before they will have the capability to thereby locate individuals remotely and kill them.

This could provide a new means of political assassination. In September 2013, an activist from the German Pirate Party, in a "protest against the EU's use of surveillance drones," flew a small Parrot quadcopter above the stage where German chancellor Angela Merkel was giving a speech, hovering it just above her head.[122] This publicity stunt demonstrated to millions of viewers both the chancellor's vulnerability and the potential reach of any non-state group that wants to target individuals. In August 2018, an assassination attempt against Venezuelan president Nicolás Maduro was carried out with two DJI Matrice 600 aerial drones loaded with explosives, which went off above his head as he spoke at the anniversary of the Venezuelan army. The president was unhurt, but seven Venezuelan soldiers were injured.[123] Like the early uses of dynamite, the effect was more about propaganda than payload; but that may change.

More significant, perhaps, will be the ability to target otherwise ordinary people whose photographs may be collected from sites like Facebook, then collated into complex facial images that can be targeted. This could be a source of blackmail, or even murder. Political leaders are protected with elaborate security systems that make it harder to get near them; most other people are not.

DRONES AS MISSILES

Not only are drones growing more advanced in terms of surveillance and targeting, they're also increasingly affordable. With UAVs becoming so economical, the distinction between reusable craft and one-time-use cruise missiles is fast disappearing, because cheap platforms are expendable.[124] As missiles, UAVs may be loaded with explosives and driven kamikaze-style into buildings—as has been attempted on numerous occasions.[125] For example, in 2011, Rezwan Ferdaus, an al-Qaeda enthusiast, plotted to load F-86 Sabre and F-15 Phantom drone aircraft with C-4 plastic explosives and fly them into the US Capitol building and the Pentagon.[126]

UAVs are certainly not invulnerable: radio-controlled UAVs can be jammed (interfering with their GPS receiver's ability to pick up signals) or spoofed (replacing their GPS signal with a spoofer's signal), and military forces can shoot them down or grab them with nets.[127] Drone manufacturers are also installing self-regulating features: the DJI Phantom 2 drone has firmware installed that employs geofencing around key spots such as airports or Tiananmen Square, for example. Some drones have "sense and avoid" technology, software that monitors the airspace for obstacles.[128] But of course those inclined to perpetrate violence with the devices aren't likely to purchase UAVs with those features, and they can also be hacked or disabled.

When it comes to defending against attacks, individuals, terrorist groups, or insurgents have the advantage of surprise, especially if aerial IEDs are directed against vulnerable civilian targets. Plus, when law enforcement agencies and military forces shoot explosive-laden flying drones out of the sky, they might still be dangerous to people on the ground. The psychological effects could be disproportionate to the potential lethality of the drone—harassing and distracting military members from above, or intimidating civilian populations and undermining governments whose top priority is to protect their citizens.

Responding to the drone could also kill civilians on the ground. If radar systems were set to the sensitivity and altitude required to detect a drone, they could react to every bird, kite, or swaying tree in the area, evacuating or locking down sensitive buildings, scrambling F-16s to respond to false alarms, or simply shooting at them from the ground.[129] All of this could spread public fear and lead to pressure for ill-considered and poorly targeted retaliatory responses. In terrorism, the important thing is usually the political effect on a population and how governments react to the attacks, not necessarily the level of destruction.

DEMOCRATIZED PRECISION STRIKE CAPABILITY

With the commercial availability of imagery tools and access to GPS, individuals have the potential for precision-strike capability comparable to what the United States enjoyed in the 1990s. The same kind of technology that is being used to develop the artificial intelligence software required for long-range package delivery to individual homes can be easily translated into finding individual targets for IEDs.[130] Autonomous drones could potentially be programmed to target a chemical plant, hydroelectric dam, or fuel depot. With sophisticated guidance capabilities, small UAVs will be able to home in on the weaknesses, using sensors and guidance systems to go after the most vulnerable point—just as heat-seeking missiles can target tank exhaust or the infrared emissions of jet engines. Or they can target places that are inherently vulnerable or dangerous: individual drones have already been found near nuclear power stations in France, and over a nuclear submarine facility in the United Kingdom, for example. They are approaching a point where they can develop their own version of "global precision strike" using tools such as drones, artificial intelligence, and robots, without the accountability or traceability to a state government.

In some cases, drones may not have to have to destroy their targets with their own weaponry; they may be able to damage targets enough that they suffer structural or mechanical failure. For example, on 22 September 2017, two Army Blackhawk helicopters on UN surveillance duty over New York City were flying low (just above the 400-foot limit for private drones) along the east shore of Staten Island, when a drone slammed into one of them. The helicopter limped to a nearby airport, and the Army pilot found that the drone had shattered into pieces, with one landing on the bottom of the main rotor system. A few inches' difference and the helicopter would have been downed.[131] In October 2017, a drone collided with a commercial aircraft about three miles out of Quebec City's international airport, as the plane prepared to land. The plane sustained minor damage and was able to land; with precision targeting, the damage could well have been much worse.[132] Both these incidents were probably accidents (which can be plenty deadly), but intentional attacks of this kind are possible.

The differences between military autonomous systems and those that are commercially available are narrowing. For example, the Aerovel Flexrotor, which is shaped like a small aircraft, has a range of more than 2,000 miles (3,400 km), a gasoline engine, and a price tag of $200,000. It can be assembled in ten minutes and configured to take off vertically. Advertised accessories include day and night imaging, radar and multispectral imaging, and an impressive communications relay.[133]

Small autonomous unmanned systems will have the ability to navigate small spaces and to pinpoint targets wherever they are. Individual entrepreneurs and inventors are making remarkable progress in building highly capable small drones that can fly almost anywhere. For example, in September 2017, Paul Kurkkala built a drone that could fly over, onto, under, and inside a moving train. The video he shot, titled "Flight of the Year," was taken from a first-person point of view. Kurkkala used goggles to look through a GoPro HER05 Session action camera that was mounted on the drone, guiding it as if he were actually under, over, and onboard the train.[134] This is incredibly cool, until you think about someone without Kurkkala's benign purpose or judgment flying a drone equipped with an IED or chemical or biological weapon under or inside the same train—or truck, or convoy, or motorcade.[135] In competitions, other individual drone operators have demonstrated remarkable piloting skills, flying drones through dense forests, along cluttered streets, into windows, and between high buildings.

Some can even pilot UAVs underground. Australian researchers are using quadcopters, developed by the Commonwealth Scientific and Industrial Research Organisation (Australia's counterpart to DARPA), to map abandoned mines. With a software program called VoxelNet, they have moved beyond one-dimensional first-person views through a camera, to 3D imagery sent to a laptop. The Australians are also exploring the use of sonar, to bounce sound waves off walls and produce 3D imagery more quickly and efficiently. The challenge in this case is that drones operating so deep in the ground cannot use GPS, so the plan is for newer generations to operate autonomously.[136] This could also be a handy capability for malicious groups that want to send drones into underground garages, dark tunnels, or urban sewer systems, for example.

Small drones launched individually or in groups will challenge the ability of individuals to seek sanctuary. Cover and concealment will be much harder because drone operators will benefit from the ability to be invisible. Individuals who are able to fly drones through first-person viewpoint cameras can be anywhere: they do not need to keep their crafts within line of sight. Operators can hide underground or in caves or buildings, making them difficult to see or to find.[137]

EVERYONE MANUFACTURES EVERYTHING WITH 3D PRINTING

Another area of innovation extending the reach of non-state actors is 3D printing (also known as "additive manufacturing"). As with drones, there are many positive uses for this technology. Soldiers in the field are using 3D printing to create on-the-spot spare parts and avoid risky, complex re-supply logistics.[138] High-end car owners are already creating replacement parts for their vehicles. The clothes industry plans to use 3D printing to manufacture perfectly fitting shoes and clothing, and cosmetic companies hope to use 3D-printed skin to test new products, avoiding animal testing. Much work is underway in creating specialty prosthetics, which will be more lifelike and perfectly matched to the patient.[139] That could be a wonderful development that promises to improve the lives of millions of people. Some scientists think that further ahead it might be possible to use an additive process called "integrated tissue-organ printers" to produce transplantable organs and other body parts. In 2002, engineers at Wake Forest Institute for Regenerative Medicine were able to 3D print an artificial kidney, created by building a printed scaffold and then coating it with the patient's own cells.[140] Implanting 3D-printed organs into human beings is not yet feasible, because scientists haven't cracked the code for how to provide them adequate blood supply. In the nearer term, many scientists think bioprinted human skin already holds tremendous promise, for skin grafts or for perfectly fitted facial parts like noses or ears.[141]

Predictably, however, potentially malicious uses of the technology have also emerged. It already enables anyone to produce simple weapons and other devices, like drones, by themselves at low cost. For some years now, very basic 3D printers have cost no more than about $200 and are easily purchasable online.[142] At this writing, the files to print small planes or UAVs like Razor, for example, cost about $20 to $40 each, and the material with which to print them costs about $20 to $30. These are pretty rudimentary capabilities: more-advanced materials (like titanium) and more-complex 3D-printed products (like aerospace parts) are more expensive and require much better printers. But printer prices are dropping year-to-year as capabilities rise.

Though new to most of us, 3D printing dates back at least to 1984, with engineer Charles Hull's patented invention of stereolithography, a printing process that enables a 3D object to be manufactured from digital data. Scientists and engineers have used 3D printing to produce one-off prototype models since the 1990s. The process has also been used in manufacturing for some time. But combined with a *cluster* of new developments today, 3D printing is poised to have more significant impact on worldwide manufacturing and logistics in the future. Married with artificial intelligence and robots, additive manufacturing can sharply reduce shipping and inventory costs, and enable companies to make things cheaply in small batches (so at an affordable per unit cost), on the spot where they are needed.

Large companies are at the forefront of 3D printing, especially the aerospace industry, which is already producing highly advanced spare parts for aircraft, for example, and the medical technology sector, which is heavily investing in bioprinting. But 3D printing also gives small operators the ability to produce all manner of items, including car parts, toys, jewelry, various types of food, and perfectly fitted clothing so far. They can do so without government oversight, producing guns, for example, with no background checks or purchase records required.

Cody Wilson, a law student at the University of Texas, became infamous for creating a 3D-printed gun in 2012. Using a crowdfunding website to collect donations from opponents of gun regulation, he and his friends created an organization called Defense Distributed and launched the Wiki Weapon Project, to make files for creating firearms available for anyone to download for free. To demonstrate the process, they used a low-cost, open-source 3D printer known as RepRap, and produced a plastic gun called "The Liberator," capable of firing a .38 caliber bullet, which they demonstrated in a YouTube video. After they released the blueprint file online, it was downloaded 100,000 times in forty-eight hours. Demonstrating the danger posed, two journalists smuggled a homemade 3D-printed Liberator onto a train traveling between the United Kingdom and France,[143] and two other journalists snuck a 3D-printed Liberator gun into the Israeli Knesset.[144]

The US State Department stepped in and had the file taken down, labeling the release a form of international arms trafficking.[145] Defense Distributed filed suit against the US government, claiming unconstitutional infringement of the First Amendment provision of free speech and the Second

Amendment's right to bear arms.[146] On 29 June 2018, the Trump admin-istration settled the suit, stating that 3D-printed guns were approved "for public release (i.e., unlimited distribution) in any form." Then, hours before they were to be published online, a federal judge, responding to a lawsuit filed by attorneys general from nineteen states and Washington, DC, blocked their release on national security grounds.[147]

The FBI has been investigating the potential for malicious actors to print improvised explosive devices for years. While Defense Distributed has been one of the most vocal and high-profile groups pursuing freely available 3D-printed capabilities for firearms, it is not the most advanced. A large number of 3D-printing gun designs are now freely available on the Web. Yet with approximately 270 million civilian-owned firearms accessible in the United States, this is hardly the easiest way to acquire a gun.[148]

Concern about the current ability of individuals to print weapons within their homes must also be tempered with a few important caveats about the state of the technology thus far. There is a big difference between what firearms manufacturers are already doing with 3D printers (Solid Concepts Inc. already 3D-prints some gun parts, for example) and what hobbyists have thus far been able to do.[149] At this point most home 3D printers are not very sophisticated. They yield products that have small defects that, especially in a firearm, render it more dangerous to the user than to the target, because it might blow up in his or her hands when firing. The guns also tend not to be accurate beyond very close range. In addition, the various polymers that individuals typically use to make them are inferior to metal in many ways, such as tensile strength, elasticity, long-term stability, and temperature re-sistance. High-quality metals and advanced polymers (such as those rein-forced with glass, carbon, or aramid) continue to be expensive for hobbyists and small businesses, who still tend to outsource their harder projects to bigger companies.[150]

Not all 3D printing is alike, and what's known as Direct Metal Laser Sintering (DMLS) is done in printers that cost somewhere between $600,000 and $1 million—well beyond the reach of hobbyists—though the firearms industry is already using them to produce functional weapons.[151] The strength and wear of additive manufactured parts made by amateurs tends to be greatly inferior to that of parts produced by industry: by now we should have learned that you cannot make gold out of lead.

That said, even if most firearms made this way are unreliable for firing, because they are such accurate representations of weaponry, they don't need to be fired to have an effect. Plastic guns do not set off metal detectors (al-though their metal firing pins and bullets do). They can be used to intimidate victims, who have no idea how fallible a 3D-printed gun may be. Would you want to call that bluff?

More to the point, the technology is improving and getting cheaper. MIT's Computer Science and Artificial Intelligence Lab has produced a multi-material ("Multi-Fab") printer that can employ up to ten different materials, including metals, which will sell for about $7,000.[152] In late 2013, Solid Concepts produced the world's first, openly announced, 3D-printed metal firearm, a pistol modeled after the Colt US government model 1911, which successfully fired more than fifty rounds.[153] As of yet, the equipment to work with metal 3D printing is quite expensive, most models costing tens of thousands more than the $1,500 computerized milling machine, and the process requires direct laser metal sintering, which is a quite specialized skill.

The military is well aware of the potential of 3D printing, which could, in fact, be important to solving logistics challenges in future conventional wars. For example, the US Marine Corps hopes to train maintenance crews, intelligence, infantry special operations, and even tank battalions to use 3D printers routinely to produce replacement parts and other items down range. They have developed an expeditionary fabrication lab (X-FAB), which is a deployable 20×20-foot shelter equipped with four 3D printers, a scanner, and computer-aided design software system.[154] Creative solutions to expensive problems are emerging, such as replacing broken small UAVs like the hand-tossed RQ-11 Raven, with new unmanned Scout or Nibbler drones that can be printed in the field, making them cheaper, faster, and more accessible.[155]

It's not that individuals and small groups are *exceeding* the capabilities of state military organizations through additive manufacturing. But the elimination of the need for some of the advanced logistics that supported Industrial Revolution–era armies is narrowing the technological gap between them. In addition, a looming question regarding the military adoption of sophisticated programs and sensitive data, especially the computer-assisted design (CAD) programs that are at the heart of 3D-printed field-deployed systems, is whether they can be sufficiently protected from hackers and thieves. In 2016, a group of Johns Hopkins master's degree students easily inserted malicious code into the blueprint for a 3D-printed quadcopter: one of its propellers disintegrated in flight, and it dropped like a stone.[156] Imagine if that had been a critical replacement part for an aircraft. Will sensitive electronic files be widely shared? Could there be a future Edward Snowden who decides to take it upon himself to make these new tools of war "transparent"?

Leveraging Power

Individually these threats are not more lethal than the widespread firearms and explosives that terrorists, insurgents, and individuals already use to deadly effect, and we can develop countermeasures. What makes them challenging and different, however, is their low cost, unpredictability, and ability to be grouped together.

The major cost imbalance between twentieth-century military technologies and twenty-first-century drones, loitering munitions, IEDS, and commercial robots leaves states vulnerable to warfare of economic attrition. Using a cheap capability that can be easily replaced can undermine the economic viability of even the most advanced military force. In March 2017, for example, the Israeli Defense Forces used a $3 million Patriot interceptor (part of their Iron Drone defenses) to shoot down a small $200 quadcopter drone that had flown over Israeli air space.[157] And mini-UAVs can be easily repaired with inexpensive parts that are simple to acquire. Even for states, certain technologies are cheap equalizers that put at risk billion-dollar platforms and scramble the balance of power equation. US Special Operations have to practice using camouflage netting to avoid letting inexpensive reconnaissance drones see them, for example.[158] Aircraft carriers and fighter jets currently have few countermeasures for swarming drone attacks that cost a tiny fraction of the price of these large, highly advanced systems.

In January 2018, Syrian rebel forces sent thirteen cheap, homemade, armed drones against Russian forces at the Hmeimim airbase and the Russian naval facility at Tartus. The Russians responded with jammers, guns, and missiles, and said no serious damage had occurred.[159] This was the first known example of an aerial swarming (or at least mass UAV) attack used in conflict.[160] What will happen when there are 300 such drones carrying larger payloads, aimed at vulnerable spots, like an aerial defense system or a command headquarters? And what if defensive jammers don't work because the drones are autonomous and therefore not reliant on GPS or other signals from the ground? What if defensive systems shoot down 250 of them, or even 295, but a few get through and find their targets?

Autonomy and artificial intelligence (AI) are changing the future calculus of major powers like the United States and China, which are vigorously competing to be at the forefront of using these capabilities for military purposes. As we will see in the next chapter, autonomy and AI are also broadening the scope for weaker non-state actors to carry out surprise attacks.

CHAPTER 9

An Army of One Launches
Many: Autonomy and Artificial
Intelligence

We are wiser than the computers. We created them.[1]

> Stanislav Petrov, Soviet lieutenant colonel who in 1983 disregarded
> a false computer alert of incoming US nuclear missiles and averted
> nuclear cataclysm. He said he "had a funny feeling in his gut."[2]

AUTONOMY REFERS TO the ability of a device or system to perform a function without direct human input.[3] Autonomous capability may be a minor or major part of a weapon or weapons system, ranging, for example, from a simple thermostat that turns on a fan in a tank to a complex algorithmically programmed missile system that can independently identify and shoot down targets. A degree of autonomy has been built into some weapons for decades, especially defensive systems, such as the US Navy's Aegis combat system, dating to the 1980s. Others include the Russian Arena system for protecting ground vehicles like tanks, dating to about 1995, and the Patriot missile system, first used by the US Army in the 1991 Persian Gulf War and currently part of Israel's Iron Dome.

The key to understanding the various degrees of autonomy is to consider the world from a machine's point of view, and then envision its relative capacity to sense the environment, process what is happening, and act—with or without the direct involvement of human beings.[4] That is its autonomy, and the ability of machines to do these three things independently—sense-process-act—has dramatically improved in the past decade. The cluster of recent technological advances in remote sensing, enormous sources of data ("big data"), computer processing power, and machine learning together yield the potential for powerful artificial intelligence driven autonomy today and tomorrow.

A Spectrum of Autonomy

To distinguish sustaining innovations in autonomy from disruptive ones, it is useful to categorize degrees and types of autonomy along a spectrum, starting with simpler "automatic" or "automated" weapons, where humans play an enormous role at time of use, and progressing gradually up to the most sophisticated AI-enabled systems, where humans have little or no role at the time of use.[5] The involvement at the time of use is key: at a certain level, all machines, and whatever programming may control them, start out as human creations. The machines that we are discussing use a range of ways, from primitive to extremely advanced, to gather or interpret data and travel through the sense-process-act sequence.

Beginning with the simplest category, an automatic system is one that gets a direct signal from the environment, such as through a human operator pressing a button, does little or no processing, and just carries out what its built-in capabilities are designed to do. In other words, it operates with preset, built-in instructions that are fed to it either mechanically, as with the computers at Bletchley Park that broke the Enigma code during World War II and were programmed mechanically by plugs and switches, or in code. Some writers distinguish between automatic weapons, by which they mean simple mechanical ones, like machine guns, and automated weapons, meaning they follow preset rules, like a thermostat. But those distinctions are fuzzy and debated, so we will call them both automatic.[6]

All automatic weapons rely on direct human action for deployment or activation, and they carry out only specific predetermined actions in response.[7] You press a button, pull a trigger, or step on a trip wire, for example, and, assuming the machine does not malfunction, the preset response happens, the same way, every time. As we have seen in our study of the machine gun in chapter 5, automatic weapons date to the Ager and Gatling guns of the US Civil War. Automatic weapons systems go back to built-in, rules-based systems like the Norden Bombsight, invented in the 1920s, which incorporated an analog computer and autopilot to deliver bombs more accurately. Another early such system was the V-1 buzz bomb, used by the Germans during World War II, which was equipped with an autopilot guidance system to regulate altitude and airspeed.[8] Bombs with timed detonation, as well as IEDs and landmines, also fall into this broad category, because the trigger for their explosion is built into the device in advance.

The history of modern terrorism demonstrates that terrorists love automatic weapons. They allow attacks to be carried out from a safe distance and anonymously. Many of them are also quite cheap to obtain or build, and the fact that these weapons are generally imprecise in targeting victims is of little to no concern to terrorists. Indeed, indiscriminate killing has been integral to the psychological impact of terrorism. We can expect that terrorists will quickly embrace any new forms of automatic weapons that are relatively inexpensive and accessible.

The next step on the spectrum toward more sophisticated autonomy is semi-autonomous weapons, which the US Department of Defense defines as "a weapon system that, once activated, is intended to only engage individual targets or specific target groups that have been selected by a human operator."[9] Or in loose military parlance, a human is "in the loop," meaning that while a machine can sense the environment, process to some degree, and suggest a course of action, it cannot act unless a human being gives approval.[10]

Many semi-autonomous weapons are first fired by humans, then sense or detect the enemy and make self-adjustments to hit the target. They include guided munitions, or "smart bombs" such as cruise missiles and torpedoes, and have elements of autonomy built into them, like homing devices. But a human being actively decides which targets the weapon should hit. The hellfire missiles shot from Reaper drones and used to target terrorists are semi-autonomous, for example. After firing, semi-autonomous weapons can correct human mistakes in aim or trajectory, but they still usually deploy when a human launches them and hit something that a human has selected. Radar, lasers, acoustics, or satellites typically guide their homing systems, and they can also track a moving target. Most are dependent on GPS, but to prepare for environments where GPS is taken out, the United States has developed homing munitions that can also be guided by imagery. Many of these weapons can be targeted with great precision, at particular buildings or even floors within a building, so they can be very discriminate—assuming the human beings deploying them intend to discriminate and have accurate intelligence about the target.[11]

Semi-autonomous systems encompass a huge category of weapons. Most of the unmanned systems we discussed in chapter 8 are semi-autonomous.

As we have seen, to determine whether non-state actors such as terrorists and insurgents would use them, much depends on size, cost, and capabilities. Armed quadcopters and small drones are well within their reach; semi-autonomous "smart bombs" are not. It's notable that both short- and long-range precision-guided weapons have not spread from state to state to the degree that American defense experts predicted in the 1990s they would.[12] This is partly because other states haven't had to engage in global war fighting as the United States has, especially since 2001. In addition, major powers such as China and Russia have skipped full development of highly expensive, semi-autonomous precision weapons and instead gone directly for those with greater autonomy.

The next step in the spectrum is human-supervised autonomous operation, where in normal operations the machine senses, decides, and acts on its own, unless a human being actively intervenes to *stop* it. Using our military parlance again, this is having a human "on the loop." It is also referred to as "narrow artificial intelligence." According to the US Department of Defense, human-supervised autonomy technologies are autonomous weapon systems that are "designed to provide human operators with the ability to intervene and terminate engagement, including in the event of a weapon system failure, before unacceptable levels of damage occur."[13] Supervised autonomy gives human beings a role, but it leans much more heavily on the speed and capabilities that advanced computerized weapons have.

For decades, advanced militaries have been waging an arms race in developing both offensive and defensive human-supervised autonomous systems, in which innovations are designed for use against an adversary's improved capabilities.[14] More than thirty states deploy human-supervised defensive systems now, or soon will.[15] Examples include Israel's Trophy tank protection system (which the United States is buying for its M1 Abrams tanks); the US Centurion Counter Rocket, Artillery, and Mortar (C-RAM) system (like a land-based Phalanx system); Germany's AMAP-ADS vehicle protection system; Russia's DROZD-2 (another active defense system for tanks); US surface ship anti-torpedo defense systems (ATTDS); and the US Terminal High Altitude Area Defense anti-ballistic missile system (THADD), which is deployed in Guam, the United Arab Emirates, and South Korea.[16] The aim is to match or exceed the speed of an attacker's system.

Another example of human-supervised autonomy is Israel's Harpy loitering munition, which is a kind of self-destructing tactical (Group 3) UAV that can loiter over an area for hours after it is launched, looking for a particular type of enemy radar signal. It is designed to target radars, not people.[17] When it detects that signal, it selects and engages the target on its own, without further human involvement. According to the Center for the Study of the Drone at Bard College, eight other countries have already purchased Israeli Harpy loitering munitions (Azerbaijan, China, Germany, India, Kazakhstan, South Korea, Turkey, and Uzbekistan), and China has reverse-engineered a version of its own (the ASN-301 anti-radiation loitering munition system).[18] On the spectrum from human-supervised to fully autonomous, this one starts to edge pretty close toward fully autonomous, because it does not have to ask anyone's permission before choosing a target and killing it.[19]

These systems are generally used for defense but as the technology evolves, the line between offense and defense is becoming increasingly blurred. Depending on a weapon's technical specifications like speed and range (which are often kept secret so that enemies cannot develop countermeasures), it may be just as easy to send an autonomous weapon into enemy territory as it is to use it strictly to protect oneself. These systems are extremely expensive. At the low end, for example, the Trophy tank protection system costs $350,000 per unit. For this reason they are most likely to be used only by highly advanced states in the future.[20] Many experts believe that human-supervised autonomy is becoming compulsory for major powers, because future attackers will have the ability to use computer-assisted weapon systems that exceed the ability of human beings to react to them.

Fully autonomous weapons, operated by artificial intelligence (AI) without any meaningful human control,[21] often referred to as "full AI," do not yet exist, but on our current trajectory they will not be long in coming. This is due to major developments in five areas: 1) the exponential growth of computing performance since the mid-twentieth century, 2) an enormous increase in private investment in AI, 3) better and more capable sensors to produce real-time feedback, 4) the availability of huge datasets that can "teach" powerful computers new skills, and 5) dramatic ongoing improvements in machine learning capabilities that allow computers to harvest and use the data in the enormous new datasets.[22] These technological developments together mean that autonomous systems of all types are getting cheaper and more widely dispersed, year-to-year. The result could be the kinds of lethal autonomous weapons (LAWS) that prominent figures such as the late physicist Stephen Hawking and Tesla and SpaceX chief's Elon Musk have warned against, and that the United Nations' Group of Governmental Experts has been arguing over since 2017.[23]

The private technology industry, often funded by defense ministries, is driving progress in artificial intelligence. Rapid development of driverless cars is being joined by innovation of AI capabilities in land-based robots and small UAVs, such as glider aircraft that can make navigational decisions autonomously on the fly.[24] They learn from on-board sensors that feed information about the environment into algorithms designed to predict air patterns, and then they independently plan a route forward.[25] They are equipped with deep learning neural networks, meaning complex sets of algorithms modeled after the human brain that can recognize objects and conduct language translation. These machines may be getting smarter at a faster rate than human intelligence will be able to keep up with.[26] We'll talk more about that a little later in the chapter.

This kind of cutting-edge artificial intelligence is not the sort of cheap technology that terrorist groups or insurgents typically have the resources to develop; however, some of it is openly available or open-sourced, so the barriers to entry will become lower. Looking a few decades into the future, rogue states or state-sponsored proxy groups might gain access to this technology either through proliferation or by hacking into commercial sources, and divert it to violent uses.

Cruder forms of autonomous technology are already available commercially. Teenaged amateurs have used Knowledge Enhanced Electronic Logic (KEEL), a technology created by NATO that allows non-technically trained people to generate code for targeting and to create hobbyist drones that progressively adapt to their environments.[27] An autonomous aerial drone that is programmed to kill random human beings at a certain location, identified merely by their thermal or infrared signatures, is within reach and could be much more lethal than a truck bomb or an assault rifle is now.[28] It would be difficult for anyone in the targeted area to escape.

Rogue states and malicious non-state actors will find it easier to gain access to a range of autonomous weapons, such as fully autonomous UAVs, than it was for them to build biological, chemical, or nuclear weapons, because tomorrow's UAVs will be cheaper and use widely dispersed technology. They will give weaker actors dramatically increased reach and power. Conventional deterrence may be ineffective in combating this threat, because autonomous drones may be difficult to attribute to their source. Their use for political violence may be internationally destabilizing.

Weapons equipped with such capability will be a leap into the unknown, and experts bitterly disagree about the implications. Looking into the future, Figure 9.1 (opposite page) summarizes how our theory of lethal empowerment applies to the spectrum of autonomous weapons systems, their targeting, and their most likely new users.

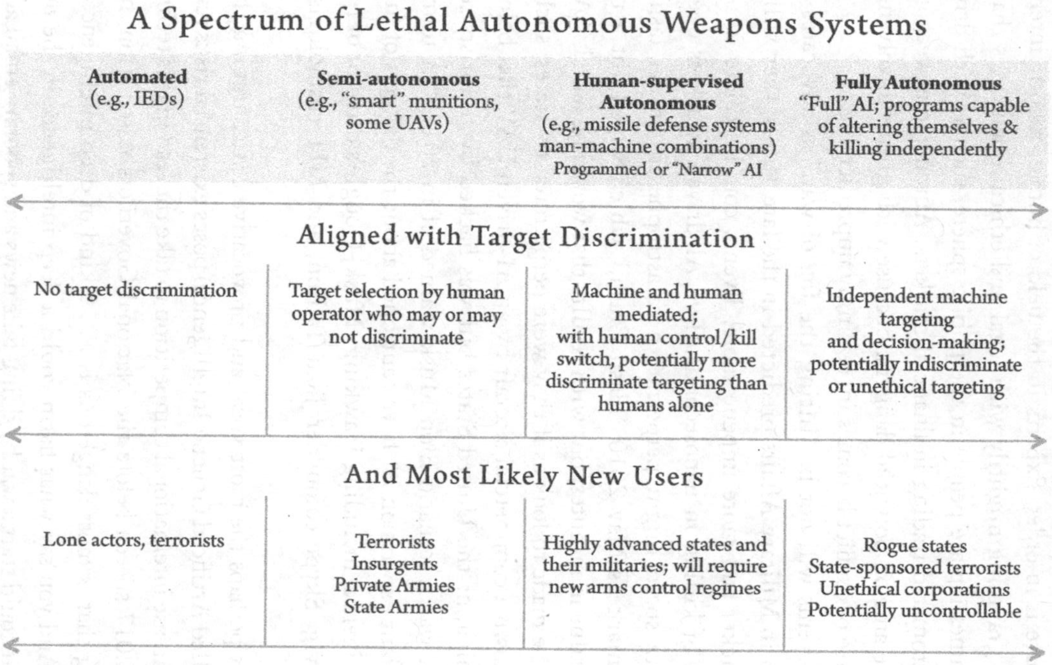

A Spectrum of Lethal Autonomous Weapons Systems

Automated (e.g., IEDs)	Semi-autonomous (e.g., "smart" munitions, some UAVs)	Human-supervised Autonomous (e.g., missile defense systems man-machine combinations) Programmed or "Narrow" AI	Fully Autonomous "Full" AI; programs capable of altering themselves & killing independently
Aligned with Target Discrimination			
No target discrimination	Target selection by human operator who may or may not discriminate	Machine and human mediated; with human control/kill switch, potentially more discriminate targeting than humans alone	Independent machine targeting and decision-making; potentially indiscriminate or unethical targeting
And Most Likely New Users			
Lone actors, terrorists	Terrorists Insurgents Private Armies State Armies	Highly advanced states and their militaries; will require new arms control regimes	Rogue states State-sponsored terrorists Unethical corporations Potentially uncontrollable

FIGURE 9.1

The Perils of Full Artificial Intelligence

Most of the public debate over autonomous weapons has focused on ultra-advanced technologies not yet in operation that have been featured in futuristic movies, such as *Ex Machina*, the *Terminator* series, and *Avengers: Age of Ultron*, in which robots run amok. Sensationalized as these depictions are, serious debate is in order. Experts in the field of human-robotic interaction are already grappling mightily with legal and ethical quandaries that arise when machines replace people in the public sphere, such as with driverless cars and automated trading in financial markets. Autonomous weapons systems up the ante on potential dangers, because with greater involvement of AI the course of conflict becomes increasingly unpredictable.

Such systems will not be "lifting the fog of war," as advocates of the Revolution in Military Affairs predicted in the late 1990s; they will make the fog thicker and more impenetrable. Events could escalate to unforeseen levels of killing in nanoseconds. Two AI-driven attackers operating at gigahertz speed might perpetrate a catastrophe in similar fashion to the stock market's May 2010 "Flash Crash," which wiped out a trillion dollars of value in minutes, but with results that would be lethal.[29] Attacks could also be made to look as if they were perpetrated by states, such as by shooting down a commercial aircraft by an AI-driven UAV that belongs to Russia, China, or the United States but was hacked by terrorists seeking to set off a war, a scenario Elon Musk warned of in 2017.[30] A number of other brilliant experts on AI have cautioned about the dangers of full artificial intelligence, including Hawking, Musk, Google's director of research Peter Norvig, Skype cofounder Jaan Tallinn, and MIT cosmologist Max Tegmark.[31]

Musk is perhaps the most vocal and provocative critic, arguing that full AI (also called Artificial General Intelligence) poses existential risks resulting from the intense international competition to take the lead in this technology. In a July 2017 speech before the National Governors Association, he said, "The thing that is most dangerous is . . . a kind of deep intelligence in the network. And you say what harm could a deep intelligence in the network do? Well, it could start a war by doing fake news and spoofing email accounts and fake press releases and just by manipulating information."[32] That same month, China unveiled plans to become the world's predominant power in artificial intelligence by 2030. A couple of months later, at a televised event with Russian students, President Vladimir Putin told them that whichever country becomes the best at artificial intelligence will become "the ruler of the world."[33] Musk believes that full artificial intelligence, which he says is far more dangerous than nuclear weapons, will cause the outbreak of World War III.

But rather than cite fantastical, fictional movies or use alarmist language, let's look at what, exactly, artificial intelligence has thus far been able to do. DeepMind, a London-based artificial intelligence lab owned by Alphabet's Google, used neural networks in combination with other technology to build an AI system called AlphaGo that beat the world champion player of the game Go (a ancient Chinese strategy game) in 2017.[34] AlphaGo learned by playing against the best human players, starting out pretty poorly, then absorbing and adopting their tactics, and soon outsmarting them.

Google then went even further and developed a program called AlphaGo Zero, which strictly played against itself, using what they called reinforcement learning, meaning there was no human guidance or data involved, just knowledge of the game's rules. AlphaGo Zero began to recognize patterns in the game, and soon developed new approaches and strategies that no human being had ever seen.[35] David Silver, its lead programmer, said at an October 2017 press conference, "By not using human data—by not using human expertise in any fashion—we've actually removed the constraints of human knowledge. It's therefore able to create knowledge itself from first principles."[36] In other words, AlphaGo was fully autonomous in its ability to play Go.

The reason this is important is that critics have long argued that artificial intelligence would be limited and governed by the human-created algorithms that drive it. This achievement demonstrated that, in fact, machines can learn on their own, unconstrained by the inputs human beings feed into them, and achieve outputs that human beings may not be able to predict or even understand. Proponents of artificial intelligence are quick to point out that this does not necessarily make AlphaGo, or any comparable future program, dangerous. AlphaGo only knows how to do one thing—play Go.[37] But this achievement should nevertheless give us pause to consider what the future prospects of such machine-based, self-generating, unbounded cognitive power could be. Human-developed algorithms are becoming so complex, building one upon another with machine learning woven in, that even their creators admit that they are losing the ability to understand them.[38]

The dangers of completely autonomous artificial intelligence must be weighed seriously. The complex and thorny issues behind full artificial intelligence deserve in-depth analysis that will not be fully engaged here, as that would require another book, and this capability is not likely to be obtained by non-state actors in the near term. On this matter, I will defer to the expertise of Stephen Hawking who warned in a 2017 speech, "Unless we learn how to prepare for, and avoid, the potential risks, AI could be the worst event in the history of our civilization. It brings dangers, like powerful autonomous weapons, or new ways for the few to oppress the many. . . . I am an optimist and I believe that we can create AI for the good of the world. That it can work in harmony with us. We simply need to be aware of the dangers, identify them, employ the best possible practice and management, and prepare for its consequences well in advance."[39]

The Predictions of Lethal Empowerment Theory

The theory of lethal empowerment laid out in earlier chapters predicts that it's not fully autonomous, full-AI systems that non-state actors will seek to acquire, at least not as a priority. It's the more accessible, less expensive, less difficult to operate automatic and semi-autonomous systems they'll jump at first. Indeed, their adoption of them is well underway.

Non-state actors already have what they need to produce simple autonomous weapons by employing open platforms and smartphone technologies. They have been innovating for years. In 2011, for example, Ansar al-Islam in Iraq produced a video demonstrating its ability to build driverless cars and remote-controlled machine guns, both crude robotic devices operated with wireless technology.[40] In 2016, ISIS released an instructional video that demonstrated how to turn a car into a remote-controlled weapon, including remote steering and technology to depress the foot pedals. The vehicle was "driven" by a mannequin that was coated in wire, wrapped in foil, and had an internal heating unit and thermostat to maintain body temperature and fool infrared cameras.[41] Apparently, ISIS never used this type of unit in battle. The appeal of deploying them is, however, obvious. Being able to create robotic suicide bombers would obviate the need to indoctrinate martyrs, one of the most labor-intensive and difficult aspects of terrorist activity.[42]

To assess the nearer-term prospects of lone wolf actors, terrorists groups, and insurgents applying autonomous capabilities, we'll survey the current state of the art in a range of technologies. In some of these areas, innovation is being led by the private technology industry, while in others military organizations are driving progress, though often in partnership with private companies and universities. Some of these technologies are already available to any comers, as dynamite was, while others are exclusive to military organizations but may become accessible to non-state actors, as the Kalashnikov quickly did.

Autonomous Reach

The far-reaching implications of robotics, autonomy, and artificial intelligence for the future of power projection by non-state actors are only now beginning to be understood. While major states, primarily China, Russia, Iran, Israel, and the United States, are developing capabilities with a focus on the relative power balances between them, they may inadvertently narrow the power differential between states and non-state actors in the process.

The industrial nation-state was defined in part by its capacity to field substantial armed forces. If you can close the gap between conventional and unconventional forces through the use of relatively cheap, autonomous robotics, you can wreak havoc with a small army and undermine the power of nation-states. Less sophisticated systems than those developed by major powers will suit this purpose just fine. Whereas advanced states use autonomous systems and robotics for greater accuracy and better discrimination of targets—as the United States has attempted to do in its high-profile drone campaign against al-Qaeda—sparing civilians is not the priority of terrorist groups. For their purposes, lower-grade robotics are perfectly adequate.[43] But highly capable autonomous robots will also likely become commercially available at relatively low cost in the near future, and individuals, terrorist groups, and insurgencies will surely seek to use them to carry out destabilizing violence not only against armies but also publics.[44] Particularly if deployed in large numbers, robotic systems would be very difficult to defend against, and their attacks difficult to attribute to perpetrators, enabling a rash of violence and destabilizing assassinations akin to that of the late nineteenth century.

The development of robotic systems for use in combat is being spearheaded by advanced militaries, which are already teaming soldiers with autonomous vehicles used for reconnaissance, supply convoys, or to increase fighting force. The Israeli Defense Forces (IDF) deploy unmanned ground vehicles along their borders to detect intruders without exposing soldiers to snipers. The IDF vehicles are operated remotely by an all-female unit but may soon transition to independent autonomous operation. The Russians have developed a self-driving, machine-gun firing mini tank called Nerekhta. Australia, which maintains a relatively small army of 30,000 troops that must defend a vast territory, sees unmanned systems as a vital way to increase its fighting mass and capabilities. The British Army is also testing unmanned systems, especially for reconnaissance, more advanced mine clearing, and breaching walls or doors.[45]

The further development of robotics for use in combat situations is inevitable, and with good reason. Transferring dirty, dull, or dangerous tasks that waste human time and energy and risk lives is to be applauded. The danger is that while today the most effective autonomous systems are well beyond the reach of non-state actors, as this technology develops, it will become cheaper and more accessible.

Even today, the basic sub-systems and sensors in civilian and military autonomous systems are the same. For example, originally funded by DARPA, iRobot's man portable Packbot series military robots can perform everything from surveillance, to building clearance, to IED detection. They were first used to dig through debris of the World Trade Center in the aftermath of the 9/11 attacks, seeking victims and assessing structural integrity in places where no humans or sniffer dogs could go.[46] Made by the same company that makes Roomba vacuum cleaners, PackBots are controlled by a joystick that looks like the one used in Microsoft's Xbox video games. Depending upon their sensor packages, Packbots cost between $100,000 and $200,000 apiece, but they are becoming cheaper and more useful to a larger range of both military and civilian actors.[47]

Military robotics have already been transferred to a broad range of civilian law enforcement organizations in the United States. Since 2003, as the result of a federal program to transfer excess US military equipment to law enforcement agencies throughout the country, some 280 US law enforcement agencies have acquired at least 1,000 military bomb disposal and reconnaissance robots.[48] In July 2016, Dallas police used a Northrop Grumman remote-controlled robot, a four-wheeled platform obtained from the military, armed with a 1-pound C4 plastic explosive, to kill Micah Xavier Johnson, a man who had allegedly shot five Dallas police officers, wounded seven others (including two civilians), and was holed up on the second floor of the El Centro Community College building in downtown Dallas.[49] Police remotely detonated the explosive using a cord.[50] This is the first known lethal use of a robot by law enforcement, and it was controversial.[51] Specific legal and policy guidelines for how and when law enforcement may use ground-based robotic technology to kill a suspect are weak to nonexistent in the United States.

A few future possibilities for use of automated and autonomous technology by violent actors are particularly troubling and merit closer examination.

SELF-DRIVING TRUCK BOMBS

We've earlier discussed the enormous commercial investment in self-driving vehicles by Google, Apple, Microsoft, Alibaba, and Baidu, as well as traditional car companies including Ford, universities like MIT, partnerships like Toyota-Uber (among many others), and a wide range of some 263 startups tackling various parts of the autonomy challenge (sensors, guidance systems, etc.). Looking only at data accessible in English, the Brookings Institution in 2017 found more than $80 billion had been invested in autonomous vehicle technology—which clearly underestimates global outlays.[52] Developments in autonomy will soon be applicable to an enormous range of other purposes, including image recognition, target recognition, delivering supplies, driving convoys, and a wide range of kinetic operations in extremely hazardous conditions, such as those involving radiation or chemical weapons, where humans are too slow, physically ill-suited, incapacitated, or indisposed.

The US Army has long been keenly interested in integrating autonomous and semi-autonomous tactical wheeled vehicles into their ground forces. Self-driving military systems have been under development for at least a decade, in response to the high rate of deadly attacks on supply convoys by both ambush and roadside bombs in the war in Iraq. Eventually the idea is to use them not just for hauling supplies, fuel, and water, but to control tanks and artillery. One robotic wingman concept, for example, envisions a manned command-and-control vehicle that directs unmanned combat vehicles on the battlefield.

In 2000, Congress mandated that one-third of Army vehicles be robotic by the end of the first decade of the twenty-first century, but the Army failed to achieve that. Autonomy kits that retrofit vehicles to allow for autonomous and remote control driving are already available, however, and have been tested on American highways. They have GPS navigation and can be used in leader-follower setups, where a lead vehicle is driven by a human and unmanned vehicles follow it and receive data and commands from the lead vehicle. Thus far, this capability is prohibitively expensive for the Army to deploy in mass, costing $200,000 per truck. Even teamed with the Marines, the two Services cannot achieve the economies of scale needed to make adoption viable. Also, the requirements for military vehicles are much more exacting than those for self-driving cars, because military vehicles operate in much more unpredictable environments—not necessarily sticking to roads and highways, for example.[53] Meanwhile, access to comparatively simple autonomous vehicle technology is likely to increase in the future, making it available to criminals, terrorists, and insurgent organizations. As noted earlier, both Ansar al-Islam and ISIS have already shown that they can develop rudimentary self-driving cars that they have weaponized.

Even if they're not loaded with explosives, self-driving vehicles have been proven to be relatively easy to hack, through either traditional cyberattacks or a new generation of adversarial machine learning, meaning that tomorrow's terrorists or criminals might be able to remotely drive them over cliffs or into waterways.[54] They could also use easier low-tech methods: for example, a group of computer scientists from four US research universities put a few black-and-white stickers on a stop sign, slightly altering that familiar octagonal image, and the deep neural network in self-driving cars had great difficulty recognizing the sign. The cars were fooled 85 percent of the time.[55] Rogue states, criminals, or terrorists could attack network systems that regulate driverless vehicle traffic in a major city, say New York, London, Sydney, Seoul, or Beijing, causing simultaneous failures and vehicular crashes on an unprecedented scale.[56] Or they could simply send a self-driving car into crowds, creating a driverless version of the 2016–2017 Nice, Berlin, London, Stockholm, Barcelona, and New York City vehicular terrorist attacks.[57]

HIJACKING THE INTERNET OF THINGS

The Internet of Things (IoT) is the interconnection of millions of computing devices via the Internet, equipped with sensors that directly receive and transfer data without human involvement. As the IoT grows to encompass more cars, kitchen appliances, thermostats, door locks, voice-activated assistants, and even hospital infusion pumps and heart monitors, it provides malevolent actors plentiful opportunities for hacking into systems and wreaking havoc.[58]

A great danger is that because private sector companies compete furiously to get their products to market cheaply and quickly, software engineers routinely fail to incorporate security into their designs. Release of new products takes priority over implementing security features, and since competitors' security is just as lax, properly securing these consumer products, which would lead to delays of months, would be a serious competitive disadvantage. What has resulted is a kind of race to the bottom: according to one estimate, 70 percent of all IoT devices have flaws such as unsecured software and unencrypted communication systems.[59] Thus far, companies are usually not held legally responsible for hacks that break through lax security in consumer devices. What's more, the companies themselves have little incentive to secure or encrypt these data sources, because easy access affords them a wealth of information about users. Openness and accessibility are valuable; for those who want to sell to us, having information on what millions of people do is very lucrative. But profiles of our behavior also offer extremely valuable intelligence for those who want to attack us.

Consumers have little to no control over what information is gathered through these devices because they do not own the software that runs them, or have control over that software. The Internet of Things is changing the nature of buying and owning items. According to law professor Joshua

Fairfield, a fundamental shift in property rights is underway and we're entering an era of digital serfdom, loosely resembling feudalism. Whereas serfs did not own their own land, homes, or even farm tools, we generally own the hardware of our smart devices, but the companies who produce them own the software and the information about us they gather. With some smart products, even the hardware is not owned outright, but rather rented. John Deere, for example, has told farmers that they don't really own the tractors they purchase from the company because they are licensing the software that runs them. Farmers cannot fix the vehicles themselves or take them to independent repair shops.[60]

Worse, data are collected on individuals without their knowledge, and their private information can then be sold or shared so others can exploit it. A Roomba vacuum cleaner connected via Wi-Fi to the Internet and with the (often naively given) approval of the owner can share usage and other data with the manufacturer iRobot about home layout and navigation data with third parties via the iRobot HOME app.[61] The WeVibe app, which was a smartphone-controlled erotic massage device, was actively collecting and sharing data with the manufacturer, Standard Innovation, about how often, when, and with what settings its owners used it. In 2017, the manufacturer of the "spying vibrator" settled a lawsuit for $3.7 million, and also agreed to cease collecting personal information from device users.[62] The examples are legion.

Because IoT devices are connected to the Internet, they can also be hacked, and intrusions are already widespread. Would you leave your front door wide open? In August 2017, hundreds of Internet-connected locks became inoperable because of a faulty software update by LockState.[63] It left hundreds of owners unable to lock or unlock their homes for a week. Hackers have moved from taking remote control of your PC to taking control of your smart TV or your city's CCTV cameras instead. They have hacked cars (repeated attacks on Jeep Cherokees in 2015 and 2016), power plants (malware took down Ukraine's power plants in 2016), smart bulbs (researchers showed they could hack thousands of Philips Hue smart bulbs in 2017), and voting machines (a Princeton professor hacked into one in seven minutes).[64] Relatively inexpensive IoT hacking tools are widely and cheaply available to non-state actors. Why bother planting an explosive device under a car if you can hack into a vehicle's navigation system and make it accelerate into a wall or off a bridge?[65] No need for assassination if hackers can deliver a fatal dose of insulin through the unencrypted radio communication system of the insulin pump.[66] No need to take physical hostages; just tamper with a hospital's computer-connected infusion pump to overdose a patient—then threaten to do the same to others.

According to American cryptographer and computer security expert Bruce Schneier, IoT devices are more vulnerable than your laptop or your phone, for a number of reasons. The first is that huge corporations like Apple, Samsung, and Microsoft can afford to hire large teams of engineers devoted to security, while the smaller companies that are making smart locks and thermostats, for example, cannot. Second, whereas people replace their smartphones and laptops every few years, that is not the case for smart refrigerators, pacemakers, or cars, which they will keep for five or ten years or more. Nefarious actors have much more time to discover their vulnerabilities and, because the software is rarely updated, those vulnerabilities persist year after year, just waiting to be exploited.[67] To make matters worse, a vulnerability in one Internet-enabled device, like your home router, can be used as a launching pad for attacks against a range of other connected devices you might own.[68] One small flaw and your whole computer-assisted life can be hijacked.

Much attention has been paid to the threat of espionage and cyberattacks by states, and in February 2016, US Director of National Intelligence James Clapper warned that the Internet of Things will further empower state-sponsored espionage, enabling better monitoring, tracking, and targeting of individuals. The threat of attacks by non-state actors is also high. For terrorists, a key question now, as always, is which avenues of attack are most easily available? Enormous collections of data are enticing targets, at scales of magnitude that non-state malicious actors could never dream of amassing themselves. States and corporations are focused on the potential fruits of big data rather than on the criminals and terrorists who can hack into it.

By connecting everything from home defense systems to medical devices to utility companies to hydroelectric dams to the Internet, we have made a new means of attack highly accessible. Absent better security measures, well-established processes of the diffusion of lethal empowerment will kick in. In the mid-twentieth century, airline hijackings evolved from airplane flight diversions to Cuba to the downing of airliners with hundreds of innocent people aboard. Exploiting the Internet of Things to hold people hostage or attack them will spawn increasingly violent copycat attacks.[69] Putting better defensive measures in place is essential.

AUTONOMOUS SWARMS

Autonomous capabilities allow devices to coordinate their movements and move in formations, like flocks of birds or schools of fish. This is called cooperative autonomy, or swarming.[70] The militaries of major states have for some years been developing swarming systems for deployment on land, in the sea, or in the air. Swarming may disrupt the power of major states because it is a way for weaker actors to offset a stronger power's advantages in weapons platforms like aircraft carriers or advanced fighter jets—high-value assets that would be costly and time-consuming to replace. It also compensates for small numbers: it enables weaker powers or groups to enlarge their footprint by giving a single human being control over hundreds of armed drones.

A strategy of swarming is akin to that of the thirteenth-century Mongols, who overwhelmed the better-armed European fighting forces by using decentralized, highly mobile horsemen firing arrows long distances, pretending to retreat so as to entice their opponents to break formation and then engulfing them in a swarm of riders—as bees, ants, or locusts do.[71] Since the 1980s, the Iranian Revolutionary Guard Corps (IRGC) has been mounting machine guns on Swedish Boghammer speedboats and using them to swarm US warships in the Persian Gulf. In 2017, for example, the destroyer USS *Mahan* fired warning shots and changed course when a group of small armed Iranian speedboats converged on it, testing its response. It's not that there are no defenses for swarming: the US Navy has close-in weapons systems like the Phalanx, and powerful guns like the M61 Vulcan cannon that can easily destroy any individual speedboat.[72] The danger lies either in overreacting to them, setting off a war, or being potentially overwhelmed by scores of them.[73] And with increasingly cheap, autonomous vehicles, weaker actors no longer need large numbers of willing human beings to pull that off.

Autonomous technology is advancing so quickly that a single operator can give a simple command, like "destroy this object," and a drone swarm, acting as an intelligent collective, gathering environmental information from its sensors and processing it in real time, can autonomously carry out the task. It's a form of command similar to calling a play in sports, where a coach gives the order from the sidelines but the players must together work out how to execute—except there is not necessarily a team captain or quarterback on the field.[74]

Drones can either be under the control of a human being or operate autonomously. To understand how swarming works and what swarms will be able to do in the future, we must first identify which type of coordination or autonomy is in play. Paul Scharre, who helped draft the US Department of Defense's first autonomy policy in 2011, has laid out four models of command-and-control for drones: 1) a human operator controls a drone swarm (centralized coordination); 2) a human operator controls a few "squad level" machines, which tell others in the swarm what to do (hierarchical coordination); 3) the members of the swarm interact instantaneously by voting or converging on a solution on its own (coordination by consensus); or 4) the swarm acts as a single autonomous unit where individual elements react to one another, as fish, bees, and herding animals do (emergent coordination).[75] The third and fourth types of swarming are more robust and resistant to countermeasures—mainly because there's no human operator to target or cut off. It's like the difference between a remote-controlled vehicle and an autonomous vehicle.

In numerous tests, swarms of small aerial craft have already operated as autonomous systems. A human being launches the drones and gives the group a general instruction, such as "wait" or "attack," then presses the start button, and the swarm independently chooses how to carry it out.[76] Because no one unit is acting as the "brain" of the swarm, it can be very hard to stop. If a defensive missile is shot into the middle of a swarm, for example, a few individual drones might be hit and the swarm might even temporarily disperse; but, as with swatting a few bees that are heading at you in an angry swarm, the remaining collective could just quickly regroup and attack the target again, in a different way or from a new direction.

While competing militaries are keen on deploying swarms, they will also be of great appeal to non-state actors. Swarming can employ commercial off-the-shelf capabilities that are well suited to spread to non-state actors, especially as the technology matures and they have access to human-directed swarms. In fact, one of the largest static formations of drones flown in a swarm thus far was launched by Intel, a 500-drone lightshow staged in 2016, all controlled by a single pilot using a laptop. Located above Krailling, Germany, the demonstration used Intel's Shooting Star drones, a type of UAV equipped with LED lights. At the close of the Global Fortune Forum in Guangzhou, on 7 December 2017, the Intel display was bested by an 1,180-drone light swarm flying in formations shaped like a map of China, a ship, and a kapok tree flower.[77] Intel then produced yet another light show, over its Folsom, California, facility, with 2,018 drones.[78] By the time this book is published, someone else will have upped the ante.

As of yet, military innovation is well ahead of adoption by other actors. Indeed, in airspace as well as aerospace, an arms race in flying military swarms is well underway. The United States took the lead with the Pentagon's Strategic Capabilities Office (SCO—pronounced "Skoh") driving development. SCO was the brainchild of former secretary of defense Ashton Carter, who also drove the establishment of the Defense Innovation Advisory Board, chaired by former Google CEO Eric Schmidt, based in Silicon Valley and with the purpose of building stronger ties between the Defense Department and the tech companies. Launched in 2012, SCO was run by William Roper, a thirty-something-year-old physicist with a reputation as a whiz kid for finding solutions to urgent technological problems. According to Roper, the office had only six full-time employees and about twenty contractors.[79] It operated in the same building that houses DARPA, but DARPA and SCO had different missions. While DARPA focused on new, high-end technology, SCO used creativity and engineering to utilize existing technology in new ways for more immediate needs, often drawing from commercial sources.[80]

The centerpiece of SCO's swarming work was a little snub-nosed autonomous drone about the size of a human hand, named Perdix, after a bird found in Greek mythology, which was designed to operate in large groups. Created by MIT's Lincoln Labs, Perdix was first tested by SCO in about 2014.[81] These cheap, fat little one-pound, four-winged biplanes were made of carbon fiber and manufactured using 3D printing. They look a bit like flying potatoes. A lithium polymer battery drove an inch-wide propeller at the rear of the plane.[82] Perdix drones have been shown to be capable of making decisions independently. They can be launched from the ground, with a slingshot, or by just throwing them. For swarming exercises, they were released from several fighter planes, such as F/A-18 Super Hornets, which took off with them mounted on hardpoints on both wings, then released them along a flight line. Talking to each other via radio, the Perdix drones formed up at a preselected place and set out on their mission—something like "patrol this area," for example. How they did it was up to them. In tests, they have shown themselves capable of collective decision-making, flying in adaptable formations, and correcting themselves when things go wrong, all by employing sensors and processing, without the direct involvement of a human being.

In total, the Pentagon is spending $3 billion a year on autonomous systems. Another major player is the US Navy, which is building swarming drones in their Low-Cost UAV Swarming Technology (LOCUST) program, unveiled in 2015. These small UAVs can be launched from ships, vehicles, or aircraft. For several years, the Navy has been testing the concept of using its own swarm of small boats to protect high-value ships against individual terrorist threats, like al-Qaeda's October 2000 attack on the USS *Cole* when it was docked in Yemen. They envision sending swarms of unmanned boats out to meet enemy swarms, like the IRGC speedboats in the Persian Gulf.[83] A key unresolved question in these scenarios is whether the unmanned armed boats would actually fire at and kill IRGC members, without any human direction—an escalation scenario that might spark an old-fashioned human-against-human war. In the United States, as in most advanced countries, military doctrine for autonomous swarming ranges from embryonic to nonexistent.

Other states have rapidly caught up. The China Electronics Technology Group Corporation (CETC), a state-owned enterprise, has exhibited aerial swarms of autonomous micro drones. In an undated video published in 2016, CETC used long, plastic fixed-wing UAVs that resemble classic model airplanes. Unlike the American Perdix, the 119 CETC drones were launched with a rubber band–style catapult from the ground—so no fighter jets required. China's micro craft demonstrated autonomous group control, sensing and avoiding collisions, coordinating their activities like a flock of birds. According to the narrator on the video, their purpose is to enhance intelligence, surveillance, and reconnaissance (ISR) capabilities, aid in targeting, provide even coverage over wide areas, and overwhelm defenses with their numbers.[84] Xinhua news agency quoted one CETC engineer claiming that China's UAV swarms would become a "disruptive force" that would "change the rules of the game."

Other states are following suit. In 2018, South Korea announced that it would have the capability to deploy a swarm of weaponized drones against an attack by North Korea.[85] Israel has also developed advanced capability. And in May 2018, China increased the number in their swarm to 200. State-to-state competition in swarming drones is heated.

Thus far deployment of these systems is hampered by logistics of battery power and the underdeveloped sophistication of the algorithms that drive their autonomy, as well as the lack of military doctrines for using them. But all of these factors are evolving quickly. For example, Boeing patented a new drone, "Vehicle Base Station," in 2016. When a drone runs out of battery power it returns to an inconspicuously placed base station, which physically replaces its batteries. [86] As deployment increases, it is conceivable that in the future, given individual drones' simplicity and affordability, swarming technology will become available to non-state actors.

Small drones launched by rogue actors, whether individually or in swarms, would play a disruptive role regarding cover and concealment in military operations. Soldiers who do not know the terrain would find it more difficult to seek sanctuary, and insurgent drone operators would benefit from the ability to operate systems while being effectively invisible. The physical vulnerabilities of major systems might also be targeted—as the Houthis apparently did in Yemen in March 2017, using seven low-cost Iranian-made small drones in kamikaze attacks to disable a United Arab Emirates–owned US Pac-3 Patriot missile system.[87]

UAVs may even the odds by providing a "poor man's air force." A range of UAVs will have the ability to provide a loitering presence overhead in situations where aircraft cannot fly. The cheapness of these platforms will mean that they can be sent out in tag teams, to replace one another when the batteries run down. With advances in concentrated battery power, especially solar-renewed batteries, insurgents may soon no longer even need a tag team to hover in the air. When they cannot gain air supremacy, insurgents might seek to disrupt air operations by sending aloft groups of small, cheap UAVs to be sucked into jet engines like birds. Their effects could be just as deadly as highly advanced precision missiles. Groups of autonomous drones might be programmed to hit the back of a tank or to fly into a helicopter rotor, for example. Swarms could also conceivably attack larger weapons systems, like fighter jets (or commercial airliners) by overwhelming them with numbers.[88]

SMALL AUTONOMOUS KILLER ROBOTS

Rogue non-state actors could also level the playing field by using aerial UAVs that are very small, driven by sensors, powered by fully autonomous artificial intelligence, and armed with an explosively formed penetrator (a small explosive that can drive a small piece of metal through a solid surface—such as a person's skull). This dystopian idea was introduced right before the 2017 UN meeting on autonomy. But the possibility that attack squads of so-called slaughter bots could be deployed by rogue non-state actors in the future is both technologically feasible and consistent with classic patterns of diffusion.

In early December 2017, Stuart Russell, a professor at UC Berkeley and long-standing British expert on AI, posted a seven-minute film, which immediately went viral, that vividly demonstrated how easy it would be for anyone to target individuals with an army of these tiny killers. The film depicts thousands of cheap autonomous microdrones chasing after thousands of innocent victims who cannot hide or protect themselves from the drones, which land on their heads and shoot a shaped charge into their skulls. It is a horrifying glimpse into a possible future, presented by the Campaign to Stop Killer Robots, released in advance of the 2017 UN meeting to prompt discussion of their risks.

As sensationalized as the film is, the technology depicted is within reach of a range of malicious actors. There are vast amounts of biodata being collected, especially facial recognition technology that terrorists could access. The small UAVs in the film would not be difficult to acquire or arm, though gaining access to shaped charges would stretch the capabilities of many current groups. Still, a range of actors from Aum Shinrikyo to Hamas to ISIS has already demonstrated the ability to deliver lethal explosives or substances via drone, and increasingly capable, autonomous UAVs are becoming easy for anyone to acquire. The industrial scale of coordinated, global, mass killing that the film presents is unrealistic;[89] but regarding smaller operations, terrorists and insurgents already have access to the technology required to deploy such tiny autonomous robotic killers.

Lethal empowerment theory indicates that these systems, if they are built, will easily diffuse. According to computer scientist Nicholas Weaver at the International Computer Science Institute in Berkeley, these systems would be easy to scale up and mass produce. In late 2017, he wrote, "Some back-of-the-envelope design suggests that I could build the computer necessary to run a slaughterbot, including all the inertial sensors and communications, with two outrider cellphone cameras—in a package the size of a sugar cube. Give me a $10 million budget and I could produce slaughterbots with a manufacturing cost of roughly $200 per unit (if I'm producing enough of them). After all, the base airframes and camera without the computer cost about $45. Further amplifying the concern is that everything involved—from the silicon chips to the machines needed to crank out tens of thousands of these little nightmares—are available off-the-shelf."[90]

The tiny robots already developed purportedly have reaction times up to 100 times the speed at which a human could react. State armies are developing physical countermeasures that may protect them from this type of swarm, but it's civilians who would be most vulnerable to attacks by them.

Tailored for Terrorism

Autonomous weapons will appeal to those perpetrating political violence not only because of their tactical advantages but due to their psychological effects.[91] Just as with dynamite in the 1870s, the impact of robotically conducted terrorist attacks would likely far exceed the actual devastation inflicted. After attacks have been successfully carried out, individuals may be terrified by the mere sound of small aerial drones, not knowing whether they are friend or foe, operated by police or terrorists. There is no system in place yet to distinguish them, and figuring out how law enforcement and military services can defend against such attacks will be complicated.

Some analysts emphasize the enormous advantages of states in mounting defenses. Greg Allen and Taniel Chan argue that artificial intelligence, machine learning, and sensors could make surveillance so effective that terrorists and insurgents would be unable to plan and execute operations without being discovered. Nation-state intelligence organizations armed with the latest technological tools would be able to detect and track their activity in real time. They write:

> Imagine, for instance, if the United States could have placed low-cost digital cameras with facial recognition and the robotic equivalent of a bomb-sniffing dog's nose every 200 yards on every road in Iraq during the height of US operations. If robotics and data processing continue their current exponential price declines and capability growth, this sort of AI-enhanced threat detection system might be possible. If it did exist, guerrilla warfare and insurgency as we know it today might be impossible.[92]

This is a thought-provoking scenario, and it's conceivable that it could unfold in territories that the United States or other major powers occupy militarily. That degree of surveillance might also be possible for authoritarian states such as Russia or China, which will likely not hesitate to use these tools domestically to protect the power of their regimes and tamp down domestic unrest. But within democratic states, legal frameworks supporting individual rights will prohibit such ubiquitous monitoring.

At the same time, the technological advantage of states is also being eroded. Obviously the nation-state will always have greater financial resources and manpower than individuals do, and the major powers are pouring vast resources into developing AI and associated technologies. In July 2017, China published the New Generation Artificial Intelligence Development Plan, which laid out a systematic plan to lead the world in Artificial Intelligence by 2030.[93] It also dramatically increased funding for new AI research. The United States is fighting to maintain its lead in these technologies. It increased Department of Defense research funding in artificial intelligence by about $200 million between 2016 and 2017, introduced congressional legislation to speed the development of autonomous vehicles, and blocked efforts to buy key American technology firms like chip-producer Qualcomm.[94] But costs are falling, and rogue actors may also be able to acquire hacked or pirated AI software that has been developed at great expense by states or corporations. The relative advantage is shrinking. Malevolent hackers, insurgents, and terrorists may become as capable of tracking soldiers or law enforcement officers as those state actors are of tracking them.[95]

Automation may fuel disruption in another way. Robots and computer-assisted automation are being rapidly adopted by companies, and they are widening the chasm between those who benefit from technological advances and those who suffer the economic consequences of resulting job loss. Economists Daron Acemoglu, of MIT, and Pasual Restrepo, of Boston University, examined the effects of the increase in industrial robot usage on the US labor market between 1990 and 2007. Their 2017 National Bureau of Economic Research study was the first to actually quantify the impact of robots, uncovering large and robust negative impacts on both employment and wages. They were also able to distinguish the effects of robots from other factors that politicians and pundits blame, such as imports from China or Mexico, and offshoring of US jobs to other countries.[96] Politicians and economists have argued about the positive and negative implications of increased automation, and now the evidence is becoming clear.

What some call the "robot apocalypse" is displacing human workers in large numbers and placing a strain on labor markets throughout the world. One effect is further widening economic inequality that has grown over the past several decades. According to the Organization for Economic Cooperation and Development (OECD), whose thirty-five member-states are all market-oriented, democratic, and prosperous, the gap between rich and poor is at its highest level in thirty years. In 2015, the richest 10 percent of OECD citizens had almost ten times the income of the poorest 10 percent.[97] Regarding job loss, one study by the Oxford Martin School concludes that over the next twenty years, about 47 percent of total US employment is at risk from computerization, and that the rate at which jobs are being destroyed by technological change exceeds any previous global economic adjustment.[98] This displacement is likely to accelerate in the next major recession, when companies will be pressured to replace the costs of human workers.[99] Meanwhile, workers in many fields and industries are not retraining or adapting nearly fast enough, and few corporations or governments are helping them adjust.

Modern terrorism took off during the late nineteenth century's Gilded Age, during which workers suffered from a rapid shift to mass production. In their anger over job losses, brutal working conditions, and a dramatic rise of income inequality, some of them embraced dynamite as their ideal means to protest. Those left behind in today's rapid shift to digital automation likewise have mounting political and economic grievances. Lethal empowerment theory predicts that the accessibility and potentially devastating power of the emerging cluster of technologies may induce some of them to lash out violently.

The history of innovation and diffusion of technology by terrorists indicates that we are foolishly overlooking the full implications of autonomy and artificial intelligence, not just for states but also for non-state actors. On their current trajectories of development, states and private corporations are providing technologies of unprecedented power to rogue actors. We must temper our techno-optimism with a better understanding of how, especially when offered new technological means, human beings often respond violently to disorienting change. The lessons of the past must be applied to sharper analysis of tomorrow's risks.

CONCLUSION | Strategy for Democracies in an Age of Lethal Empowerment

THE CAPABILITIES IN mobilization, reach, and command and control that emerging technologies hand to individuals and small groups have already significantly undermined the power of democratic states and changed the character of war. The even more powerful capabilities that are likely to become accessible in the future may be as destabilizing to world order as both dynamite and the AK-47 were, perhaps more so.

The trajectory of today's open innovation revolution is being shaped not just by industry titans like Jeff Bezos, Bill Gates, and Jack Ma, and visionary military leaders such as former US defense secretary Ashton Carter, but also by edgy experimenters like Cody Wilson with his 3D-printed Liberator gun, Alex Jones with his social media megaphone, and Frank Wang with his ubiquitous DJI quadcopter. Tomorrow's technological breakthrough could emerge from a prosumer or hobbyist, working in a garage or backyard. We should keep in mind that the radio and the airplane were both invented in that way, and both dramatically changed the nature of military conflict.

The degree to which military innovation took the lead over private innovation in developing lethal technologies during the twentieth century may have been anomalous. There was no clear dividing line between professional and amateur scientific communities in the late 1800s, and that is becoming true again today.

In order to begin more effectively combating the use of new technologies for disruptive violence, we must consider a number of core insights from our analysis of the history and current adoption of potentially lethal disruptive technologies by non-state actors.

Powerful Economic Incentives for Diffusion

It is vital to recognize just how powerful the market forces driving the development and diffusion of emerging technologies are. As was the case with dynamite, the positive social advances these technologies have already delivered, and those they promise for the future, are extraordinary. The demand for them, combined with the enormous profits to be made by commercializing new and emerging technologies for the widest possible adoption, are extremely strong headwinds. The lessons from the diffusion of dynamite could not be more pertinent.

Despite Alfred Nobel's vigorous efforts to obtain control over the manufacture of dynamite through constant patent battles and the formation of his trust, he had to yield to the reality that demand was so strong that other producers would continue to crop up and would apply downward pressure on pricing. His best business option was to expand his own production as rapidly as possible.

Economic imperatives also drove the genie out of the bottle of diffusion of the AK-47. Of course, anti-capitalist ideology played a role. If the Soviets had respected the international patent process, they might have retained control over the production of the original model. But even so, because the gun was so simple in design and cheap to manufacture, they would almost surely have been no more capable of preventing copycat production than Nobel was.

In both of these historical cases of diffusion and in the analysis of how today's emerging innovations have already been harnessed by malicious actors, we have seen that, as Lethal Empowerment Theory predicts, new technologies will be rapidly adopted and then innovated with by violent non-state actors when they are:

- Accessible
- Cheap
- Simple to use
- Transportable

- Concealable
- Effective (i.e., providing leverage and more "bang for the buck")
- "Multi-use" (i.e., beyond "dual use"; suitable for a wide range of contexts)
- Not cutting-edge—usually in the second or third wave of innovation
- Bought off the shelf (or otherwise easily acquired)
- Part of a cluster of other emerging technologies (which are combined to magnify overall effects)
- Symbolically resonant (which makes them more potent than just their tactical effectiveness)
- Given to unexpected uses

Market forces driving the current wave of innovation incentivize the creators of new technologies to endow them with many of these qualities. Consumers respond powerfully to fresh products that offer them exciting new capabilities, like home thermostats that can be adjusted remotely, and tractors that can monitor their own operation and communicate with the manufacturer to enhance their performance. They do so with particular alacrity when those products also offer ease of use, which is why creating increasingly simple interfaces and devices has for years been the gold standard at the forefront of commercial innovation. Competition to capture consumer dollars and on-going product loyalty also puts powerful downward pressure on price. So does the high value of gathering consumers' personal data, an unprecedented cash cow that easily offsets low prices for products sold. Hence, relatively cheap and extremely easy-to-use devices with extraordinary capabilities are reaching the broad global public with unprecedented speed.

Technological Optimism and a Boom in Tinkering

Hobbyists and would-be garage inventors are driving the provision of increasingly powerful component technologies, which they can use to make their own devices and can combine in novel ways. Groups of interactive technologies like small UAVs, 3D printers, smartphones, and simple CRISPR gene-editing tools facilitate creative experimentation. Not only are the components so readily available, but they operate on ready-made platforms that advanced scientists and engineers have already devised and delivered. Easy-to-follow instructions for their use and how they can be combined are a mouse click away. Some scientists argue that momentous changes in synthetic biology or autonomous robotics, for example, make it imperative to draw more people into amateur tinkering and experimenting.[1] The maker movement, which celebrates hands-on creativity by hobbyists of all types, along with community labs like hackerspaces, where amateurs can access technology tools and share ideas, are thriving.

With few restrictions on the use of the new technologies in place, these basement and backyard tinkerers can produce both exciting innovations and destabilizing discoveries that rival innovations by professional private laboratories and government facilities—not because they are more advanced but because they diffuse more rapidly. When today's tinkerers want to share or sell their products with virtually anyone, they do so more easily than ever before. The phenotype of this age of experimentation is Alfred Nobel, working alone and then amassing enormous personal wealth, not J. Robert Oppenheimer leading a team of scientists in the service of their country in a government facility at Los Alamos.

The technological enthusiasm and hobbyism of today is strikingly similar to that at the turn of the nineteenth century, and it has led to a comparable degree of willful blindness about the risks. In 1903, the manager of a big New York dynamite manufacturing company begged publicly in the *New York Times* for new laws to abide by: "It's one of the easiest things in the world to buy dynamite enough in this city to blow up half of lower Broadway. . . . A total stranger could go into any powder company in this city and buy all the dynamite he had money to pay for, and not a question would be asked as to what use the explosive was to be put. . . . I have often talked to other powder men about selling explosives to every one willing to buy, whether he would be able to give a satisfactory account of himself or not. But the law is at fault, not the powder men. Give us a law which we all must obey, and we shall be only too willing to follow it."[2] Those who sell dangerous products sometimes appreciate their risks the most.

Even as horrifying dynamite-fueled violence took its toll, the energies of a range of inventors and experimenters never flagged, from professional scientists to prosumers, hobbyists, and ordinary consumers. Indeed, each gory attack only seemed to heighten their enthusiasm in an atmosphere of giddy techno-optimism and get-rich-quick financial breakthroughs. Individuals experimented with explosives in their basements, workshops, and sheds, the vast majority focusing on things like blowing up tree stumps or seeing if they could make it rain, not remotely interested in violence. But those who were, wreaked havoc.

Today we have a tendency to focus exclusively upon doctrines or motivations for terrorist attacks, assassinations, and other shocking violence, seeing radicalization or ideology or mental illness or injustice as the roots of the problem. But attackers must also have the means.

New Communications Technologies Are Powerful Incentives to Violence

Facebook has said it was blindsided by the use of its platform by Russian meddlers in the US 2016 elections. Study of the history of political violence shows that the harnessing of powerful new means of communication to the public by nefarious actors should be expected.

We've seen that dynamite and the AK-47 were seized on to launch waves of attacks not only due to their nature as accessible and powerful technologies, but also because of their symbolic power. That symbolic resonance was greatly heightened by the advent of new means of communicating about attacks. In both cases, drawing attention to attacks, developing a global reputation for effectiveness, and building a twisted heroic narrative were crucial to the spread of violence. For dynamiters, the means was sensationalist global newspaper coverage, and for the insurgents who took power with the AK-47, it was carefully crafted revolutionary propaganda delivered in vivid images on the nightly network news.

The attempt to undermine the democratic process today with "fake news" is deeply troubling, but other uses of the new communications platforms by malicious actors are equally problematic. Less attention has been paid to how ISIS harnessed social media for recruitment, and the extraordinary effectiveness with which it built a massive following must be much more seriously processed because others will be even better at it. The rash of live-streamed violence also deserves focused attention. Live-streaming video technology combined with first-person filmmaking, millions of automated bots to spread the word through a broad range of social media platforms, and the manipulative tools of the attention economy are adding incendiary motivation to carry out ever-more lurid and powerful attacks.

Arguing today that social media companies are not responsible for the content they facilitate is like absolving nineteenth century hyperbolic newspapers for their shoddy professional practices. Both enabled fake news, and both accelerated terrorism. The *New York Times* developed much higher standards; so must Facebook and Twitter.

Militaries Are Facing the Innovator's Dilemma

The balance of military power among the United States, China, and Russia is a vital focus for all analysts seeking to anticipate future threats to the world order. The United States and its allies are facing serious sustaining threats. China's assertiveness in the South China Sea, Russia's intervention into the Syrian civil war, and the North Korean nuclear threat are all high-technology challenges they must attend to. They must not lose sight of these many threats or fail to plan for them and innovate to deter or meet them. But just as dominant companies can be blindsided by market disruptors, militaries can be blindsided by non-state actors.

Military planners tend to think of lethal use by private actors, even insurgents, as tactical nuisances against which they can readily defend. And that has proven largely true; superior lethal technology, force, and command capabilities marshaled by military forces have effectively countered many threats. ISIS's use of armed quadcopters on the battlefield killed or wounded dozens of unfortunate Iraqis, but the primitive drones were soon neutralized or disabled. The online targeted recruitment of teenagers in East London, North Carolina, or Sydney worried law enforcement and military planners, but the comeback was the military defeat of ISIS in Iraq and Syria. Armed quadcopters may easily attack civilians or political officials in the future, but many analysts argue that defenses—from chicken wire to nets—can easily fend them off. Swarms of autonomous drones with facial recognition capability could target specific people, but terrorists would be unlikely to be able to use them in numbers large enough to threaten an army.[3]

Yet other strategists assert that the existing threat of commercial drones is significant, even for conventional armies. A March 2018 report from the US National Academy of Sciences concludes, "[C]onsumer [quadcopters] are easy to buy. Their performance is improving dramatically, their cost has dropped significantly, and there are millions of them around the world. [T]hey pose a significant and growing threat to US warfighting forces when used for nefarious means rather than as intended."[4] During his September 2017 reappointment hearing, US Chairman of the Joint Chiefs of Staff General Joseph F. Dunford, USMC, said that commercially available drones were "at the top of our list for current emerging threats in the current fight [in Iraq]."[5] If dominant militaries struggle to defend against consumer drones, even those that don't arrive in swarms, then civilians are far less likely to be protected from them.

Right now it is possible to walk into Best Buy or Verizon (or just go on-line) and purchase the technology necessary to recruit, arm, and field your own private army. That force won't match the kinetic capabilities of the US Army or its partners and allies, but it doesn't need to. Insurgents who are more patient, more committed, or more dedicated to a local cause or territory will always have an advantage over foreign forces whose goal is to get in, solve a problem, and get out. We've seen this with insurgencies from Algeria to Vietnam to the current resurgence of the Taliban in Afghanistan.

Truly disruptive technological innovations change the nature of the market by lowering the barriers to entry and ballooning the number of participants. Accessible lethal technologies give a much broader range of actors the power to start wars, and major powers are losing the capacity to end them.

Disruptive Private Armies and the ISIS Precedent

When ISIS first appeared, many terrorism experts thought it was just al-Qaeda rebranded with a black flag. They were caught off guard by the group's recruitment of 40,000 followers from throughout the world and the scale of conventional operations it was capable of. They had not counted on ISIS capitalizing on second- and third-generation systems by jury-rigging them. The UAVs and other primitive robotics the group made use of were laughable compared to those used by the United States, United Kingdom, France, Israel, and other advanced states. But they enabled some highly effective operations. By the end of its fighting in Iraq, for example, ISIS was employing UAVs in small aerial formations to maintain local electronic networks so that other systems could operate despite efforts to jam them. The Islamic State's territorial conquest was short-lived largely because the group offered the United States and its allies clear targets to attack conventionally—an example that future insurgent groups will learn from and avoid. Those with much more sophisticated, less anachronistic grievances will improve upon the ISIS model. Copycats may not endorse so-called jihadist terrorism but resurgent right-wing, left-wing, or ethno-nationalist causes—or even new ideologies that we have yet to anticipate—will follow.

The conditions are in place for mobilizing additional powerful non-state groups to act more effectively. We are already contending with a range of untraditional actors in so-called hybrid conflicts, as between Russia and Ukraine and with Syria and its many anti-Assad factions. China's efforts to construct military and commercial outposts on artificial islands in disputed territory in the South China Sea, threatening free navigation and obstructing sea lines of communication, are another case in point. The potential for an armed clash, either directly between China and a neighboring state or indirectly through the use of Chinese proxies, like civilian fishing vessels or swarms of unmanned drone ships, are high.[6]

The wave of refugees and displaced people from the conflicts in Syria and Iraq—the greatest movement of humanity since the Second World War—is exacerbating anger and a sense of division and dissolution from the Middle

East to Europe to the United States. There is increasing political fraction-alization, from Iraqi Kurdistan, to East Turkestan, to Kashmir, to Tibet, to Eastern Ukraine, to countless independence or self-government movements that are gaining fresh impetus or traction. Crisis situations with long-term local roots are joined now by a global sense of disarray, combined with the new technological means to act on proliferating grievances.

Tomorrow's battles will be defined not just by China, Russia, the United States, and other states with enormous budgets and impressive technolog-ical programs, but also by private armies exploiting the new technologies. Rather than attempting to win a war, the new goal of some groups is to keep clashes below the level of conventional war and achieve incremental gains in territory, economic resources, intellectual property, and political backing, using the media to build long-term support from sympathizers or project *faits accomplis*. New technologies are perfectly suited for a strategy of raids, meaning quick irregular skirmishes directed at distracting, confusing, or undermining a stronger enemy. Predominant twentieth-century military concepts like deterrence, coercion, and compellence are being sidestepped by clandestine attacks, incremental small actions ("salami-slicing"), theft of intellectual property, and the hacking of elections. The US Defense Science Board calls non-state actors' capacity for persistent cyber attacks and intrusions "death by 1,000 hacks."[7] From cyberspace to Crimea, to Ukraine, to the South China, today's most effective players are keeping their activity at a level where they gain the spoils of war without having to fight or be "coerced" at all. They are fighting a long-game war of attrition.

Asymmetrical warfare can be just as devastating over time as conven-tional state-to-state war, especially if it drives governments to neglect their long-term interests. Taking an indirect approach—in this case undermining powerful enemies before they become fully aware of it–is not a new idea. "To subdue the enemy without fighting is the acme of skill," wrote Sun Zi in the late sixth century B.C. In the words of a 2011 Russian manual on psycholog-ical warfare: "The population doesn't even feel it is being acted upon. So the state doesn't switch on its self-defence mechanisms. . . . [I]nformation war is supple, you can never predict the angle or instruments of an attack."[8]

These actors would not be able to defeat major state armies in direct com-petition on the battlefield, but they're well aware they don't need to. Newly diffused technologies offer them means to actively foment and exploit do-mestic and regional discord, dissatisfaction, and unrest and to pursue what well-known American strategist and former undersecretary of defense for policy Fred Iklé dubbed annihilation from *within*.[9]

Responding to the Threat

The most effective way to respond to the fast-moving changes of an open revolution is to align all the participants, including government, industry, and individual citizens, around incentives for developing protections.

THE PROFIT MOTIVE FOR PROTECTIONS

Inventors and the companies that promote their innovations bear some responsibility for mitigating foreseeable risks and for educating policymakers about what they are. The argument that private companies are not responsible for policy doesn't wash. Policy has always lagged behind innovation because innovators are the ones who understand the full potential and limitations of their creations. They have a duty to their fellow citizens to take responsibility for helping to identify and mitigate the risks.

Of course, that hasn't proven a compelling argument to the companies thus far. A more effective approach to persuading them may be to stress the profit to be made from protecting the public from the risks—and the perils to their businesses from failing to do so. The broader public can be encouraged to take greater responsibility for convincing them, insisting that effective safety and privacy measures be integrated into our appliances, cars, and other consumer products.

Companies must be required to inform the public much better about the risk they are exposing themselves to in making a purchase. The development of technologies with which we can be spied on and targeted has raced ahead of our awareness of those capabilities. Just try to buy a new automobile that is not computer-networked to its producer, collecting an average of 25 gigabytes of data every hour you drive.[10] Beginning in 2014, nearly 100 percent of new cars were tracking, recording, and wirelessly transmitting your driving habits and history to a cloud or a third-party vendor, many with dubious security protections. This included external-facing sensors to detect obstacles, but also internal sensors to study you. Some new, high-end cars, like Tesla's Model 3, GM's 2018 Cadillac CT6, and the Subaru 2019 Forester, have internal-facing cameras that monitor drivers for signs of distraction.[11] How many members of the public are aware of these intrusions?

We must insist on better protection against the risks, knowing what data are being collected, who controls it, who owns it, and how well safeguarded it is. There will be no change until consumers demand, and become willing to pay for, products that protect their privacy, are harder to hack, and cannot be used for lethal purposes.

We must not hold back innovation, just refocus it on enhancing protections and reducing risk. There are practical solutions to many of the safety problems we have identified here, and especially over the long term, developing them—and paying for them—will be much cheaper than ignoring the threats. We must begin vigorously demanding safety.

Defensive measures against weaponized small drones, for example, are poorly developed and well behind where they should be technologically. Shooting nets at drones, jamming communications, employing lasers, or sending trained eagles to snatch small drones out of the air are all primitive answers to a problem that will get worse. Making geofencing around all sensitive sites standard government practice is long overdue. To capture market forces in the direction of safety, we could allow individuals to purchase anti-drone geofencing around their own homes, if they so wish.

REGULATION IS NOT NECESSARILY STRANGULATION

Elected representatives and government policymakers must face their responsibility to enact reasonable laws that reduce risks, encouraging innovation within frameworks that serve the public interest and protect national security. We have accepted regulation or strict industry guidelines for every other major technological innovation, from building codes to automobile standards to airline regulations to nuclear plants and well beyond. Such measures protect citizens, which is at the heart of what governments are meant to do. They make democracies stronger, more resilient, and more resistant to attacks.

In this respect, it is valuable to reflect on the different approaches taken by governments in Europe and in the United States to end the first wave of terrorism by dynamitings. In the United Kingdom, a regulatory system for dynamite's sale and safe transport was instituted early on, which Alfred Nobel was able to adhere to in building his business there. In Europe broadly, once there was widespread awareness of the threat, governments passed regulations limiting access to explosives and also developed better international police cooperation, notably with the establishment of Interpol. This seemed to tamp down the numbers of attacks, though the arrival of a greater cataclysm, the First World War, also played a vital role.

In the United States, with its sparse, complicated patchwork of state and local regulation, federal government focus fell sharply on perpetrators, with vigorous anti-immigration laws passed even when the majority of dynamiters were Americans associated with the labor movement. These laws had little impact upon the rate of attacks. The railroad industry stepped forward to regulate dynamite's transportation across state lines but otherwise the explosive remained easy to acquire.

Compared to Europe, US bombings built to a crescendo much later into the twentieth century, ending with Roosevelt's New Deal and the 1935 passage of the National Labor Relations Act. Industry self-regulation and addressing the causes of popular unrest played an important role in ending dynamite violence in the United States, where legitimate political activity gradually supplanted bombings.

Today European lawmakers have been legislating a patchwork of data and privacy regulations concerning how personal information may be collected and used, while in the United States industry self-regulation by the companies is still favored.[12] Thus far, the Silicon Valley giants such as Amazon, Google, Twitter, and Facebook have fallen far short of providing the protections needed. They must recognize that they have powerful incentives to do much better. On 19 July 2018, after the Cambridge Analytica dataprivacy scandal and the implementation of privacy regulations in Europe (the General Data Protect Regulation or GDPR), Facebook lost $120 billion in market value in one day, the largest ever one-day loss in value for a US traded company.[13] Young users fled from Facebook to YouTube, Snapchat, and Instagram (which Facebook owns); but all of these platforms are vulnerable to future debacles and business instability.[14] In March 2019, Facebook owner Mark Zuckerberg tried to get ahead of the curve by proposing regulations in the *Washington Post* that would effectively squeeze out upstart competition and eliminate the platform's responsibility to police itself.[15] Admitting the problem was a step in the right direction; but the only way to build smarter, fairer oversight is for lawmakers and the public to seriously educate themselves about the risks and demand better solutions.

Policymakers bear responsibility for learning more about new technologies and analyzing their implications proactively. Our politicians are largely ignorant about the risks and opportunities inherent in everything from bioprinters to UAVs to the Internet of Things to autonomy to artificial intelligence—and this prevents them from protecting ordinary citizens from attacks on life and property. In democracies, elected representatives, judges, prime ministers, and presidents are all answerable for safeguarding the public interest, and they have a lot of catching up to do.

BUILDING STRONGER NATIONAL SECURITY

In the United States, the 11 September attacks led to vigorous efforts to increase homeland security, through enhanced airline security, better intelligence, more cooperation with allies, and a vast increase in institutional capacity for counterterrorism. Similar measures were put in place in hundreds of allied and partner countries throughout the world. As a result, the capacity for traditional counterterrorism methods such as intelligence sharing, direct action, and engaging local partners in conflicts from Mali to Yemen to Pakistan greatly increased. But there is a huge hole in that big, global protection network. The attack surface for our domestic vulnerability has vastly increased at the same time that new actors have potent new means to attack it.

We have worried about terrorists targeting infrastructure in cyber attacks, and that threat remains. But it is joined now by the threat of weaponizing aerial drones, targeting individuals through widely available facial recognition software, or hacking vulnerable Internet-connected devices like cars, medical equipment, and individual homes through the Internet of Things. The answer to these security challenges is not necessarily more traditional security. Just as important is to shrink the big, juicy targets we've made available. Countries, such as Sweden, Finland, Norway, and the Baltic states on the periphery of Russia, are well accustomed to defending themselves against cyberattacks, loss of cell service, jamming of GPS signals, or a flood of polarizing disinformation and propaganda designed to inflame ethnic tensions. They are decades ahead of the United States and other democratic states in reducing their vulnerabilities by increasing civic education and public resilience. Finland has a robust educational program in online literacy, for example, and Latvia is developing a similar approach.[16] In 1957, Russia's launching of Sputnik sparked a comprehensive societal response to the challenge in the United States. Today's threat is even more urgent.

One thing about the current open revolution is sure: the integration of new technologies into unexpected combinations will surprise us. The degree of systems integration and command-and-control that emerging technologies are providing has never before been within reach of individual actors or small groups. Swarming, self-driving truck bombs, and robots are just the beginning. We must also be mindful of the scale and breadth of vulnerability we have built into our societies. The Internet of Things provides an avenue of access into millions of Internet-connected devices and appliances. With artificial intelligence, single individuals will have a shot at building armies without needing to collect large numbers of human beings. Semi-autonomous and autonomous weapons systems will enable small forces to hold their own against vastly superior forces. On our current trajectory, without both better defensive measures and greater regulation of risk, the result will be wars of attrition that democracies cannot win.

The United States and its allies have repeatedly found themselves unprepared for new threats but then responded with extraordinary unity, vigor, and ingenuity. The United States most glaringly lagged in countering the buildup of Nazi power and recognizing the threat from Imperial Japan, but after Pearl Harbor the pace of response was astonishing. We failed again in the 1990s to really step up to the threat from al-Qaeda until the 9/11 attacks, but thereafter instituted extensive security measures that have thwarted a host of attempted follow-ups. We have the innovative, economic, and strategic capacity to devise and deploy protections to meet the current and future challenges. But if we are to do so, we must harness our unparalleled capabilities now, with all the talent, energy, and drive we have shown we can muster. We must not wait for another unanticipated attack to shock us into action.

EPILOGUE | Power to the People

Audrey Kurth Cronin

'Power to the people' may have been a 1960s radical slogan of the Left; but in the United States, at least, the concept has since emerged from the Right. A dramatic illustration unfolded on January 6th, 2021, when a group of some 800 right-wing, anti-government, accelerationist, conspiracy-minded American citizens, mobilized by groups such as the Oath Keepers, Boogaloo Bois, Proud Boys, Three Percenters, and QAnon, stormed the US Capitol seeking to overturn the 2020 presidential election. Goaded by President Trump, who told them to "march to the Capitol," "fight like hell," and "take back our country," these violent extremists overwhelmed US Capitol police, breached the building, and sought to kill or kidnap Members of Congress. Five people died after a vital underpinning of democracy—a democratic election and peaceful transfer of power—was jeopardized for the first time in American history. None of this would have happened without the deliberate mass mobilization of demonstrators into the Capitol, outnumbering and overwhelming law enforcement.

It was not just digital technology that caused the Capitol insurrection. The effects of the pandemic, alongside riots and protests in the aftermath of live-streamed, racist police killings of American Black people, especially George Floyd, brought popular passions to a boil. In 2020 and 2021, US left-wing anarchists and Antifa members also engaged in violence, clashing with alt-right demonstrators at rallies and causing widespread destruction in places like Portland, Oregon and Minneapolis, Minnesota, where a police precinct was burned to the ground.[1] But far-left incidents were fewer and less lethal than far-right attacks, which had been gradually building in the United States since 2009 and claimed 91 victims (compared to 19 by far-left groups) between 2015 and early 2021.[2] Regardless of ideological slant, today the vast majority of terrorist attacks and plots in the United States arise through online channels, and when one platform is shut down, followers migrate to another.

What we have in much of the developed world is a weaponized internet. Right-wing extremist groups are born online and thrive through shared symbols, memes, slogans, and unhinged doctrines about the need for a race war and the desire to act as "soldiers." They mimic ISIS, whose online radicalization (especially via Twitter) drew converts who had little understanding of Islam, knew nothing about Syria, held different ideological viewpoints, but believed they would find paradise through apocalyptic violence. ISIS gathered followers in one place, slapped together a proto-state, and, after sweeping through Iraq in July 2014, offered conventional armies a target for counterattack. Today's online extremists typically have never met, do not share the same ideology, have no prior record, and recognize no formal leaders; yet they drive each other toward apocalyptic terrorism or mass shootings via Facebook, 8chan, Gab, Parler, Signal or Telegram.

Many right-wing terrorists worship their predecessors online, calling them "saints" or "chads" (slang for Alpha males), then seek to copy them in a global digital tag-team. For example, white supremacist Robert Bowers killed eleven worshippers in an October 2018 attack on the Tree of Life synagogue in Pittsburgh, Pennsylvania. Reading about it in New Zealand, Brenton Tarrant then killed 51 people at three mosques in Christchurch, New Zealand five months later (15 March 2019). The next month, stirred by the actions of "Saint Tarrant" (as posters on 8chan called him), John Earnest killed 1 and wounded three in a synagogue near San Diego, California (27 April 2019). Patrick Crusius, seeking to join them, then killed 23 people in El Paso, Texas a few months later (3 August 2019). Philip Manshaus mentioned Crusius when he killed two at the al-Noor Islamic Center near Oslo, Norway a week later (August 10, 2019). They were all white supremacists but had never met and did not agree on their specific aims: Bowers and Earnest were anti-Semitic, Tarrants and Manshaus were anti-Muslim, and Crusius was anti-immigrant. What united them was a desire to bring on a "race war"—and to earn fame and immortality online.

Terrorists and extremists are joined by governments who are redesigning state sponsorship for the digital age. In an environment tailor-made for subterfuge, weak states launch foreign influence operations or amplify domestic campaigns. Often it is impossible to distinguish principal from proxy—even by the proxy. Russia, long master of this sort of tradecraft, has used bots, fake accounts, stolen information, and trolling to amplify messages and aggravate existing divisions, especially in democratic societies. Russian intelligence services and their stooges have purchased commercial bot networks, hired Western marketing firms, and supported local nationals to manage extremist pages and websites. They have helped polarize politics in democratic states and communities, for example by supporting both the Black Lives Matter movement and the White Lives Matter countermovement in the US, the Brexit referendum in the UK, and the far-right Lega party in Italy.[3] China, Iran, Saudi Arabia, and the United Arab Emirates have followed Russia's lead, sowing political tensions in Australia, Hong Kong, Israel, and Jordan, among others. Disinformation has always played upon existing fault lines.

All of this makes one wonder what exactly advanced Western and allied military forces are fighting to preserve, as enemies cut the legs out from under them at home. Effective strategies must always reflect the conditions in which they arise. Because of the properties, good and bad, of technologies described in this book, the historical context has changed, not just in the technological dimension but also in the *social* dimension of warfare.[4] Yet our policymakers have not adapted. Instead of confronting a changed strategic environment, public and private leaders respond to symptoms by prevaricating, profiteering, and pandering for money and attention from their base. Above all, pundits and platforms insist, we must protect freedom of speech. Boundlessly? Even when doing so supports foreign or domestic influence operations against us?

Democracy is more at risk than at any time since World War II. In addition to the easier-to-see technological advances of China's Communist Party and the PLA, our military and security services must develop more integrated, clear-eyed strategic approaches to bottom-up threats at home that build upon themselves. The best way to defeat democracies is to exploit their weaknesses. Intelligence services have always played upon the faults of their targets—to blackmail, manipulate, or confuse them. Now they have the perfect tools to do so, aimed at juicy targets eager to fight over fault lines, with plausible deniability easier than ever. For Western democracies, particularly the United States, this is annihilation from within.

If the United States and its democratic allies are to survive, they will need to be more alert to the challenges of digital technologies that are threatening fundamental norms of democratic governance internally. In the United States, that means Americans must return to the U.S. Constitution, continuing the long, often fractious conversation that established the country more than two centuries ago and gave it a touchpoint for all to reference.[5] When the US Constitution was written, only white male landowners could vote; but ten Constitutional amendments arrived in the first six months, seventeen more in the next 203 years. Slowly the document has adapted and evolved, interpretations updated or extended as the historical context has changed.

Today Americans must continue that conversation in a new technological context, not with weapons but with words. What exactly does it mean now to defend the Constitution from all enemies, foreign and domestic, when they're attacking through digital channels, and it is hard to know who and where they are? How should the legal structure adapt to technological tests of what it means to have freedom of speech, freedom of the press or freedom from search and seizure? The answers are not just for Constitutional scholars but for ordinary Americans—including the misguided "soldiers" who stormed the Capitol on January 6, 2021, forgetting that members of the American military swear not to protect the president (*any* president) but the Constitution. Americans do not have to agree on everything; but if we do not reaffirm the principles that unite us, our quarrels will tear us apart.

Meanwhile democracy's enemies, from individual criminals to state-supported syndicates, are launching ransomware attacks against infrastructure, crippling private industries, undermining government operations, and echoing the damage of conventional bombing eighty years ago—but with secrecy, impunity, and the ability to siphon off data now. Why fight the United States or its allies directly? No armed drone, Patriot missile, F-35 or nuclear weapon can answer this attack.

Our institutions and economy are at risk. Yet the chasm between technology and strategy is wider than ever. To respond, the United States and its allies must confront the role of disinformation, shore up our defenses against internal and external attack, and educate our populations about the risks of digital technologies—then adapt our legal structures to deal with them. All these remedies have worked before and can work now, as detailed in this book. Securing the future of democracy requires not just countering overt efforts by autocracies such as China, Russia, North Korea, and Iran to harm us—indeed, that's the easy part. Competing with China on artificial intelligence or quantum computing makes no difference if democratic institutions lose their integrity and democratic publics no longer support their own governments.

External and internal security are seamlessly linked through the global internet. One of the lessons of the Cold War is that when the United States and its allies misdiagnose threats that are militarily weaker than we are yet more focused and determined, we lose wars. Right now, we are losing a war of attrition at home. To restore real power to the people, we must do what has worked in the past: unify around core democratic values, regulate risks to secure order and preserve liberty, and lower our vulnerability to the darker sides of new technologies.

| Methodology

Power to the People Political Violence Innovation Database (P2P-PVID)

The Power to the People Terrorist Innovation Database (P2P-PVID) has been compiled by Audrey Kurth Cronin and a team of graduate research assistants from George Mason University and American University for the purpose of analyzing how non-state groups and individuals over time have been empowered by "new" accessible innovative technologies, including dynamite, Kalashnikov assault rifles, drones, and social media to organize, plan, and execute political violence. This methodology appendix describes the P2P-PVID collection and coding rules including definitions and inclusion criteria, primary and secondary data sources, data collection process, as well as description of variables and possible values.

The P2P-PVID currently consists of three data sets looking at:

1) The manufacture, adoption, and use of dynamite for political violence during the "First Wave" of Modern Terrorism (1867 to 1934);
2) The manufacture, dissemination, and use of AK-47 and Kalashnikov variants for political violence (1947 to present);
3) The manufacture, dissemination, and use of drones and other autonomous systems for political violence.

All data will be made available at audreykurthcronin.com, as soon as feasible following the publication of the book. A description of the methodology for collecting and coding information for each data set follows.

Data Set on Manufacture, Adoption, and Use of Dynamite for Political Violence from 1867 to 1934 (P2P-Dynamite-PVID)

The P2P-Dynamite-PVID consists of five separate data subsets covering the period 1867 (the year dynamite was first patented by Alfred Nobel) through 1934 (a year when anarchist acts became blurred with the rise of fascism and the prelude to the Spanish Civil War). The first data subset consists of "anarchist bombing incidents" (defined below) that occurred throughout the world during the "First Wave" of modern terrorism. The second data subset contains the locations and founding dates of factories that manufactured dynamite and other high explosives during this period. The third data subset includes historical data on dynamite pricing in the United States and Europe. The fourth data subset contains the name, location, and circulation information for anarchist newspapers and other anarchist periodicals published and distributed during this time. The fifth data subset then reviews the dates and countries where anti-terrorist and explosive control laws were enacted.

ANARCHIST BOMBING INCIDENT DATA SET (1867–1934)

For the purposes of this research project, the term "anarchist" is broadly defined to include those individuals and groups that rebel against authority and government, and often use violent means (propaganda of the deed) to try to overthrow the existing order. Under this broad rubric, anarchists include self-identified individuals, as well as Irish nationalists, nihilists, insurgents, revolutionaries, insurrectionists, and social/political agitators who used violence in an attempt to implement political and social change during the late nineteenth and early twentieth centuries.

An "anarchist bombing incident" is an event, as identified in a primary or secondary source, where an anarchist (as defined above) commits an explosive act (e.g., attack with a bomb) or engages in an activity with explosives that is thwarted or otherwise fails. Included in this latter category are bombs that were discovered but fail to explode, bombing plots that were discovered before they were implemented, and anarchists who accidently blew themselves up. Non-anarchist bombings, such as those associated with purely criminal activities or domestic disputes, thus not politically motivated, are not included. In a few cases, "bombings" originally attributed to anarchists were later found to be accidents (e.g., natural gas or munitions explosions), and removed from the data set.

Digitized historical newspaper articles were the primary source of anarchist bombing incident data. Primary newspaper sources included historic editions of the *New York Times* and *Times* of London, and over eighty additional English-language newspapers from the United States, Great Britain, Australia, and Ireland. In addition, many incidents were identified from French-, Spanish-, Italian-, Portuguese-, and Russian-language primary newspaper sources. We made every possible effort to be comprehensive; however, available newspaper databases in other languages varied in their scope and quality. Parliamentary documents, especially from the annual report of Her Majesty's Inspectors of Explosives, provide details on incidents primarily in the United Kingdom and Ireland, but also some foreign "outrages" in other countries. Secondary sources of incident data included scholarly books and articles by noted anarchist historians including Paul Avrich, Barbara Gage, Anna Geifman, Richard Bach Jensen; and Irish nationalist historians Joseph McKenna and Niall Whelehan. Additional Italian anarchist bombings incidents were also provided via a French-language anarchist database "Chronologie de l'anarchisme et des mouvements et activités utopiques et libertaires italiens." Where possible, secondary source data were verified using primary source historic newspaper articles.

Digitized primary and secondary sources were searched for bombing incidents using Boolean terms including "Terrorist AND Dynamite," "Terrorist AND Bomb," "Anarchist AND Dynamite," "Anarchist AND Bomb," and "Anarchist AND Explosion." As of 1 September 2018 a total of 1,291 unique anarchist bombing incidents were identified and recorded in an Excel spreadsheet. Variables collected for all incidents included incident date, incident location (city and country), target type, perpetrator/group name, explosive used (if known), and the number of casualties including both wounded and killed. In all incidents, the reference source was recorded and an image of the reference was stored for future review. In cases where four or more casualties occurred, two or more source articles were collected and reviewed confirming the incident's authenticity and casualty values. In some cases, news reports varied in the actual number of casualties, so we reported a range of casualties for both wounded and killed. All incidents were then geocoded using latitude and longitude so that they could be plotted and analyzed for geographic patterns in relation to the locations of explosive factories and anarchist publication offices.

DYNAMITE/HIGH EXPLOSIVE FACTORY DATA SET (1867–1934)

Following the invention of dynamite in 1867, Alfred Nobel quickly secured patents in Sweden, France, Great Britain, and the United States, and established dynamite factories in twelve European countries including Norway, Sweden, Finland, France, Great Britain, Austria-Hungary, Germany, Poland, Switzerland, Italy, Spain, and Portugal. The locations of the European factories are documented on the Nobelprize.org website[1] and in a biography of Alfred Nobel by Kenne Fant.[2] Nobel also set up factories in the United States, and negotiated deals with American chemical manufacturers, most notably the E. I. du Pont de Nemours Company, to control the manufacturing and sale of dynamite in North America under what became known as the Powder Trust. The best record of US dynamite and high-explosive factory locations was *The Report of the Chief Inspector of the Bureau of Safe Transportation of Explosives and Other Dangerous Materials* published in February 1908.[3] The report, published on the eve of the breakup of the Powder Trust, provides the factory name, ownership, location, and explosive products produced at every plant in the United States and Canada. Many of these locations are confirmed in a subsequent scholarly article on the breakup of the Powder Trust by William Stevens in the *Quarterly Journal of Economics* published in 1912. Additional explosive factory locations and initial production dates were also identified using other secondary scholarly articles and information from explosive manufacturer websites. All factory locations were then geocoded using latitude and longitude so that they could be plotted and analyzed for geographic patterns in relation to the anarchist bombing incidents. Through 1 September 2018 a total of 161 factories producing dynamite and other high explosives in over twenty countries between 1867 and 1934 were identified and plotted. In a few instances, factories ceased explosive manufacturing due to closure or industrial accident.

HISTORICAL DYNAMITE PRICING DATA SET (1867–1934)

For much of the First Wave of Modern Terrorism period, the manufacture and sale of dynamite was controlled by two cartels—the Powder Trust in the Western Hemisphere, and the Nobel Trust in Europe. The purpose of these trusts was to control the price of dynamite and eliminate competition within each trust's sales domain. In the United States, in order to shroud the Powder Trust's anti-competitive activities, pricing decisions were made in closed board cartel meetings. No printed price catalogs appeared until

1907—near the time when judicial proceedings began until the eventual breakup of the Powder Trust in 1912. Fortunately, the Hagley Museum and Library in Wilmington, Delaware, has board meeting minutes of the Gunpowder Trade Association and DuPont Companies with dynamite pricing information during the Powder Trust period in both the United States and Europe. The museum also houses a collection of E. I. du Pont de Nemours price lists beginning with Price List No. 1 (1 December 1907),[4] with the price per pound and transport costs of various grades and brands of dynamite delivered in various DuPont sales territories throughout the United States. The data set contains general dynamite pricing (adjusted for inflation) for various periods between 1867 and 1906, and specific detailed DuPont dynamite pricing (adjusted for inflation) in the United States after 1906. Since nearly 70 percent of the P2P-Dynamite-PVID bombing incidents in the United States happened after 1906, the data set also allows analysis of the relationship between the price of dynamite and bombing incidents at various geographic locations around the United States for the period 1907 to 1934.

ANARCHIST NEWSPAPER AND OTHER PERIODICAL PUBLICATION DATA SET (1867–TO 1934)

Anarchist newspapers and other publications were the nineteenth-century equivalent of the Internet websites of today. Through these publications, anarchists were able to disseminate their ideology and methods, including information on how to manufacture and detonate explosive devices, and carry out anarchist bombing attacks. The primary sources of anarchist publications, their publishing location, circulation, and publication dates can be found on a number of specialty academic websites including the *Anarchy Archives, Mapping American Social Movements: Anarchist Newspapers and Periodicals 1872–1940* (University of Washington), and the *Bibliothek der Freien* (Germany). As of 1 September 2018 a total of 644 anarchist publications produced between 1867 and 1934 were identified and documented. Variables collected and coded include publication name and type, production location (city and country), dates of publication, language of publication, and when available, circulation numbers and publication interval (e.g., daily, weekly, monthly, etc.). All anarchist publication locations were then geocoded using latitude and longitude so that they could be plotted and analyzed for geographic patterns in relation to the anarchist bombing incidents.

The final data set contains the country, provisions, and enactment date of anti-anarchist and explosive control legislation passed and implemented during the period, along with country and dates for other key legal events such as the filing of patents, formation of explosive companies and legal trusts, arrest and prosecution of key bombing suspects, and the signing of anarchist-related international agreements. This data was gathered from historical newspaper articles primarily from the *New York Times* and the *Times* (London), as well as references from key scholarly secondary sources. These data are used as key markers to examine the before and after impact that a range of laws and events had on anarchist bombing.

Data Set on the Manufacture, Adoption, and Use of Kalashnikov Rifles for Political Violence from 1947 to Present (P2P-Kalashnikov-PVID)

The P2P-Kalashnikov-PVID consists of three separate data subsets covering the period 1947 (the year the AK-47 was invented by Mikhail Kalashnikov and first manufactured by the Soviet Union) through the present day. The first data subset consists of the locations and founding dates of factories that manufactured Kalashnikov rifles and variants (as defined below) during this period. The second data subset lists the countries whose militaries have used the Kalashnikov and variants as a primary weapon at some point during this period. The third data subset is a chronology of conflicts where Kalashnikovs and variants were a primary weapon for one or more conflict combatants.

KALASHNIKOV AND VARIANT FACTORY DATA SET (1947 TO PRESENT)

Following Mikhail Kalashnikov's invention of the assault rifle that became known as the AK-47, the Soviet Union quickly established manufacturing facilities at a factory location in Izhevsk. During the 1950s, the USSR began to export the technology and license the gun's manufacture to Soviet Warsaw Pact allies, and key Communist proxies and trading partners including China, North Korea, and Egypt. During the 1960s and beyond, the technology further disseminated via additional authorized licensing agreements, and unlicensed copying of the design and unauthorized manufacture by other countries including Finland, Israel, India, Pakistan, and even the United States. Along the way, refinements were made to the initial AK-47 design by the Soviet Union, its authorized agents, and the unlicensed manufacturers, resulting in a wide variety of Kalashnikov variants. In this project, variants encompass Soviet models including the AK-47, AKM, AK-74, and AK-103; Chinese Type-56, Type-63, Type-68, and Type-81; North Korean Type-58 and Type-88; and variants made in East Germany (AKS, MPi-K), Hungary (AK-55, AKM-63, NGM-81), Poland (PMK, M-88), Romania (PM md. 63/65, PM md. 80, PM md. 90), Bulgaria (AKK, AKS), Albania (AKMS), Finland (RK-62, Valmet M76, RK-95), Israel (IMI Galil), India (INSAS), Iraq (Tabuk), Pakistan (PK-10), Ethiopia (Gafat Et-97/1), Iran (KLS, KLF, KLT), Sudan (MAZ), Nigeria (OBJ-006), Croatia (APS-95), Kosovo (AK Sopmod), Armenia (K-3), Vietnam (AKM-1, AKM-VN, TUL-1), and Serbia (M64, M70, M72, M76, M77, M80, M82, M85, M90, M91, M92, M99, M21)—all told, over fifty variants in nearly forty countries.

The locations of factories that manufactured the Kalashnikov and Kalashnikov variants came from a number of books and online sources, as well as a visit to the Kalashnikov museum in Izhevsk, Russia (including an interview of Jan Ismailovich Landau, museum historian). Books included *The Gun* by C. J. Chivers, *Rifles of the World,* 3rd ed. by John Walter, and *The AK47 Story: Evolution of the Kalashnikov Weapons and Small Arms of the World* by Edward Ezell. Online sources included *Kalashnikov (AK 47) Factory and Manufacturer Location* website by theakforum.net, aftermathgunclub.com, revolvy.com, and globalsecurity.org. As of 1 September 2018 a total of sixty-two factories in thirty-six countries were identified and recorded in an Excel spreadsheet. Data collected included factory name, location (city, country), year founded, types of variants manufactured, and estimated number of rifles manufactured (if known). All Kalashnikov factory locations were then geocoded using latitude and longitude so that they could be plotted and analyzed for geographic patterns.

COUNTRIES WHOSE MILITARIES USE KALASHNIKOV RIFLES AND VARIANTS DATA SET (1947 TO PRESENT)

There are over 130 countries whose militaries have adopted Kalashnikovs or Kalashnikov variants over the past seventy years. Information on national military Kalashnikov usage comes primarily from the *Small Arms Survey* (Oslo, Norway), *Jane's Directory of Military Small Arms Ammunition, Military*

Small Arms of the 20th Century, 7th ed. by Ian Hogg, and *Tools of Modern Terror: How the AK-47 and AR-15 Evolved into Rifles of Choice for Mass Shooting* by C. J. Chivers (*New York Times*, 2016). Data collected included country of military usage and type of Kalashnikovs used.

CHRONOLOGY OF KALASHNIKOV CONFLICTS DATA SET (1947 TO PRESENT)

This dataset chronicles over seventy-five different conflicts where the Kalashnikov or a variant was the primary weapon of one or more of the combatants. Data sources included *Kalashnikov Culture: Small Arms Proliferation and Irregular Warfare* by Christopher Carr, *The Gun* by C. J. Chivers, *The AK-47 and AK74 Kalashnikov Rifles and Their Variations* by Joe Poyer, and *The AK-47: Kalashnikov-series Assault Rifles (Weapon)* by Gordon Rottman. Data collected included name of conflict, countr(ies) of usage, combatants, years of conflict, Kalashnikov variants used (if known), and Kalashnikov country of origin/manufacture (if known). Oftentimes, Kalashnikovs manufactured in one country were bought and traded in key Kalashnikov marketplaces such as Bakaaraha (Somalia), Cox's Bazaar (Bangladesh), Darra (Pakistan), Souks (Yemen), and the Iguazu Triangle (South America). Via legal and illegal sales and trades, Kalashnikovs have migrated from conflict to conflict.

Data Set on Manufacture, Adoption, and Use of Drones (P2P-Drones-PVID)

The P2P-Drones-PVID data set organizes military UAVs or "drones" into five categories based on size (takeoff weight) and ceiling (maximum altitude). (See Figure 8.1.) The five categories are 1) micro (≤ 20 lbs. takeoff weight), 2) small (or mini) (21–55 lbs. takeoff weight), 3) tactical (56–1,320 lbs. takeoff weight and <18,000 feet ceiling), 4) medium altitude (1,320–18,000 lbs. takeoff weight and 15,000–18,000 feet ceiling), and 5) high altitude (1,320–32,000 lbs. takeoff weight and 18,000–65,000 feet ceiling). The current data set consists of over 150 drone models manufactured and operated by the militaries in over forty countries.

Data collected for each drone model include drone name/model, country of origin, company/manufacturer, takeoff weight (approx. lbs.), ceiling (approx. feet), speed (approx. knots), payload (approx. lbs.), mission (e.g., ISR, strike, transport), and price (if known) in US dollars. Data was collected from a number of online sources including the Center for a New American Security (CNAS) Drone Database (http://drones.cnas.org/drones/), Center for the Study of the Drone at Bard College (https://dronecenter. bard.edu), Department of Defense, *Unmanned Systems Integrated Roadmap, FY 2013–2038* (among other DoD sites), and corporate websites for military drone manufacturers. Classification and additional drone specifications were also collected from scholarly publications by Michael Horowitz, Matthew Fuhrmann, and Peter Singer; and the *Springer Handbook of Unmanned Aerial Vehicles* (2015), edited by Kimon Valavanis and George Vachtsevanos.

NOTES

Introduction

1. Eric Schmitt, "Pentagon Tests Lasers and Nets to Combat a Vexing Foe: ISIS Drones," *New York Times*, 23 September 2017; and Don Rassler, *The Islamic State and Drones: Supply, Scale, and Future Threats*, Combating Terrorism Center at West Point, US Military Academy, July 2018, pp. 1–5.

2. See, for example, Brian Jenkins, *High Technology Terrorism and Surrogate War: The Impact of New Technology on Low-level Violence*, RAND Report #P-5339 (January 1975); Paul Wilkinson, ed., *Technology and Terrorism* (London: Frank Cass, 1993); Brian A. Jackson, "Technology Acquisition by Terrorist Groups: Threat Assessment Informed by Lessons from Private Sector Technology Adoption," *Studies in Conflict & Terrorism* 24 (2001): 183–213; Adam Dolnik, *Understanding Terrorist Innovation: Technology, Tactics, and Global Trends* (London: Routledge, 2007); and Brian A. Jackson and David R. Frelinger, "Rifling through the Terrorists' Arsenal: Exploring Groups' Weapon Choices and Technology Strategies," *Studies in Conflict and Terrorism* 31 (2008): 183–213.

3. Elsa B. Kania and John Costello, *Quantum Hegemony? China's Ambitions and the Challenge to US Innovation Leadership*, Center for a New American Security, 12 September 2018, at https://www.cnas.org/publications/reports/quantum-hegemony.

4. Global Risks Report 2017, World Economic Forum, 12th ed., at http://reports.weforum.org/global-risks-2017/.

5. Committee on Forecasting Future Disruptive Technologies and National Research Council, *Persistent Forecasting of Disruptive Technologies* (Washington, DC: National Academies Press, 2010), xv. As the committee noted, "The term was first coined by Bower and Christensen in 1995 to refer to a type of

technology that brings about a sudden change to established technologies *and markets.*" [Emphasis added.] To understand whether or not something is "disruptive," the technology cannot be separated from how it is taken up and used by humans (i.e., the market). The original definition of disruption contrasts to that used in military innovation studies. For example, Terry Pierce "adapts the Christensen model for military use" yet describes disruptive innovations as "those that resulted in an improved performance along a warfighting trajectory that traditionally had not been valued." Terry C. Pierce, *Warfighting and Disruptive Technologies: Disguising Innovation* (London: Frank Cass, 2004), 25. That is a perfect definition of a sustaining technology or innovation.

6. For example, J. F. C. Fuller wrote in 1919, "Tools, or weapons, if only the right ones can be discovered, form 99 per cent. of victory." "The Secret of Victory," Weekly Tank Notes, January 25, 1919; cited by J. F. C Fuller, *Armament & History* (New York: Da Capo Press, 1998), 31.

7. Joseph L. Bower and Clayton M. Christensen, "Disruptive Technologies: Catching the Wave," *Harvard Business Review* 73, no. 1 (January–February 1995): 43–53.

8. The concept was fully developed in Clayton M. Christensen, *The Innovator's Dilemma* (New York: Harper Business, 2011; original edition Harvard Business School, 1997).

9. Ibid., xviii.

10. See, for example: Jill Lepore, "The Disruption Machine: What the Gospel of Innovation Gets Wrong," *New Yorker*, 23 June 2014; Erwin Danneels, "Disruptive Technology Reconsidered: A Critique and Research Agenda," *Journal of Product Innovation Management* 21, no. 4 (2004): 246–58; *Journal of Product Innovation Management* 23, no. 1 (2006) (entire issue is devoted to Christensen's theory); and Andrew A. King and Baljir Baatartogtokh, "How Useful Is the Theory of Disruptive Innovation?," *MIT Sloan Management Review* 57, no. 1 (fall 2015): 77–90. See also Peter Dombrowski and Eugene Gholz, "Identifying Disruptive Innovation: Innovation Theory and the Defense Industry," *Innovations* (Spring 2009): 101–17.

11. Benz received a patent on 29 January 1886. Working nearby, Gottleib Daimler also deserves credit for the internal combustion engine, having designed the first motorcycle in 1885.

12. The model T initially came in numerous colors but Ford famously switched to black paint in 1914 so as to save money. Later models also had doors. The first mass-produced automobile was actually the Curved Dash Oldsmobile designed by Ransom E. Olds. Robert W. Domm, *Michigan Yesterday and Today* (Minneapolis: Voyageur Press), 29; and Michael Rodriquez, *R. E. Olds and Industrial Lansing* (Chicago: Arcadia, 2004). However, Henry Ford perfected the assembly line and excelled in marketing.

13. Arthur W. Einstein Jr., *"Ask the Man Who Owns One": An Illustrated History of Packard Advertising* (Jefferson, NC: MacFarland, 2016), 52.

14. S. M. Bowden, "Demand and Supply Constraints in the Inter-War UK Car Industry: Did the Manufacturers Get It Right?" *Business History* 33, no. 2 (April 1991): 246.

15. Christensen (2011), 143–7.

16. Robert Spousta and Steve Chan, "Hold the Drones: Fostering the Development of Big Data Paradigms through Regulatory Frameworks," *Journal of Communication and Computing* 12 (2015): 137.

17. The concept of a "pro-sumer" comes from Alvin Toffler, *The Third Wave* (New York: William Morrow, 1980).

18. For example, Erik Brynjolfsson and Andrew McAfee, *The Second Machine Age: Work, Progress, and Prosperity in a Time of Brilliant Technologies* (New York: W. W. Norton, 2014); Klaus Schwab, *The Fourth Industrial Revolution* (Geneva: World Economic Forum, 2016); and Zeynep Tufekci, *Twitter and Tear Gas: The Power and Fragility of Networked Protest* (New Haven, CT: Yale University Press, 2017).

19. For much more on the strategies of terrorism see my *Ending Terrorism: Lessons for Defeating al-Qaeda*, International Institute for Strategic Studies, Adelphi Paper #394 (London: Routledge, 2008).

Chapter 1

1. "Einstein at 70," *Liberal Judaism* 16 (April–May 1949): 12. From Einstein Archive 30–1104, as sourced by Alice Calaprice, "On Peace, War, the Bomb, and the Military," *The New Quotable Einstein* (Princeton, NJ: Princeton University Press, 2005), 173.

2. As far as I know, the first person to talk about "open" and "closed" technological revolutions was James Moor, who applied the concept to major commercial technologies such as computers, automobiles, or electricity that have a major impact upon how a society functions. I am gratefully indebted to him. My framework is related to his but a little different, in that it considers only technological revolutions that affect state military power. See James H. Moor, "Why We Need Better Ethics for Emerging Technologies," *Ethics and Information Technology* (2005) 7: 111–19.

3. Daniel R. Headrick, *Power over Peoples: Technology, Environments, and Western Imperialism, 1400 to the Present* (Princeton, NJ: Princeton University Press, 2010), especially chapter 7: "Weapons and Colonial Wars, 1830–1914," 257–301. According to Headrick, European and European American advances in the late nineteenth century were the result not merely of firearms, but also of the range of means that industrialization put at their disposal, including steamboats and medical advances.

4. Winston Spencer Churchill, *The River War: An Account of the Reconquest of the Sudan*, 2nd ed. (New York and London: Longmans, Green and Co., 1902; New York: Dover Publications, republished 2006), 300.

5. Max Weber, "Politics as a Vocation"[*Politik als Beruf*], lecture given in Munich, 28 January 1919; in *Essays in Sociology*, translated and edited by Howard Garth and Cynthia Wright Mills (New York: Free Press, 1946), 26–45. Others have developed more elaborate arguments about the relationship between war and state building, most notably Charles Tilly.

6. Fred Kaplan, *The Wizards of Armageddon* (New York: Touchstone Books, 1983), 9–10.

7. Bernard Brodie, ed., *The Absolute Weapon: Atomic Power and World Order* (New York: Harcourt, Brace and Company, 1946). See also Lawrence Freedman, *The Evolution of Nuclear Strategy*, 3rd ed. (New York: Palgrave Macmillan, 2003); and Fred Kaplan, *The Wizards of Armageddon* (New York: Touchstone Books, 1983).

8. Bernard Brodie and Fawn M. Brodie, *From Crossbow to H-Bomb: The Evolution of the Weapons and Tactics of Warfare* (Bloomington: Indiana University Press, 1962; rev. and enl. edition, 1973), 7.

9. The story is related in Herodotus's *Histories*, although many details remain ambiguous.

10. Brodie and Brodie (1973), 8.

11. Ibid., 261–2.

12. Ibid., 280ff.

13. Lynn White Jr., *Medieval Technology and Social Change* (Oxford, UK: Oxford University Press, 1962). D. J. A. Ross, H. Brunner, and Stanislav Andreski also argued the stirrup was essential. The thesis has been disputed by historians of the feudal era, however, including J. R. Strayer, Philippe Contamine, and John Beeler. See Robert L. O'Connell, *Of Arms and Men: A History of War, Weapons, and Aggression* (Oxford, UK: Oxford University Press, 1989), 322, notes 12 and 13.

14. Clifford J. Rogers, "The Military Revolutions of the Hundred Years War," *Journal of Military History* 57 (1993): 241–78. This thesis is contested by medieval historian Kelly DeVries, *Medieval Military Technology* (Peterborough, Ontario, 1994); and "Catapults Are Not Atom Bombs: Towards a Redefinition of 'Effectiveness' in Premodern Military Technology," *War in History* 4 (1997): 454–70; and Rogers responds in Clifford J. Rogers, "The Efficacy of the English Longbow: A Reply to Kelly DeVries," *War in History* 5, no. 2 (1998): 233–42; and Clifford J. Rogers, "The Development of the Longbow in Late Medieval England and 'Technological Determinism'," *Journal of Medieval History* 37 (2011): 321–41.

15. Another important argument was offered by Caro M. Cipolla in his 1985 book *Guns, Sails, and Empires*, in which he demonstrated how, beginning in the fifteenth century, the European powers put massive firepower on ocean-going ships, built global commercial empires, and soon ruled the world. Caro M. Cipolla, *Guns, Sails, and Empires: Technology Innovation and the Early Phases of European Expansion, 1400–1700* (Manhattan, KS: Sunflower University Press, 1985).

16. Michael Roberts, *The Military Revolution, 1560–1660*, inaugural lecture at Queen's University, Belfast, 1956; reprinted in *Essays in Swedish History* (London, 1967). Roberts argued that the period 1550–1650 witnessed a military revolution in warfare in Europe that was crucial to state formation. His original conception was not about technology at all, but about changes in tactics, strategy, scale, and social impact. He was followed by Geoffrey Parker, who expanded on the thesis and focused on military innovation. Geoffrey Parker, "The 'Military Revolution,' 1560–1660—A Myth?," *Journal of Modern History* 48, no. 2 (1976): 195–214; and Geoffrey Parker, *The Military Revolution: Military Innovation and the Rise of the West, 1500–1800* (Cambridge, UK: Cambridge University Press, 1988), 43. An influential set of arguments about the "Revolution in Military Affairs" predominated in the United States at the end of the century. See, inter alia, Andrew Krepinevich, "Cavalry to Computer: The Pattern of Military Revolutions," *National Interest* 37 (fall 1994): 30–42; and Clifford J. Rogers, "The Military Revolutions of the Hundred Years' War," *Journal of Military History* 57 (April 1993): 241–73.

17. See, for example, David Parrott, *The Business of War: Military Enterprise and Military Revolution in Early Modern Europe* (Cambridge, UK: Cambridge University Press, 2012).

18. In 2016, Duke University professor Alex Roland summed up the core thesis, "From the Stone Age to the Nuclear Age, Technology Has Driven the Evolution of Warfare." Alex Roland, *War and Technology: A Very Short Introduction* (Oxford, UK: Oxford University Press, 2016). See also Alex Roland, "Technology and War: The Historiographical Revolution of the 1980s," *Technology and Culture* 34, no. 1 (January 1993): 117–34.

19. Martin Van Creveld, *Technology and War: From 2000 B.C. to the Present* (New York: Free Press, 1989), 1.

20. For criticisms of too great a focus on technology at the expense of strategy, see Michael Howard, "The Forgotten Dimensions of Strategy," *Foreign Affairs* 57, no. 5 (summer 1979): 975–86 and "How Much Can Technology Change Warfare?" in *Two Historians in Technology and War*, Strategic Studies Institute, US Army War College, Carlisle, Pennsylvania, 1994; Hew Strachan, *The Direction of War: Contemporary Strategy in Historical Perspective* (New York: Cambridge University Press, 2013); Richard Betts, "Should Strategic Studies Survive?," *World Politics* 50, no. 1 (October 1997); Colin Gray, *Modern Strategy* (Oxford, UK: Oxford University Press, 2000); Beatrice Heuser, *The Evolution of Strategy: Thinking War from Antiquity to the Present* (Cambridge, UK: Cambridge University Press, 2010); and Lawrence Freedman, *The Revolution in Strategic Affairs*, The Adelphi Papers, No. 318 (1998), International Institute for Strategic Studies (London).

21. Michael Howard, *War in European History* (Oxford, UK: Oxford University Press, 1976), 135.

22. Military historian Dennis Showalter led the way in this with his brilliant 1975 book about the importance of railways to modern war. Dennis Showalter, *Railroads and Rifles: Soldiers, Technology, and the Unification of Germany* (Hamden, CT: Archon Books, 1975).

23. Foundational works include Max Weber's 1922 seminal work on bureaucracy, later expanded upon by James Q. Wilson and many others. Max Weber, *Economy and Society* (Berkeley: University of California Press, 1978); Frederick Winslow Taylor, *The Principles of Scientific Management* (New York: Dover Publications, 1997); and James Q. Wilson, *Bureaucracy* (New York: Basic Books, 1991).

24. Williamson R. Murray and Allan R. Millett, eds., *Military Innovation in the Interwar Period* (Cambridge, UK: Cambridge University Press, 1998). See also "Preserving Primacy: A Defense Strategy for the New Administration," *Foreign Affairs.com*, 3 August 2016.

25. The British called it "Radio Direction Finding" or "R.D.F.," later known as radar. Winston Churchill, *The Gathering Storm*, Vol. 1 of *The Second World War* (New York: Houghton Mifflin Company, 1948; RosettaBooks edition, 2002), 138, 140.

26. Stephen Peter Rosen, *Winning the Next War: Innovation and the Modern Military* (Ithaca, NY: Cornell University Press, 1991), 136.

27. Ibid., chapter 5: "New Blood for the Submarine Force," 130–47.

28. A select number of serving US military officers go to civilian universities to earn doctorates. Recent prominent examples include: David Petraeus, PhD, International Relations and Economics, Princeton University; H. R. McMaster, PhD, History, University of North Carolina at Chapel Hill; John Nagl, DPhil, International Relations, Oxford University (Rhodes Scholar); Suzanne Nielsen, PhD, Harvard University, chair of the Social Sciences department, West Point; and Isaiah Wilson, PhD, Government, Cornell University.

29. Samuel Huntington spearheaded this analysis at Harvard beginning in the late 1960s, including work by his students Vincent Davis, *The Politics of Innovation: Patterns in Navy Cases* (Denver, CO: University of Denver Press, 1967); and Ronald J. Kurth, *The Politics of Innovation in the Navy*, unpublished Harvard dissertation, 1970. Later examples of Intra-Service analysis include Gregory A. Engel, "Cruise Missiles and the Tomahawk," in B. Hayes and D. Smith, eds., *The Politics of Naval Innovation* (Newport, RI: US Naval War College, 1994), 18–22; W. Blair Haworth, *The Bradley and How It Got That Way: Technology, Institutions, and the Problem of Mechanized Infantry in the United States Army* (Westport, CT: Greenwood Press, 1999); David E. Johnson, *Fast Tanks and Heavy Bombers: Innovation in the US Army 1917–1945* (Ithaca, NY: Cornell University Press, 1998); Suzanne Nielsen, "Preparing for War: The Dynamics of Peacetime Military Reform," Harvard dissertation, 2003; and *An Army Transformed: The US Army's Post-Vietnam Recovery and the Dynamics of Change in Military Organizations* (Carlisle, PA: US Army War College, 2010); and John Nagl, *Learning to Eat Soup with a Knife* (Chicago: University of Chicago Press, 2002).

30. Charlie Miller, *Serving Two Masters: Doctrinal Evolution in the 20th Century US Army*, unpublished dissertation, Columbia University, 2002; Benjamin M. Jensen, *Forging the Sword: Doctrinal Change in the US Army* (Stanford, CA: Stanford University Press, 2016); Andrew A. Gallo, *Understanding Military Doctrinal Change during Peacetime*, unpublished dissertation, Columbia University, 2018.

31. A prominent early study of an inter-Service rivalry was political scientist Harvey Sapolski's 1972 book *Polaris System Development*, the story of the 1950s fight among the US Navy, Army, and Air Force over Intercontinental Ballistic Missiles. Originally the Navy was part of the Army's Jupiter medium-range ballistic missile program; but, fearing a loss of mission, the Navy began to develop its own solid-fueled Polaris missile program. The obstacles were not just other Services: Sapolsky detailed resistance within the traditional Navy, among both surface fleet admirals and the scrappy submarine force. To outflank them, Admiral Arleigh Burke, chief of naval operations, and Charles Thomas, secretary of the Navy, created the Navy's Special Projects Office, gave it resources and talented officers, and had it report directly to them. In 1960, the Polaris missile was the first nuclear-armed, submarine-launched ballistic missile deployed in the US Navy fleet. In an ironic twist, Secretary of Defense Charles Wilson assigned the Jupiter (which still had a squat shape so as to fit into submarines) to the Air Force in 1956, limiting the Army to short-range surface-to-surface missiles. But the Air Force still preferred its own Thor missile to the shorter-range, more accurate Jupiter missile. The Jupiters were eventually deployed nonetheless, sent to Italy and Turkey, though they were removed under secret agreement with the Soviet Union during the Cuban Missile Crisis a year later. See Harvey Sapolsky, *Polaris System Development: Bureaucratic and Programmatic Success in Government* (Cambridge, MA: Harvard University Press, 1972). Other examples of inter-Service analysis include Michael H. Armacost, *The Politics of Weapons Innovation: The Thor-Jupiter Controversy* (New York: Columbia University Press, 1969); Andrew J. Bacevich, *The Pentomic Era: The US Army between Korea and Vietnam* (Washington, DC: National Defense University Press, 1986); Owen R. Cote, "The Politics of Innovative Military Doctrine: The US Navy and Fleet Ballistic Missiles" (Cambridge, MA: PhD dissertation, MIT 1998); and Frederic A. Bergerson, *The Army Gets an Air Force* (Baltimore: Johns Hopkins University Press, 1980).

32. Examples include Barry Posen, *The Sources of Military Doctrine: France, Britain, and Germany between the World Wars* (Ithaca, NY: Cornell University Press, 1984); Kimberly Marten Zisk, *Engaging the Enemy: Organization Theory and Soviet Military Innovation, 1955–1991* (Princeton, NJ: Princeton University Press, 1993); and Deborah Avant, *Political Institutions and Military Change: Lessons from Peripheral Wars* (Ithaca, NY: Cornell University Press, 1994).

33. Examples include Elizabeth Kier, *Imagining War: French and British Military Doctrine between the Wars* (Princeton, NJ: Princeton University Press, 1997); Theo

Farrell and Terry Terriff, eds., *The Sources of Military Change: Culture, Politics, Technology* (London: Lynne Rienner, 2002); Theo Farrell, "Culture and Military Power," *Review of International Studies* 24, no. 3 (July 1998): 407–16; Dima Adamsky, *The Culture of Military Innovation: The Impact of Cultural Factors on the Revolution in Military Affairs in Russia, the US, and Israel* (Stanford, CA: Stanford University Press, 2010); and Thomas G. Mahnken, *Technology and the American Way of War since 1945* (New York: Columbia University Press, 2008).

34. Some important work argued for a widening focus. In 2006, for example, the RAND Corporation's Adam Grissom surveyed the field and highlighted less well-known examples of bottom-up technological innovation in the ranks. These included the study of junior and mid-ranking US Marines in Haiti, Dominican Republic, Nicaragua from 1915 to 1940, whose experiences led to the development of Small Wars doctrine; analysis of assault tactics developed by German storm troopers in World War I, which Captain Willy Rohr shared with other units in a one-week training course beginning in December 1915; and an examination of Nazi soldiers' repurposing of the 88 mm Flak anti-aircraft cannon, which was intended for air defense but instead turned very effectively against tanks and troop tactics. Adam Grissom, "The Future of Military Innovation Studies," *Journal of Strategic Studies* 29, no. 5 (October 2006): 905–34. Grissom in turn cites the following sources: Keith B. Bickel, *Mars Learning: The Marine Corps Development of Small Wars Doctrine, 1915–1940* (Boulder, CO: Westview Press, 2000); Bruce I. Gundmundsson, *Stormtroop Tactics: Innovation in the German Army, 1914–1918* (Westport, CT: Praeger, 1995); Timothy T. Lupfer, *The Dynamics of Doctrine: The Change in German Tactical Doctrine During the First World War* (Leavenworth, KS: US Army Combat Studies Institute, 1981); and Thomas L. Jentz, *Dreaded Threat: The 8.8 cm Flak 18/36/37 in the Anti-Tank Role* (Boyds, MD: Panzer Tracts, 2001).

35. The origin of the American RMA framework was a reworking of the Soviet Military-Technical Revolution of the 1970s. See Dima P. Adamsky, "Through the Looking Glass: The Soviet Military-Technical Revolution and the American Revolution in Military Affairs," *Journal of Strategic Studies* 31, no. 2 (April 2008): 257–94.

36. Bill Owens with Ed Offley, *Lifting the Fog of War* (Baltimore: Johns Hopkins University Press, 2000), 100.

37. Roger Trinquier, *Modern Warfare: A French View of Counterinsurgency* (New York: Praeger, 2006); David Galula, *Counterinsurgency Warfare: Theory and Practice* (New York: Praeger, 2006); Frank Kitson, *Low Intensity Operations: Subversion, Insurgency, and Peacekeeping* (London: Faber and Faber, 2011).

38. Excellent books include those by Australian-American former army officer and high-level US adviser David Kilcullen's *The Accidental Guerrilla* (2011), former US State Department political officer Carter Malkasian's *War Comes to Gramser* (2013), and former British Gurkha Rifles officer Emile

Simpson's *War from the Ground Up* (2012). David Kilcullen, *The Accidental Guerrilla: Fighting Small Wars in the Midst of a Big One* (Oxford, UK: Oxford University Press, 2011); Carter Malkasian, *War Comes to Garmser: Thirty Years of Conflict on the Afghan Frontier* (Oxford, UK: Oxford University Press, 2013); and Emile Simpson, *War from the Ground Up: Twenty-First Century Combat as Politics* (Oxford, UK: Oxford University Press, 2013).

39. This is a shortening of Rogers's definition: "An innovation is an idea, practice or object that is perceived as new by an individual or other unit of adoption." Everett M. Rogers, *Diffusion of Innovations* (New York: Free Press, 1962; 5th ed., 2003), 12. Joseph Schumpeter defined innovation as "doing things differently in the realm of economic life." J. A. Schumpeter, *Business Cycles* (New York, 1939), Vol. 1, 84.

40. Benoît Godin, *"Meddle Not with Them That Are Given to Change": Innovation as Evil,* Project on the Intellectual History of Innovation, Working Paper No. 6 (2010), accessed at http://www.csiic.ca/PDF/IntellectualNo6.pdf. See also Emma Green, "Innovation: The History of a Buzzword," *The Atlantic*, 20 June 2013, at https://www.theatlantic.com/business/archive/2013/06/innovation-the-history-of-a-buzzword/277067/.

41. This was not the origin of the concept of innovation, which according to Benoît Godin dates to Ancient Greece. He argues that before the twentieth century, innovation was considered a vice and had nothing to do with creativity. See Benoît Godin, *Innovation: A Conceptual History of an Anonymous Concept*, Working Paper No. 21, Project on the Intellectual History of Innovation, Montreal, Quebec, Canada, at http://www.csiic.ca/PDF/WorkingPaper21.pdf.

42. Joseph Schumpeter, *Capitalism, Socialism, and Democracy*, chapter 7, "The Process of Creative Destruction" (New York: Harper & Brothers, 1942; First Harper Perennial Modern Thought Edition, 2008), 81–86.

43. See, for example, Barry Posen, *The Sources of Military Doctrine: France, Britain, and Germany between the World Wars* (Ithaca, NY: Cornell University Press, 1984), 29–30. In a parenthetical comment, Posen admits that "stability" might be a less loaded term than "stagnation"; however, he continues to use stagnation as the converse of innovation.

44. Rogers (2003), 5. There are numerous updated editions of the original 1962 book. While he notes diffusion research on innovations as diverse as cell phones, medical drugs, farming innovations, smoking cessation, and teaching practices, he makes no mention of the diffusion of military innovations.

45. Rogers (2003) includes an extended bibliography of hundreds of books and articles on diffusion. See pp. 477–535.

46. The first post–Cold War analysis of military diffusion was by Emily Goldman and Leslie Eliason, eds. *The Diffusion of Military Technology and Ideas* (Stanford, CA: Stanford University Press, 2003). In general, military diffusion studies have followed in the tradition of Cold War studies of

technology transfer and proliferation. An exception to the state-to-state framework is the excellent study of the global spread of small arms by the Small Arms Survey in Geneva, Switzerland [http://www.smallarmssurvey. org]. For more, see chapter 6.

47. John Lynn, "Heart of the Sepoy: The Adoption and Adaptation of European Military Practice in South Asia, 1740–1805," chapter 2 of Goldman and Eliason (2003), 33–62. For a more in-depth analysis, see John Lynn, *Battle: A History of Combat and Culture* (Oxford, UK: Westview Press, 2003).

48. Horowitz introduced a model of diffusion he called "adoption-capacity theory," which held that financial and organizational requirements affected how military innovations spread in the international system and which actors would adopt them. In his study of modern suicide terrorism, Horowitz argued that traditional models of military innovation could also apply to terrorist organizations like al-Qaeda, the Tamil Tigers (LTTE), and Hamas. Although he made little mention of the explosives technologies behind suicide IEDs, his was the first effort by a military innovation theorist to step outside traditional state-to-state frameworks and consider the global impact of the diffusion of suicide tactics between terrorist groups. Michael Horowitz, *The Diffusion of Military Power: Causes and Consequences for International Politics* (Princeton, NJ: Princeton University Press, 2010).

49. Karen Buckley has recently completed a brilliant dissertation on the topic of the spread of suicide tactics, technology, and practices. Karen Buckley, *Deadly Contagion: The Tactical Use and Migration of Suicide Bombings*, George Mason University dissertation, 2018.

50. According to Emily Goldman this was mainly because the Royal Navy lost control of the Fleet Air Arm (and its air power advocates), the Germans and Italians established independent air forces (who didn't care about the navy), while the US Navy maintained control over its aviators. Emily O. Goldman, "Receptivity to Revolution: Carrier Air Power in Peace and War," chapter 10, pp. 267–303; and Emily O. Goldman and Andrew L. Ross, "Conclusion: The Diffusion of Military Technology and Ideas—Theory and Practice," 385; both in Goldman and Eliason (2003).

51. Andrade qualifies the argument, saying that this was not the *only* consideration for China's decline, but an important factor. Tonio Andrade, *The Gunpowder Age: China, Military Innovation, and the Rise of the West in World History* (Princeton, NJ: Princeton University Press, 2016), 7. See also Andrade, "An Accelerating Divergence? The Revisionist Model of World History and the Question of Eurasian Military Parity: Data from East Asia," *Canadian Journal of Sociology* 36, no. 2 (2011): 185–208; Jack Goldstone, "Comment on 'An Accelerating Divergence?'," *Canadian Journal of Sociology* 36, no. 2 (2011): 209–12. This is a dynamic field of research that focuses on new sources and analyses emerging out of Asia, correcting European military histories that fail to appreciate the innovativeness of the Chinese and

Japanese in such things as cannon, firearms, organizational integration, and military drilling before the eighteenth century. For a broader argument and history, see also Ian Morris, *Why the West Rules—for Now* (New York: Farrar, Straus, and Giroux, 2010).

52. Charles O. Hucker, "Hu Tsung-Hsien's Campaign against Hsü Hai, 1556," in *Chinese Ways in Warfare*, edited by Frank A. Kierman Jr. and John K. Fairbank (Cambridge, MA: Harvard University Press, 1974), 73–74.

53. Horowitz (2010), 5.

54. Victor Lefebure, *Riddle of the Rhine: Chemical Strategy in Peace and War* (London: W. Collins Sons, 1921), 21–22. Chemical weapons were banned under the Hague Convention of 1899.

55. Many of these examples are discussed in George Raudzens, "War-Winning Weapons: The Measurement of Technological Determinism in Military History," *Journal of Military History* 54, no. 4 (1990): 403–33. Fuller's Plan 1919 was never fully carried out, but it laid the groundwork for the German blitzkrieg style of warfare. Other historians arguing against technological determinism include Martin Biddle, Colin Gray, and Williamson Murray, all of whom stress the need to place technology into the broader context of its use. Likewise Stephen Biddle and Bruce Gudmundsson emphasize the importance of military doctrine and decision-making over technology.

56. Recalling his view at the time, Nixon wrote in memoirs, "We have the power to destroy his war-making ability. The only question is whether we have the will to use that power." Richard M. Nixon, *RN: The Memoirs of Richard Nixon* (New York: Warner Books, 1978), Vol. 2, pp. 85–86; cited by Joseph R. Cerami, "Presidential Decisionmaking and Vietnam: Lessons for Strategists," *Parameters* (winter 1996–1997): 66–80.

57. Elisabeth Bumiller, "Soldier, Thinker, Hunter, Spy: Drawing a Bead on Al Qaeda," *New York Times*, 3 September 2011, at https://www.nytimes.com/2011/09/04/world/04vickers.html. See also Audrey Kurth Cronin, "Why Drones Fail: When Tactics Drive Strategy," *Foreign Affairs* (July/August 2013).

58. Ian Castle, *The First Blitz: Bombing London in the First World War* (Oxford, UK: Osprey Publishing, 2015), 15.

59. According to the 1947 US Strategic Bombing Survey, V-1s injured more than 60,000 Londoners during the latter half of the war, and V-2s injured some 25,000 more in 1944–1945. "Aircraft Division Industry Report," The United States Strategic Bombing Survey (Aircraft Division, January 1947), 112–16.

60. For a compelling and provocative discussion, see Sean McFate, *The New Rules of War: Victory in the Age of Durable Disorder* (New York: William Morrow, 2019).

61. John Stone, "Technology, Society, and the Infantry Revolution of the Fourteenth Century," *Journal of Military History* 68, no. 2 (April 2004): 361–80.

62. Napoleon massed enormous armies, and his opponents were forced to match his numbers. European battles went from a height of 60,000 to 80,000 men on the

battlefield in the mid-seventeenth century, to 460,000 at Leipzig in the early nineteenth century. Martin van Creveld, *Technology and War: From 2000 B.C. to the Present* (New York: Free Press, 1989), 113. See also my "Cyber-Mobilization: The New *Levée en Masse*," *Parameters* 36, no. 2 (summer 2006): 77–87.

63. This is not to ignore the important role of artillery in the war. The point is that Clausewitz wrote about other, broader factors changing war's character.

64. Hew Strachan, *Carl von Clausewitz's* On War: *A Biography* (London: Atlantic Books, 2008).

65. Carl von Clausewitz, *On War*, edited and translated by Michael Howard and Peter Paret (Princeton, NJ: Princeton University Press, 1976), 100.

66. For a lively overview of twentieth-century military technologies that were crucial to current civilian technologies, see David Hambling, *Weapons Grade: How Modern Warfare Gave Birth to Our High-Tech World* (New York: Carroll & Graf, 2005).

67. "Google Leads the Race to Dominate Artificial Intelligence," *The Economist*, 7 December 2017, at https://www.economist.com/news/business/21732125-tech-giants-are-investing-billions-transformative-technology-google-leads-race.

68. Govini, *Department of Defense Artificial Intelligence, Big Data, and Cloud Taxonomy*, 2017, at http://www.govini.com/home/insights/.

69. US Defense Innovation Board, Open Meeting Minutes, Defense Innovation Unit Experimental Headquarters, Mountain View, CA, 12 July 2017, p. 2, at https://media.defense.gov/2017/Dec/18/2001857959/-1/-1/0/2017-2566-148525_MEETING MINUTES_(2017-09-28-08-53-26).PDF.

70. Comment made at the Carnegie Corporation workshop on "The Strategic Stability Implications of Emerging Technologies," Elliott School, George Washington University, Washington, DC, 1 June 2018.

71. James Vincent, "Google Opens Chinese AI Lab, Says 'Science Has No Borders'," *The Verge*, 13 December 2017, at https://www.theverge.com/2017/12/13/16771134/google-ai-lab-china-research-center.

Chapter 2

1. Justin Huggler, "Islamist Extremist Ricin Plot Foiled by German Police," *The Telegraph*, 14 June 2018.

2. Christian Jokinen, "Foiled Ricin Plot Raises Specter of 'More Sophisticated' IS-Inspired Attacks," *Terrorism Monitor* 16, no. 16 (10 August 2018): 1, at https://jamestown.org/program/foiled-ricin-plot-raises-specter-of-more-sophisticated-is-inspired-attacks/.

3. Richard Edwards, "Poison-tip Umbrella Assassination of Georgi Markov Reinvestigated," *The Telegraph*, 19 June 2008, at https://www.telegraph.co.uk/news/2158765/Poison-tip-umbrella-assassination-of-Georgi-Markov-reinvestigated.html.

4. Henry Samuel, "Man Charged after French Police Foil Paris Ricin Terror Plot," *The Telegraph*, 18 May 2018, at https://www.telegraph.co.uk/news/2018/05/18/french-police-foil-ricin-terror-plot-arrest-egyptian-brothers/.

5. Statement by German Public Prosecutor General, 20 June 2018; cited by Florian Flade, "The June 2018 Cologne Ricin Plot: A New Threshold in Jihadi Bio Terror," *CTC Sentinel*, Combating Terrorism Center at West Point, August 2018, pp. 1–4.

6. Interview with a Germany security official, July 2018; and Statement by German Public Prosecutor Genera, 24 July 2018; both in Flade (2018).

7. Right-wing groups have distributed ricin instruction manuals and even materials through their print-based networks for decades. See, for example, Jonathan Tucker and Jason Pate, chapter 10: "The Minnesota Patriots Council," *Toxic Terror: Chemical and Biological Weapons* (Cambridge, MA: The Belfer Center, 2000), 159–84.

8. Susan Borowski, "From Beans to Weapon: The Discovery of Ricin," *American Association for the Advancement of Science*, 11 July 2011, at https://www.aaas.org/beans-weapon-discovery-ricin.

9. Moor (2006), 112.

10. The gap between average- and best-practice technologies narrows. Joel Mokyr, *The Lever of Riches: Technological Creativity and Economic Progress* (Oxford, UK: Oxford University Press, 1990), 10.

11. "Explosives" include dynamite, grenades, mortars, and improvised explosive devices, and "firearms" include automatic weapons, shotguns, and pistols. Gary LaFree, Laura Dugan, and Erin Miller, *Putting Terrorism in Context: Lessons from the Global Terrorism Database* (London: Routledge, 2015), 100.

12. The Small Arms Survey defines small arms as revolvers and self-loading pistols, rifles and carbines, assault rifles, sub-machine guns, and light machine guns. See Small Arms Survey at http://www.smallarmssurvey.org/weapons-and-markets/definitions.html.

13. Crashing airplanes into targets has a long history. For a concise listing of incidents, see Adam Dolnik (2007), 39–40.

14. Paul Mozur, "China, Not Silicon Valley, Is Cutting Edge in Mobile Tech," *New York Times*, 2 August 2016, at http://www.nytimes.com/2016/08/03/technology/china-mobile-tech-innovation-silicon-valley.html.

15. Christopher K. Gegge, "To Ryse against Ones Kingdome: The Roots of Insurgency," *The Ancient World* 44, no. 1 (2013): 20–29. According to Gegge, the word was first used in English during the reign of Henry VIII in the United Kingdom's Ecclesiastical Appeals Act of 1532.

16. See Bruce Hoffman, *Inside Terrorism* (New York: Columbia University Press, 2017), 37; and Audrey Kurth Cronin, "ISIS Is Not a Terrorist Group," *Foreign Affairs* (March/April 2015): 87–98.

17. For many more examples, see Timothy Howe and Lee L. Brice, eds., *Brill's Companion to Insurgency and Terrorism in the Ancient Mediterranean* (Leiden, Netherlands: Koninklijke Brill, 2016).

18. See David C. Rapoport, "Fear and Trembling: Terrorism in Three Religious Traditions," *American Political Science Review* 78, no. 3 (September 1984): 658–77; James J. Bloom, *The Jewish Revolts against Rome, A.D. 66–135* (London: McFarland & Company, 2010). Some scholars disagree with the label "terrorist" for all three of these early groups, but there's no arguing that symbolic non-state violence against innocents for broader purposes, or terrorism, has ancient historical roots.

19. Rapoport (1984), 665; W. B. Bartlett, *The Assassins: The Story of Medieval Islam's Secret Sect* (London: Sutton Publishing, 2001); Gerard Chaliand and Arnaud Blin, "Zealots and Assassins," chapter 3 of *The History of Terrorism: From Antiquity to Al Qaeda* (Berkeley and Los Angeles: University of California Press, 2007), 55–78.

20. Mark Twain, *Following the Equator: A Journey around the World* (Hartford, CT: The American Publishing Company, 1897; Dover Productions, Reprint Edition, 1989), 428. See also Megan Garber, "The History of 'Thug'," *The Atlantic*, 28 April 2015.

21. David C. Rapoport, "The Four Waves of Modern Terrorism," chapter 2 of *Attacking Terrorism: Elements of a Grand Strategy*, edited by Audrey Kurth Cronin and James M. Ludes (Washington, DC: Georgetown University Press, 2004), 46–73.

22. Ibid., 47–50.

23. Ibid.; Dipak K. Gupta, "Waves of International Terrorism: An Explanation of the Process by Which Ideas Flood the World," chapter 2 of *Terrorism, Identity and Legitimacy*, edited by Jean E. Rosenfeld (London: Routledge, 2011), 30–43.

24. The phrase was first linked to the activities of rural Italian anarchists but popularized by the French anarchist Paul Brouse in 1877.

25. Lindsay Clutterbuck, "The Progenitors of Terrorism: Russian Revolutionaries or Extreme Irish Republicans?" *Terrorism and Political Violence* 16, no. 1 (spring 2004): 154–81.

26. For new insights regarding prewar fears of anarchism spreading to the Balkans, see Richard Bach Jensen, "Anarchist Terrorism and Counter-terrorism in Europe and the World, 1878–1934," in *The Routledge History of Terrorism*, edited by Randall David (London and New York: Routeldge, 2015), 111–29.

27. "Terrorism" and "insurgency" overlap, but they are different. Terrorist groups number dozens or hundreds of members, attack civilians, do not hold territory, and cannot directly confront military forces. Insurgencies are numerically larger than terrorist groups, operate as military units, seize and hold territory, and are strong enough to target military forces. Rapoport's waves include both. See Cronin (2009), 154; and Audrey Kurth Cronin,

"ISIS Is Not a Terrorist Group: Why Counterterrorism Won't Stop the Latest Jihadist Threat," *Foreign Affairs* (March/April 2015): 87–98.

28. For much more on this, see chapter 3 of Audrey Kurth Cronin, *How Terrorism Ends* (Princeton, NJ: Princeton University Press, 2009), 73–93.

29. Using the ITERATE dataset, political scientists Karen Rasler and William R. Thompson tested the waves theory and found it explained the period they analyzed (1968–2004). Rasler and Thompson, "Looking for Waves of Terrorism," *Terrorism and Political Violence* 21, no. 1 (2009): 28–41.

30. Mark Sedgwick, "Inspiration and the Origins of Global Waves of Terrorism," *Studies in Conflict & Terrorism* 30 (2007): 97–112. Sedgwick sees a German, Chinese, Afghan, and Italian wave.

31. Plastic explosives, technological descendants of dynamite, were also important to the second wave.

32. To avoid confusion, I have used the modern state names. The actual British mandate was for Mesopotamia, for example, which includes parts of Iran, Syria, Turkey, and Iraq. Margaret Macmillan, *Paris 1919: Six Months That Changed the World* (New York: Random House, 2003), chapter 8: "Mandates," pp. 98–108; and Quincy Wright, *Mandates under the League of Nations* (Chicago: University of Chicago Press, 1930).

33. The Avalon Project Documents in Law, History and Diplomacy, Yale Law School, Lillian Goldman Law Library, at http://avalon.law.yale.edu/wwii/atlantic.asp.

34. Max Boot, *Invisible Armies: An Epic History of Guerrilla Warfare from Ancient Times to the Present* (New York: W. W. Norton, 2013).

35. There have been two large spurts in UN membership: 1955–1960 (when forty-one states joined) and 1991–1993 (when twenty-six states joined). The first is the height of postwar decolonization and the second the aftermath of the Soviet Union's breakup.

36. The AK-47 (or its facsimile) appears on the flag of Mozambique and Hezbollah, and on the coats of arms of Zimbabwe, Burkina Faso (1984–1997), and East Timor.

37. The Black September operatives were told to demand the release of Palestinian prisoners and, after holding the Israelis for no more than twenty-four hours, to ask for a plane to fly them and the hostages to an Arab country. Simon Reeve, *One Day in September: The Full Story of the 1972 Munich Olympics Massacre and the Israeli Revenge Operation "Wrath of God"* (New York: Arcade Publishing, 2000). Note that Sedgwick (2007) thinks the size of the third wave is exaggerated because that is when Western terrorist statistics began to be kept.

38. In 1987, North Korean agents also planted a bomb in an overhead storage bin on Korean Air 858, killing 115 people.

39. Bags became subject to thermal neutron activation screening, which detects the level of nitrogen. Military and commercial explosives such as SEMTEX,

RDX, and PETN are high in nitrogen. M. Mitchell Waldrop, "FAA Fights Back on Plastic Explosives," *Science* 243, no. 4888 (13 January 1989): 165–6. See also Anthony Fainberg, "Explosives Detection for Aviation Security," *Science* 255, no. 5051 (20 March 1992): 1531–7.

40. Helpful overviews of changing media practices include Jason Burke, "How the Changing Media Is Changing Terrorism," *The Guardian*, 25 February 2016, at https://www.theguardian.com/world/2016/feb/25/how-changing-media-changing-terrorism; and Charlie Beckett, *Fanning the Flames: Reporting Terror in a Networked World*, Tow Center for Digital Journalism and Democracy Fund Voice, Columbia School of Journalism, September 2016, at https://www.cjr.org/tow_center_reports/coverage_terrorism_social_media.php/#problem-covering-terrorism.

41. Stephen Frederic Dale, "Religious Suicide in Islamic Asia: Anticolonial Terrorism in India, Indonesia, and the Philippines," *Journal of Conflict Resolution* 32, no. 1 (March 1988): 37–59.

42. Suicide attacks are by no means a strictly religious phenomenon, however. Secular groups such as the Tamil Tigers (LTTE) were masters of the tactic.

43. Rapoport thinks the current wave will fizzle out some time in the 2020s, to be gradually replaced by a new one. Jeffrey Kaplan argues that the fifth wave will be characterized by groups such as Sudan's Janjaweed and the Uganda's Lord's Resistance Army that begin on an international level, then turn inward toward violent tribalism. Kaplan, "Terrorism's Fifth Wave: A Theory, a Conundrum and a Dilemma," *Perspectives on Terrorism* 2, no. 2 (January 2008); and *Terrorist Groups and the New Tribalism: Terrorism's Fifth Wave* (London: Routledge, 2010). Jeffrey D. Simon sees the fifth wave as characterized by lone actors and their use of technology. Simon, "Technological and Lone Operator Terrorism: Prospects for a Fifth Wave of Global Terrorism," chapter 3 in Rosenfeld (2011), 44–65. Others counter that Rapoport is wrong—that the fourth wave is unique and will last much longer than the first three did. Rod Lyon and Stephanie Huang, "Fifth Wave Terrorism: Have Predictions Jumped the Gun?," *The Strategist*, Australian Strategic Policy Institute (ASPI), 27 July 2015, at http://www.aspistrategist.org.au/fifth-wave-terrorism-have-predictions-jumped-the-gun/.

44. This according to Martha Crenshaw, who distinguishes between strategic, tactical, and organizational innovation. See Martha Crenshaw, "Innovation: Decision Points in the Trajectory of Terrorism," Appendix II of Maria J. Rasmussen and Mohammed M. Hafez, *Terrorist Innovations in Weapons of Mass Effect: Preconditions, Causes, and Predictive Indicators*, Workshop Report, The Defense Threat Reduction Agency, 2010, at https://calhoun.nps.edu/bitstream/handle/10945/25358/2010_019_Terrorist_WME.pdf;sequence=4. Discussions of terrorist group "learning" also prevail, emphasizing dynamic reactions to counterterrorism measures. For example, soon after 9/11, the RAND corporation produced very good studies of

terrorist and insurgent learning (cited in next footnote). See also E. Ahmed, A. Elgazzar, and A. S. Hegazi, "On Complex Adaptive Systems and Terrorism," *Physics Letters* A, 337(1/2) (2005): 127–9; Nancy Hayden, "The Complexity of Terrorism: Social and Behavioral Understanding Trends for the Future," in *Mapping Terrorism Research: State of the Art, Gaps, and Future Direction*, edited by Magnus Ranstorp (London: Routledge, 2006), 33–57; and Jao Ricardo Faria, "Terrorist Innovations and Anti-Terrorist Policies," *Terrorism and Political Violence* 18, no. 1 (2006): 47–56.

45. These include: Brian A. Jackson, with John C. Baker, Kim Cragin, John Parachini, Horacio R. Trujillo, and Peter Chalk, *Aptitude for Destruction*, Vol. 1: *Organizational Learning in Terrorist Groups and Its Implications for Combating Terrorism* (Santa Monica, CA: RAND Corporation, 2005); Brian A. Jackson, John C. Baker, Kim Cragin, John Parachini, Horacio R. Trujillo, and Peter Chalk, *Aptitude for Destruction*, Vol. 2: *Case Studies of Organizational Learning in Five Terrorist Groups* (Santa Monica, CA: RAND Corporation, 2005); Kim Cragin, Peter Chalk, Sara A. Daly, Brian A. Jackson, *Sharing the Dragon's Teeth: Terrorist Groups and the Exchange of New Technologies* (Santa Monica, CA: RAND Corporation, 2007); Paul Wilkinson, ed., *Technology and Terrorism* (London: Frank Cass, 1993); and David Clarke, ed., *Technology and Terrorism* (New Brunswick, NJ: Transaction Publishers, 2004).

46. A few have begun pushing toward a broader perspective. See Gary Ackerman's interesting doctoral thesis on weapons adoption across a range of cases, "'More Bank for the Buck': Examining the Determinants of Terrorist Adoption of New Weapons Technologies," unpublished, King's College dissertation, London, 2014; Gabriel Koehler-Derrick and Daniel James Milton, "Choose Your Weapon: The Impact of Strategic Considerations and Resource Constraints on Terrorist Group Weapon Selection," *Terrorism and Political Violence* (2017), 1–20; and Rashmi Singh, "A Preliminary Typology Mapping Pathways of Learning and Innovation by Modern *Jihadist* Groups," *Studies in Conflict & Terrorism* 40, no. 7 (2017): 624–44. On processes of "creativity," see also Gina Scott Ligon, Karyn Sporer, and Douglas C. Derrick, "Violent Innovation: Creativity in the Domain of Terrorism," chapter 28 of *The Cambridge Handbook of Creativity across Domains*, edited by James C. Kaufman, Vlad P. Glaveanu, and John Baer (Cambridge, UK: Cambridge University Press, 2017), 507–22.

47. Jackson (2001); and M. Sedgwick, "Inspiration and the Origins of Global Waves of Terrorism," *Studies in Conflict and Terrorism* 30, no. 2 (2007): 102.

48. Adam Dolnik, author of a 2007 book on terrorist innovation, writes, "Terrorist innovation would simply fall into the realm of emulation and adaptation, as technologies used by terrorists have *never* been completely new." *Understanding Terrorist Innovation* (London: Routledge, 2007), 5. See also Bruce Hoffman, "Terrorist Targeting: Tactics, Trends, and Potentialities," *Terrorism and Political Violence* 5, no. 2 (summer 1993): 12–29.

49. Ariel Merari, "Terrorism as a Strategy of Struggle: Past and Future," *Terrorism and Political Violence* 11, no. 4 (1999): 54.

50. Adam Dolnik, "The Dynamics of Terrorist Innovation," chapter 4 of *Understanding Terrorism Innovation and Learning: Al Qaeda and Beyond*, edited by Magnus Ranstorp and Magnus Normark (London: Routledge, 2015), 77.

51. The Provisional IRA split off from the original Irish Republican Army (which became known as the "Official IRA") in 1969. The Palestinian Al Fatah had initially trained members of the Provisional IRA in Lebanon. J. Bowyer Bell, *The Secret Army: The IRA 1916–1979* (Cambridge, MA: MIT Press, 1980), 439–40.

52. Personal interview with law enforcement officials, March 2004; cited by Richard A. Jackson, "Provisional Irish Republican Army," chapter 5 in *Aptitude for Destruction*, RAND, 97–98.

53. Jackson et al. (2005), 99. According to the London *Independent* newspaper, about 120 PIRA members were killed by "own goals"—premature explosions or accidental shooting incidents. "Terrorists Killed by Their Own Devices," *Independent* (London), 20 February 1996, at https://www.independent.co.uk/news/terrorists-killed-by-their-own-devices-1319857.html.

54. Regarding weapons, the PIRA were pleased to discover that the US Army's M-16, in its Colt AR-15 semi-automatic version, could be bought freely as modern sporting rifles in many US states. They also bought the AR-18, nicknamed "the widow maker," which could penetrate British Army body armor and armored personnel carriers. J. Bowyer Bell (1980), 439.

55. Ibid., 438.

56. JIbid., 438–9.

57. Paul Gill makes the excellent point that PIRA bomb making innovations fell in the 1980s, probably because the Libyan weapons and explosives made them unnecessary. Gill (2017), p. 581.

58. Jackson et al. (2005), 100–3; and Gill (2017), 579.

59. The most authoritative English-language source on this group's use of technology, including many interviews with imprisoned Aum members, is Richard Danzig, Marc Sageman, Terrance Leighton, Lloyd Hough, Hidemi Yuki, Rui Kotani, and Zachary M. Hosford, *Aum Shinrikyo: Insights into How Terrorists Develop Biological and Chemical Weapons*, Center for a New American Security, July 2011, at https://www.cnas.org/publications/reports/aum-shinrikyo-insights-into-how-terrorists-develop-biological-and-chemical-weapons; and second edition, December 2012, at https://s3.amazonaws.com/files.cnas.org/documents/CNAS_AumShinrikyo_SecondEdition_English.pdf?mtime=20160906080510. This information is mainly drawn from these two sources. Other important books about the cult include Robert Jay Lifton, *Destroying the World to Save It* (New York: Henry Holt, 1999); and

Ian Reader, *Religious Violence in Contemporary Japan: The Case of Aum Shinrikyo* (Honolulu: University of Hawaii Press, 2000).

60. On the growing risks of biological terrorism, see Bob Graham and Jim Talent, *World at Risk: The Report of the Commission for the Prevention of Weapons of Mass Destruction Proliferation and Terrorism* (New York: Vintage Books, 2008); Greg Koblentz, "Biosecurity Reconsidered: Calibrating Biological Threats and Responses," *International Security* 34, no. 4 (Spring 2010): 96–132; and Sonia Ben Ouagrham-Gormley, "Barriers to Bioweapons: Intangible Obstacles to Proliferation," *International Security* 36, no. 4 (spring 2012): 80–114.

61. The operation involved a diversionary small explosive, followed by well-coordinated automatic fire that overwhelmed the sentries. The incident was not a suicide attack, as the driver and several others jumped out of the truck prior to the blast. Zadka, *Blood in Zion: How the Jewish Guerrillas Drove the British out of Palestine* (London: Brassey's, 1995), 70–72.

62. I am not including an analysis of explosively formed penetrators (EFPs), which are even more powerful and lethal than IEDs. EFPs were originally invented in the 1930s by the oil industry, to punch holes in metal pipes, so any country with an oil industry has them. There were factories for producing them in Iraq and some may have come in to the country from Iran.

63. Here I am including roadside bombs, car bombs, and smaller devices. DOD Personnel and Military Casualty Statistics, *Defense Manpower Data Center, Casualty Summary by Reason, October 7, 2001 through August 18, 2007*; cited by Wilson (2007), 1–2. As of 2008, there were 606 amputees and 5,792 who suffered traumatic brain injury as the result of a blast. The total number of military deaths was 4,122 in Iraq and 561 in Afghanistan. Hannah Fischer, *United States Military Casualty Statistics: Operation Iraqi Freedom and Operation Enduring Freedom, CRS Report for Congress, #RS22452, 9 September 2008*.

64. In 2016, it was restructured into the Joint Improvised-Threat Defeat Organization, or JIDO. JIEDDO's mission was to defeat IEDs "as a weapon of strategic influence."

65. This is a representative list. For more information, see JIEDDO's annual reports at http://www.jieddo.mil. See also US Government Accountability Office, *Warfighter Support: DoD Needs Strategic Outcome-related Goals and Visibility over its Counter-IED Efforts*, GAO-12-280, 22 February 2012, at https://www.gao.gov/products/GAO-12-280. See also Peter Cary and Nancy Youssef, "JIEDDO: The Manhattan Project That Bombed," Center for Public Integrity, 27 March 2011, at https://www.publicintegrity.org/2011/03/27/3799/jieddo-manhattan-project-bombed.

66. Exact numbers of PIRA are unknown; however, this estimate is based on the database of active members between 1970 and 1998, compiled by Paul Gill and John Horgan. See "Who Were the Volunteers? The Shifting Sociological

and Operational Profile of 1,240 Provisional Irish Republican Army Members," *Terrorism and Political Violence* 25, no. 3 (2013): 435–56.

67. "JIEDDO: The Manhattan Project That Bombed," (2014) and http://www.pica.army.mil/pmccs/CombatMunitions/Defeat/Rhino.html.

68. Noah Shachtman, "The Pentagon and Its Bogus Bomb-zapper: A Love Story," *Wired*, 28 March 2011, at https://www.wired.com/2011/03/pentagon-still-hearts-its-bogus-bomb-zapper/; and Cary and Youssef, Center for Public Integrity (2011).

69. Rick Atkinson, "With Jammers, US Targets Bombmakers," *The Argus Observer*, 8 October 2007, at https://www.argusobserver.com/news/with-jammers-u-s-targets-bombmakers/article_e57a69d5-bc9e-5e3f-b376-c28699c5cc30.html.

70. Quoted in Noah Shachtman, "Pentagon Bomb Squad Chief Looks Back," *Wired*, 16 November 2007, at https://www.wired.com/2007/11/the-other-day-i/.

71. Government Accountability Office, *Warfighter Support: Actions Needed to Improve the Joint Improvised Explosive Device Defeat Organization's System of Internal Control*, GAO-10-660, 1 July 2010, at https://www.gao.gov/products/GAO-10-660; and Andrew Tilghman, "Effectiveness of Some Counter-IED Efforts Unproved, Pentagon Report Finds," *Military Times*, 9 August 2016, at https://www.militarytimes.com/2016/08/09/effectiveness-of-some-counter-ied-efforts-unproved-pentagon-report-finds/.

72. Kelsey Atherton, "When Big Data Went to War—and Lost," *Politico*, 11 October 2017.

73. Francesco Trebbi, Eric Weese, Austin L. Wright, and Andrew Shaver, *Insurgent Learning*, National Bureau of Economic Research Working Paper No. 23475, 1 January 2018, accessed at https://www.aeaweb.org/conference/2018/preliminary/paper/2YY6H8HR.

74. JIEDDO statistics reported by Spencer Ackerman, "$265 bomb, $300 Billion War: The Economics of the 9/11 Era's Signature Weapon," *Wired*, 8 September 2011, at https://www.wired.com/2011/09/ied-cost/.

75. Ibid.; and JIEDDO Annual Report 2010, at https://www.jieddo.mil/content/docs/JIEDDO_2010_Annual_Report_U.pdf.

76. Max Boot, *Invisible Armies: An Epic History of Guerrilla Warfare from Ancient Times to the Present* (New York: W. W. Norton, 2013), 54–55.

77. Those who study terrorists tend to use the word "contagion" more often than "diffusion," probably because it implies human-to-human contact and follows in the criminology tradition. Robert T. Holden, "The Contagiousness of Aircraft Hijacking," *American Journal of Sociology*, Vol. 91, No. 4 (January 1986): 874–904; Manus I. Midlarsky, Martha Crenshaw, and Fumihiko Yoshida, "Why Violence Spreads: The Contagion of International Terrorism," *International Studies Quarterly* 24 (1980): 262–98. James Forest argues that contagion initially referred to inspiration (or motivation) and diffusion to

operational ability. James J. F. Forest, "Introduction," *Teaching Terror: Strategic and Tactical Learning in the Terrorist World* (New York: Rowman & Littlefield, 2006), 4. Sedgwick (2007) further divides operational contagion between specific technique and general strategy. See also Gary LaFree, Min Xie, and Aila M. Matanock, "The Contagious Diffusion of Worldwide Terrorism: Is It Less Common Than We Might Think?," *Studies in Conflict & Terrorism* 41, no. 4 (2018): 261–80.

78. Holden (1986).

79. Actual direct evidence—deaths—resulting from the Werther effect at the time is hard to find today. One person who killed herself was Christine von Lassbery, who drowned herself in the River Ilm in Weimar, in January 1778, after a failed love affair. She was found with a copy of the novel in her pocket. A second person, Fanni von Ickstatt, jumped from a tower in Munich, but the direct connection to Goethe was unproven. See Michael Hulse's introduction to *The Sorrows of Young Werther* (London: Penguin Classics, 1989), 12–13. The term "Werther effect" has persisted, describing a phenomenon of copycat suicides that has been rigorously proven in numerous modern cases.

80. Leonard Berkowitz and Jacqueline Macaulay, "The Contagion of Criminal Violence," *Sociometry* 34, no. 2 (June 1971): 238–60. The relationship did not hold for other types of crime such as larceny, burglary, and auto theft.

81. See Edward Heyman and Edward Mickolus, "Observations on 'Why Violence Spreads'," *International Studies Quarterly* 24, no. 2 (June 1980): 299–305; and Manus I. Midlarsky, Martha Crenshaw, and Fumihoko Yoshida, "Rejoinder to 'Observations on "Why Violence Spreads,"'" *International Studies Quarterly* 24, no. 2 (June 1980): 306–10.

82. Holden (1986).

83. Ibid., 875.

84. Ibid.; Dolnik (2007), 53–55; and Laura Dugan, Gary LaFree, and Alex R. Piquero, "Testing a Rational Choice Model of Airline Hijackings," *Criminology* 43, no. 4 (2005): 1031–65.

85. Alan Yu, "How A Gene Editing Tools Went from Labs to a Middle-School Classroom," *NPR*, 27 May 2017, at http://www.npr.org/sections/alltechconsidered/2017/05/27/530210657/how-a-gene-editing-tool-went-from-labs-to-a-middle-school-classroom. Note that these kits are very simple and not AI/autonomy enabled. The bigger issue is the ability to order reagent from pharma/research distributors, a risky prospect—although they are beginning to ask questions regarding who is ordering what. I am grateful to Elizabeth Prescott for this observation.

86. Right now this behavior is mostly self-experimentation, but that could change.

87. Everett M. Rogers, *Diffusion of Innovations* (New York: Free Press, 1962; 5th ed., 2003), 11.

88. Rogers (2003), chapter 6: "Attributes of Innovations and Their Rate of Adoption," 219–66.
89. Rogers (2003), 39–45.
90. Jared Diamond, *Guns, Germs, and Steel* (New York: Norton Press, 1999), 243.
91. Ibid., 242–4.
92. Contrast this with the Manhattan Project, conducted in secret with the clear purpose of building an atomic bomb before Nazi Germany could. It took three years and cost $2 billion. See Diamond (1999), 242.

Chapter 3

1. The identity of the bomber has never been definitively proven. In his 1991 book, Paul Avrich argues on the basis of circumstantial evidence that it was Mario Buda. See Paul Avrich, *Sacco and Vanzetti: The Anarchist Background* (Princeton, NJ: Princeton University Press, 1991), 204–7. Mario Buda had worked in construction and was familiar with the use of dynamite. Jeffrey D. Simon, "The Forgotten Terrorists: Lessons from the History of Terrorism," *Terrorism and Political Violence* 20, no. 2 (2008): 201. Mike Davis argues that the perpetrator was "Mike Boda" (obviously the same person), a supporter of the anarchist theorist Luigi Galleani. Mike Davis, *Buda's Wagon: A Brief History of the Car Bomb* (London: Verso, 2007), 1.
2. Deaths were between 38 and 40. Casualty figures vary, probably due to poor forensic tools and claims in subsequent lawsuits. Even the local *New York Times* contradicts itself, with 40 claimed at first, but lower death counts in later editions. A 15 November 1920 *New York Times* headline read "40th Wall Street Death," and a 23 November 1920 *New York Times* article cites 40 killed and upwards of 300 injured. *New York Times* articles in May 1921 and a *Washington Post* article on first anniversary both also cite 40 deaths. However, a *New York Times* article a month earlier (April 1921) cites 37 killed and over 200 injured, and *New York Times* articles a few months later (September and December 1921) list 39 killed. Meanwhile a December 1921 article in the *Chicago Tribune* lists 38 killed and over 200 wounded. The best we can do is estimate. Davis claims there were 40 dead, though he does not cite a source. See also Beverly Gage, *The Day Wall Street Exploded* (New York: Oxford University Press, 2009), 160–1, who cites 38 killed and 143 injured.
3. The building currently belongs to a Chinese global real estate investment group and sits empty. Gilbert King, "Anger and Anarchy on Wall Street," 4 October 2011, Smithsonian.com, at http://www.smithsonianmag.com/history/anger-and-anarchy-on-wall-street-96057606/?no-ist.
4. This paragraph, including the names and descriptions of the early devices, is drawn from "Fire Drug," chapter 1 of Jack Kelly, *Gunpowder: Alchemy, Bombards, & Pyrotechnics: The History of the Explosive That Changed the World* (New York: Basic Books, 2004), 1–18.

5. J. R. Partington, *A History of Greek Fire and Gunpowder* (Baltimore: Johns Hopkins University Press, 1960), 65–81.

6. Trevor N. Dupuy, *The Evolution of Weapons and Warfare* (New York: Da Capo Press, 1984), 146.

7. Brodie and Brodie (1973), 44–45; and Albert Manucy, *Artillery through the Ages*, National Park Service Interpretive History Series no. 3 (Washington, DC: US Government Printing Office, 1949), 4–5.

8. Antonia Fraser, *Faith and Treason* (New York: Random House, 1996), 121–2.

9. Some historians doubt aspects of this story, as it depended on the coerced confessions of the conspirators and government accounts. A cultural icon, Guy Fawkes is remembered in an old joke as "the last man to enter Parliament with honest intentions." James A. Sharpe, *Remember, Remember: A Cultural History of Guy Fawkes Day* (Cambridge, MA: Harvard University Press, 2005), 6.

10. Beginning in the sixteenth century, numerous tests for the strength of gunpowder were devised, including the Trauzl test and the ballistic pendulum test. See G. I. Brown, *The Big Bang: A History of Explosives* (Stroud, UK: Sutton Publishing, 1998), 38–44.

11. The exact number of Paris victims killed and wounded is disputed.

12. William S. Dutton, *One Thousand Years of Explosives: From Wildfire to H-bomb* (New York: Rinehart and Winston, 1960), 91; Brown (1998), 9, 71.

13. Showalter (1975).

14. Dutton (1960), 115.

15. John Keegan, *History of Modern Warfare* (New York: Vintage Books, 1993), 305–6; and James M. McPherson, *Battle Cry of Freedom: The Civil War Era* (Oxford, UK: Oxford University Press, 2003), 12.

16. James Ford Rhodes, "Sherman's March to the Sea," *American Historical Review* 6, no. 3 (April 1901): 466–74. Northern gunpowder was of a better quality than that available in the South. Brown (1998), 32.

17. There are no records of how many died, and the estimates vary wildly.

18. The safety fuse was invented in 1831 by William Bickford, which greatly reduced the danger to miners. Brown (1998), 72.

19. Dutton (1960), 116; Brown (1998), 77–79.

20. At the Battle of Edgehill in 1642, Oliver Cromwell reportedly yelled, "Put your trust in God, my boys, but mind to keep your powder dry." William Blacker claims it in his 1834 poem "Oliver's Advice."

21. Smokeless powder was invented in 1884.

22. "Black Powder's Soul: The Quest for the Elusive Saltpeter," chapter 2 of Stephen R. Brown, *A Most Damnable Invention: Dynamite, Nitrates, and the Making of the Modern World* (New York: St. Martin's Press, 2005), pp. 25–50.

23. On Nobel's life and work, see Kenne Fant, *Alfred Nobel: A Biography* (New York: Arcade Publishing, 1991); and Ulf Larsson, *Alfred*

Nobel: Networks of Innovation (Stockholm: Archives of the Nobel
Museum, 2008).

24. Fant (1991), 178–9.
25. G. Nagendrappa, "Alfred Bernhard Nobel," *Resonance*, June 2013, 500–13.
26. Brown (1998), 91–94.
27. Nobel never married, but he did have two romantic relationships: a mistress,
    Sofie Hess, who was unfaithful and tried to make a claim on his estate after
    his death; and Countess Bertha Kinsky, his former secretary, who became a
    peace activist and greatly influenced him late in his life.
28. Fant (1991), 46–77. Fant puts the barge in Bockholm Bay (p. 73); the
    Nobel Prize museum says Lake Mälaren. Birgitta Lemmel, "Alfred Nobel's
    Industrial Activities in Vinterviken," https://www.nobelprize.org/alfred-
    nobel/alfred-nobels-industrial-activities-in-vinterviken/.
29. The second explosion was probably the nitroglycerine being used to
    manufacture the dynamite.
30. "The Pacific Coast: The Nitro-Glycerine Explosion in San Francisco.
    Terrific Effects of the New Explosive Compound—Horrible Scene—Entire
    Demolition of the Building—Melancholy Loss of Life—Minute Description
    of the Tragic Event," *New York* Times, May 13, 1866, p. 3.
31. "Explosion in Greenwich Street: Fortunate Escape of a Hotel and Its
    Inmates—Twenty-four Persons Wounded—No One Mortally," *New York
    Times*, 6 November 1865, p. 2; and G. M. Mowbray, "Nitroglycerine,"
    lecture given at the Stevens Institute of Technology, reprinted in *Scientific
    American*, 27 May 1876, p. 341.
32. J. E. Jorpes, "Alfred Nobel," *British Medical Journal*, 3 January 1959, p. 3.
33. "Experiments with Nitro-Glycerin or Blasting Oil," *Scientific American* 14,
    no. 20 (12 May 1866): 322.
34. Fant (1991), 79.
35. In the United States, an anti-nitroglycerine law was passed on 26 July 1866.
    Fant (1991), 89. It could be used, but it had to be transported in metal
    containers and labeled "Dangerous." In 1869 the Nitroglycerine Act was
    passed in England. The British act banned the use of nitroglycerine in any
    form, making it impossible for Nobel to sell dynamite there. Fant (1991),
    141. Four years later he had received a dispensation: nitroglycerine could be
    manufactured in England, just not transported there. In 1870, the British
    dynamite company was founded in Ardeer, on Scotland's western coast.
36. Fulminate of mercury also easily explodes but is expensive and hard to store.
    This design required only a small amount.
37. It helped that dynamite was invented at about the same time that the
    pneumatic drill (1871) and the diamond drill (1863) were—both also
    important tools to cut through rock. Regarding explosives techniques, see
    Willard Young, "Dynamite as a Railroad Builder: High Explosives as One of
    the Principal Agents in the Building and Reconstruction of Our Railroads,"

Scientific American 104, no. 25 (24 June 1911): 620. Also James R. Thurman, *Practical Bomb Scene Investigation*, 2nd ed. (Boca Raton, FL: Taylor and Francis, 2011), 44–45.

38. The project used both the original dynamite and Nobel's blasting gelatin. Fant (1991), 4. The dynamite factory at Isleten, Switzerland, was built so as to supply the project with dynamite. Ulf Larsson, *Alfred Nobel: Networks of Innovation* (Stockholm, Sweden: Archives of the Nobel Museum), 88.

39. "Tunneling Market in Brazil," *Tunnel*, Issue 5 (2011), at http://www.tunnel-online.info/en/artikel/tunnel_2011-05_Tunnelling_Market_in_Brazil_1245307.html.

40. As the Russian word "Volya" can also be translated "Liberty" or "Freedom," some English accounts name this group "People's Liberty" or "People's Freedom."

41. Alexander II was also the tsar who sold Alaska to the United States for $7.2 million in a treaty signed 30 March 1867.

42. For a masterful comprehensive analysis, see Franco Venturi, *The Roots of Revolution: A History of the Populist and Socialist Movements in 19th-Century Russia* (New York: Alfred A. Knopf, 1960).

43. David Saunders, *Russia in the Age of Reaction and Reform, 1801–1881* (London: Longman, 1992), 317–21.

44. Sources disagree as to whether the dynamite was imported or completely constructed by Nikolai Kibalchich, the group's brilliant explosives expert. Some writers claim that the group sent an agent to Switzerland to buy dynamite that was smuggled into Russia. Kibalchich, who was the designer of the IEDs, stated at his trial, however, that no explosives had been imported and that the operatives themselves had made them. Lee B. Croft, *Nikolai Ivanovich Kibalchich: Terrorist Rocket Pioneer* (Tempe, AZ: Institute for Issues in the History of Science, 2006); Ronald Seth, *The Russian Terrorists* (London: Barrie and Rockcliff, 1966).

45. David Footman, *Red Prelude* (London: Barrie & Rockliff, The Cresset Press, 1944), 112–13.

46. Footman (1944), 139–40.

47. The phrase was used by minister of the court Alexander Adlerberg, reporting to the tsar. This description of the actions of the plotters is from Radzinsky, who drew it directly from original sources, including post-arrest confessions. Edvard Radzinsky, *Alexander II: The Last Great Tsar*, translated by Antonia W. Bouis (London: Free Press, 2005).

48. Laqueur claimed that the dynamite would also have been insufficient to kill the tsar. Walter Laqueur, *The Age of Terrorism* (New York: Little, Brown, 1987), 105, note 34.

49. Perovskaya had taken over leadership of the group when Andrei Zhelyabov, the group's mastermind, was arrested. Ironically, her father had lost his job as a result of the 1866 attempt on the tsar's life.

50. According to Footman, the bombs weighed five pounds and had within them two glass tubes placed crosswise. Upon striking a hard object, the glass tubes broke and set off the chemical reaction that ignited the detonator. Footman (1944), 175.

51. According to Alexander Yarmolinsky, he was an errand boy for the butcher. Avrahm Yarmolinsky, *Road to Revolution: A Century of Russian Radicalism* (New York: Macmillan, 1959), 279.

52. Kotik Grinevitski is also Ignacy Hryniewieski Grinevizky (in Russian, Игнатий Иоахимович Гриневицкий). "Kotik," which means Kitten, was his group nickname.

53. Emelianov described this during his trial. Footman (1944); and Peter Kropotkin on the assassination of Alexander II, an excerpt from James Harvey Robinson and Charles Beard, eds., *Readings in Modern European History*, Vol. 2 (Boston: Ginn and Company, 1980), 362–3.

54. There were a total of four bombers sent by Perovskaya to kill the tsar: Ryasakov; Grinevitsky; Timothy Mihailov, a factory hand, aged twenty-one; and Ivan Emilianov, a student, aged nineteen. Mikhailov was hanged with the others in 1881 and Emilianov earned himself a life of hard labor in Siberia. See Footman (1994), Radzinsky (2005); Walter G. Moss, *Alexander II and His Times: A Narrative History of Russia in the Age of Alexander II, Tolstoy, and Dostoevsky* (London: Anthem Press, 2002); and Hugh Seton-Watson, *The Decline of Imperial Russia, 1855–1914* (London: Methuen, 1952). See also Cronin (2009), 123–5.

55. Richard Pipes, *Russia under the Old Regime* (London: Penguin, 1974, 2nd ed., 1995), 298; Radzinsky puts the number at twenty-five. Radzinsky (2005), 311.

56. The number of members of Narodnaya Volya is disputed and difficult to confirm. According to one source, People's Will had some 500 full members and several thousand sympathizers behind it. This estimate is higher than most other sources. S. S. Volk, *Narodnaya Volia 1879—1882* (Moscow and Leningrad, 1966), 277; cited by David Saunders, *Russia in the Age of Reaction and Reform, 1801–1881* (London: Longman, 1992), 336.

57. Radzinsky (2005), 426.

58. Vera Figner's execution was commuted at the last minute and she died in prison many years later. During that time she wrote a popular memoir whose English translation is *Memoirs of a Revolutionist* (DeKalb: Northern Illinois University Press, 1991).

59. Nobel's regret about terrorist use of dynamite is well documented. See, for example, Michael J. Schaack, *Anarchy and Anarchists: A History of the Red Terror and the Social Revolution in America and Europe* (New York and Philadelphia: W. A. Houghton, 1889), 29: "In speaking of the invention, Adolf Houssaye the French litterateur recently said: "He [Nobel] is a strong advocate of peace and regards with upmost horror the use of dynamite by

assassins and political conspirators. . . . 'Look you,' [Nobel] exclaimed. 'I am a man of peace. But when I see these miscreants misusing my invention, do you know how it makes me feel? It make me feel like gathering the whole crowd of them into a storehouse full of dynamite and blowing them all up together!' "

60. Many Irishmen were also driven by a crippling potato famine and hope for economic opportunity. Between 1947 and 1955, 1,300,000 Irish men and women immigrated to the United States. Short (1979), 18.

61. All from Niall Whelehan, *The Dynamiters: Irish Nationalism and Political Violence in the Wider World, 1867–1900* (Cambridge, UK: Cambridge University Press, 2015), chapter 3, 138–176. The Clan na Gael even built a submarine, the "Fenian Ram," to attack British merchant ships. See "A Fenian Dynamite Ram: A Submarine Boat Built for Use against England," *New York Times*, 19 January 1896.

62. This section draws heavily on Lindsay Clutterbuck's excellent analysis of the techniques of the Irish Fenians, 1879–1885. Clutterbuck (2004), 154–81.

63. The casualty figures differ in different sources. Clutterbuck (2004), p. 160 lists 15 deaths. Kenna (2014), p. 11, and English (2006), p. 181 say 12 were killed and some 126 injured. Contemporaneous newspaper reports likewise list 12. "Many Killed as Fenians Try to Blow Up Prison," *Manchester Guardian*, 14 December 1867, at https://www.theguardian.com/theguardian/2010/dec/14/archive-fenians-blow-up-prison. The intentions of the bombers are disputed. Short (1979) argues that the killing of innocents had been unintended. Having carefully analyzed Scotland Yard records of the bombs used, Clutterbuck (2004) counters that their size and placement proves otherwise.

64. Richard English, *Irish Freedom: The History of Nationalism in Ireland* (London: Macmillan, 2006), 181.

65. Henri Le Caron, *Twenty-five Years in the Secret Service* (London: Heinemann, 1892), 109–11; cited by Clutterbuck (2004), 160.

66. Quoted by Bernard Porter, *The Origins of the Vigilant State: The London Metropolitan Police Special Branch before the First World War* (Woodbridge, UK: The Boydell Press, 1987), 8.

67. Clutterbuck (2004), pp. 160, 163. The phrase is from an article in *Irish World*, 4 December 1875, cited by Short (1979), 38.

68. Vivian Majendie to William Harcourt, 21 December 1881, Public Record Office, HO144/84/A7266B; in Clutterbuck (2004), 165, note 47.

69. The Skirmishers were a small splinter group of Clan na Gael and much less well funded. All of the attacks by Irish nationalists are available in the database explained in Appendix B.

70. *The Citizen*, 22 December 1883 TNA HO 144/1537/1; cited by Shane Kenna, *War in the Shadows: The Irish-American Fenians Who Bombed Victorian Britain* (Newbridge, Ireland: Merrion, 2014), 94 and 353, note 85.

71. Kenna (2014); K. R. M. Short, *The Dynamite War: Irish-American Bombers in Victorian Britain* (Dublin: Gill and Macmillan, 1979); and Niall Whelehan, "Skirmishing, the Irish World, and Empire, 1876–86," *Eire-Ireland* 42, no. 1/2 (spring/summer 2007), 180–200.

72. Larsson (2008), 92.

73. Clutterbuck (2004), 176.

74. Inspector Morris Moser, as told to Charles Rideal in *Stories from Scotland Yard* (London: George Routledge and Sons, 1890), 20–27, at https://archive.org/details/129929299.2300.emory.edu.

75. Fortunately, the first bomb failed to detonate, and the other three were found and rendered harmless. Clutterbuck (2004), 169.

76. Whelehan, 195–6.

77. The dynamite was manufactured in Philadelphia. Short (1979), 205.

78. Exact casualty figures are difficult to confirm. The deaths of two police officers (Cole and Cox) appear in numerous British newspaper sources, but only vague references to other civilian casualties.

79. Short (1979), 204–10; Joseph McKenna, *The Irish-American Dynamite Campaign: A History 1881–1896* (Jefferson, NC: McFarland, 2012), 91–118.

80. For specific attacks and casualty figures, see our original database of incidents from 1867 to 1934, accessible at audreykurthcronin.com.

81. Whelehan (2015), 295–6.

82. Richard Bach Jensen, *The Battle against Anarchist Terrorism: An International History, 1878–1934* (Cambridge, UK: Cambridge University Press, 2014), 72–73; Short, 229–42; and Whelehan (2015), 295–303.

83. Short (1979), 232–3.

84. William Godwin, *An Enquiry concerning Political Justice,* Oxford World's Classics, with an introduction and notes by Mark Philp (Oxford, UK: Oxford University Press, 2013). His ideas are usually referred to as "philosophical anarchism."

85. Pierre-Joseph Proudhon, *What Is Property?*, Cambridge Texts in the History of Political Thought (Cambridge, UK: Cambridge University Press, 1994).

86. They inspired or engaged directly in violent actions, such as Stephniak-Kravchinsky's stabbing of the head of the secret police in the streets of St. Petersburg in 1878. The Russian nihilist movement was crushed in the aftermath of the assassination of Tsar Alexander II, forcing most revolutionaries underground.

87. Historian Richard Bach Jensen's assesses that between 1878 and 1914, over 220 people died and over 750 were injured, including bombers who accidentally blew themselves up, in anarchist attacks outside of Russia. Richard Bach Jensen, *The Battle against Anarchist Terrorism: An International History, 1878–1934* (Cambridge, UK: Cambridge University Press, 2014), 36. Jensen's earlier article has a lower number: at least 160 people were killed and 500 wounded. Richard Bach Jensen, "The International Campaign

against Anarchist Terrorism, 1880–1930s, *Terrorism and Political Violence* 21, no. 1 (2009): 90.

88. Anna Geifman, *Thou Shalt Kill: Revolutionary Terrorism in Russia, 1894–1917* (Princeton, NJ: Princeton University Press, 1995), 125. Given the state of government collapse at the time, it is unclear in the casualty figures which were victims of anarchists, social revolutionaries, or civil war.

89. Congressional Record, 60th Congress, 1st Session, Volume XLII, Part 5 [Senate, 9 April 1908], p. 4526; as quoted and cited by R. B. Jensen, "Daggers, Rifles, and Dynamite: Anarchist Terrorism in Nineteenth-Century Europe," *Terrorism and Political Violence* 16, no. 1 (2004): 117 and note 4.

90. Jensen (2014), 23–24; Jensen, "Daggers, Rifles, and Dynamite," p. 133. Vera Zasulich was a social revolutionary, and later a Menshevik.

91. Ronald Seth, *The Russian Terrorists: The Story of the Narodniki* (London: Barrie and Rockliff, 1966), chapter 25: "A Dozen Men and Women," 164–76.

92. See, for example, Benjamin Grob-Fitzbiggon, "From the Dagger to the Bomb: Karl Heinzen and the Evolution of Political Terror," *Terrorism and Political Violence* 16, no. 1 (spring 2004): 97–115. The outcome may have more to do with the available weapon than the intentions of the killer.

93. Jensen (2004), 133.

94. Passionate debate rages over the provenance of this bomb. See, for example, Timothy Messer-Kurse, James O. Eckert Jr., Pannee Burckel, and Jeffrey Dunn, "The Haymarket Bomb: Reassessing the Evidence," *Labor: Studies in Working-Class History of the Americas* 2, no. 2 (2005): 39–51; and the response by Bryan D. Palmer, "CSI Labor History: Haymarket and the Forensics of Forgetting," *Labor: Studies in Working-Class History of the Americas* 2, no. 2 (2005): 25–36. It was constructed exactly as instructed in Joseph Most's manual (see below).

95. Gage (2009), p. 51. The exact numbers are disputed. See also Jensen (2014), p. 30.

96. Of the eight, one was sentenced to fifteen years in prison, two had their death sentences commuted, one condemned prisoner hanged himself in his cell, and the remaining four were publicly hanged.

97. The Galleanist Italian group "The Eleventh of November" was named after the date in 1887 when they were killed. See Jeffrey D. Simon, "The Forgotten Terrorists: Lessons from the History of Terrorism," *Terrorism and Political Violence* 20, no. 2 (2008): 197.

98. Jensen (2014), 44–52.

99. Dirk Hoerder, *Plutocrats and Socialists: Reports by German Diplomats and Agents on the American Labor Movement, 1878–1917* (Munich: K. G. Saur, 1981), 371 and 382–3; cited by Jensen (2014), 47.

100. Jensen (2014), 31. Jensen concedes that other periods saw the deaths of more heads of state, especially in newly independent states of the developing world, but not in stable, established countries.

101. Dynamite had catalyzed the movement through its symbolic show of strength, especially by ordinary people against monarchs, but as time passed, those who were inspired by the Dynamitards did not always use a bomb. French president Carnot and Empress Elisabeth were both stabbed; Spanish prime minister del Castillo, Italian king Umberto and US president McKinley were all shot. Protection details had by this point improved; for high-profile, well-defended targets, dynamite bombings now mainly killed bystanders.

102. Richard Bach Jensen, "The International Campaign against Anarchist Terrorism, 1880s–1930s," *Terrorism and Political Violence* 21, no. 1 (2009): 91–92.

103. "Madrid Outrage," *The Brisbane Courier*, 4 June 1906, p. 4, at http:// trove.nla.gov.au/ndp/del/page/1553330?zoomLevel=1; and Jensen (2014), 299–301.

104. These may have been more properly labeled "nationalists." Bombings were often labeled "anarchist" attacks whether or not they were connected to or even inspired by the ideological movement.

105. Jensen (2014), 295–6.

106. It is difficult to discern exactly how the Sarajevo assassins self-identified. John Zametica argues that they were aligned with Croatia, not Serbia or Bosnia. *Folly and Malice: The Habsburg Empire, the Balkans, and the Start of World War One* (London, UK: Shepheard-Walwyn, 2017). I am grateful to Professor Sir Hew Strachan for pointing this out.

107. Christopher Clark, *The Sleepwalkers: How Europe Went to War in 1914* (New York: HarperCollins, 2013), 54. The Russians may also have played a role in the plot, pursuing a long-standing pan-Slav policy of supporting Serbia. Clark (2013), 17.

108. George Woodcock, "Anarchism in Russia," chapter 13 of *Anarchism: A History of Libertarian Ideas and Movements* (Cleveland: World Publishing Company, 1962), 399–424; Joll (1980), 166; and Vladimir Iu. Chernaiev, "Anarchists," in Edward Acton, V. Iu. Cherniaev, and William G. Rosenberg, eds., *Critical Companion to the Russian Revolution, 1914–1921* (Bloomington: Indiana University Press, 1997), 225.

109. All-Russian Extraordinary (or Emergency) Secret Commission, in Russian Всероссийская Чрезвычайная Комиссия or ВЧК (Ve-Che-Ka). Members in the organization were referred to as "chekists."

110. For more information see Philip Pomper, *Lenin's Brother: The Origins of the October Revolution* (New York: W. W. Norton, 2009).

111. In 1918, Lenin himself had been shot and badly injured by Fanya Kaplan, who was a Social Revolutionary. It was a crucial turning point leading to the civil war.

112. George Woodcock, *Anarchism: A History of Libertarian Ideas and Movements* (New York: Penguin Books, 1962; 2nd ed., 1975), 355.

113. Again, the argument here is not that state power ceased to exist, but that an unanticipated kind of new regime emerged. The anarchist wave of terrorism played a key role in causing the tsarist regime to overreact, weakening it, and contributing to the emergence of the repressive Soviet regime that ultimately replaced it. Neither the social revolutionaries nor the Russian anarchists actually wanted to *rule*. Like most groups in the first wave of modern terrorism, their attacks were consciously designed as provocations, and catalysts for change.

114. James Joll, *The Anarchists* (Cambridge, MA: Harvard University Press, 1980), chapter 7, "The Revolution That Failed," 158–76.

115. See the anarchist database of attacks, compiled for this book and explained in Appendix B.

116. Beverly Gage, *The Day Wall Street Exploded: A Story of America in its First Age of Terror* (Oxford, UK: Oxford University Press, 2009), 27; and Robert K. Murray, *Red Scare: A Study in National Hysteria, 1919–1920* (Minneapolis: University of Minnesota Press, 1955), 78. According to Murray, the reason the bomb was not more deadly is that it went off prematurely when the perpetrator stumbled on the stone steps leading up to the front door. The bomb killed him.

117. Simon (2008), 203.

118. *Cleveland Plain Dealer*, 4 June 1919; cited by Paul Avrich, *Sacco and Venzetti: The Anarchist Background* (Princeton, NJ: Princeton University Press, 1991), 165.

119. Murray (1955), 72.

120. Ibid., 212–17.

121. Ibid., 266–7.

122. Linda Levine, *The Labor Market during the Depression and the Current Recession*, Report #R40655, US Congressional Research Service, 19 June 2009, p. 8.

123. US House of Representatives, Labor Committee Hearings, 29 May 1934, p. 1202; cited by Michael Goldfield, "Worker Insurgency, Radical Organization, and New Deal Labor Legislation," *American Political Science Review* 83, no. 4 (December 1989): 1273.

Chapter 4

1. W. W. Huntley and F. M. Robinson, *Catalogue of Standard List-Price of Material Used by Railroads 1900* (Richmond, VA: I. N. Jones, 1900), advertisement on p. 35.

2. Finn J. D. John, "Dynamite Used to Be a Regular Part of Oregon Life," *Offbeat Oregon*, 11 January 2015. Notably, by 1894, only the United States continued to allow and use liquid nitroglycerine, particularly for digging oil wells.

3. David McCullough, *The Wright Brothers* (New York: Simon & Schuster, 2015).

4. Sally Shuttleworth and Berris Charnley, "Science Periodicals in the Nineteenth and Twenty-first Centuries," *Notes and Records of the Royal Society* (2016), 3, at http://rsnr.royalsocietypublishing.org; and Bernard Lightman, "Popularizers, Participation, and the Transformations of Nineteenth-Century Publishing: From the 1860s to the 1880s," *Notes and Records of the Royal Society* (2016), 3, at http://rsnr.royalsocietypublishing.org.

5. Ruth Barton, "Just before Nature: The Purposes of Science and the Purposes of Popularization in Some English Popular Science Journals of the 1860s," *Annals of Science* 55, no. 1 (1998): 1–33; and Ralph O'Connor, "Reflections on Popular Science in Britain: Genres, Categories, and Historians," *Isis* 100, no. 2 (June 2009): 333–45.

6. Lightman (2016).

7. Notably, more than 150 Nobel laureates are among *Scientific American*'s authors.

8. A. Nobel, "On Dynamite, a Recent Preparation of Nitroglycerine, as a Blasting Agent," *Report of the British Association for the Advancement of Science, 38th Meeting* (1868), 194, at https://www.biodiversitylibrary.org/item/93116 - page/810/mode/1up. "Dynamite—Review of a Paper by M. Nobel, The Inventor," *Scientific American* 19, no. 14 (30 September 1868): 216. *Scientific American* is the oldest continuously published magazine in the United States. It reports on emerging scientific and technological trends, and advocates for invention. It founded the first branch of the US Patent Agency in 1850, providing technical and legal advice to inventors. The magazine claims that by 1900, some 100,000 inventions had been patented thanks to *Scientific American*. "About Scientific American," at https://www.scientificamerican.com/page/about-scientific-american/.

9. J. P. Cooke Jr., "The Atmosphere as an Anvil," *Popular Science Monthly*, 1 June 1874, pp. 220–4.

10. "Dynamite as a Stump Puller for Land Reclamation," *Scientific American* 30, no. 22 (30 May 1874): 341; "Plowing with Dynamite," *Scientific American* 36, no. 17 (28 April 1877): 262. Nobel's agents had demonstrated this use in vineyards a year earlier. See "Explosive Agriculture—Dynamite vs. Plows," *Scientific American* 35, no. 16 (14 October 1876): 244; "Uses of Dynamite in Grubbing," *Gippsland Times* (Victoria, Australia), 28 November 1881, p. 4. See also William Young, "Dynamite on the Farm: Explosives as a Substitute for Ax, Plow, and Spade," *Scientific American* 104, no. 7 (18 February 1911): 163.

11. "Killing Cattle by Dynamite," *Scientific American* 37, no. 8 (25 August 1877): 118.

12. Kate Galbraith, "Fishing with Dynamite," *Scientific American* 54, no. 20 (15 May 1886): 310; and "Destruction of Fish by Dynamite," *Scientific American* 51, no. 4 (26 July 1884): 52.

13. "The Horrors of Fishing with Dynamite," *New York Times,* 4 February 2015.

14. "New Uses for Dynamite," *Richmond River Herald* and *Northern Districts Advertiser* (Australia), 10 October 1890, p. 8.

15. "Rainmaking by Dynamite," *Scientific American* 100, no. 13 (27 March 1909): 238.

16. "The Dynamite Gun," *Scientific American* 58, no. 23 (9 June 1888): 361; "Novel Applications of Dynamite," *Scientific American* 39, no. 15 (12 October 1878): 225; "Experiments with Explosives," *Scientific American* 42, no. 26 (26 June 1880): 406; also cited by Whelehan (2012), 144–5.

17. "Dynamite—Its Uses and How to Handle It," from the *Indian Engineer,* Calcutta, India; reprinted in *Scientific American* 58, no. 17 (28 April 1888): 260.

18. "Novel Application of Dynamite," *Scientific American* 39, no. 15 (12 October 1878): 225.

19. For a rich history of the evolution, accessibility, and legal treatment of weapons manuals, especially in the United States, see Ann Larabee, *The Wrong Hands: Popular Weapons Manuals and Their Historic Challenges to a Democratic Society* (Oxford, UK: Oxford University Press, 2015).

20. Dynamite bombs were the most innovative and effective but newspaper accounts did not always specify which was used. There was a range of small explosives, including "Orsini bombs," "petards," and "mail bombs." Anarchists in Barcelona filled coffee grinders, metal boxes, and tin pans with explosives, for example. See: J. Romero Maura, "Terrorism in Barcelona 1904–1909," *Past and Present* (Dec. 1968), 148.

21. Ronald S. Coddington, "Pinned Down at Port Hudson," *New York Times*, 14 June 2013, at https://opinionator.blogs.nytimes.com/2013/06/14/pinned-down-at-port-hudson/.

22. Sydney C. Kerksis and Thomas S. Dickey, *Field Artillery of the American Civil War* (Atlanta: Phoenix Press, 1968), 459; Anthony Saunders, *Reinventing Warfare 1914–18: Novel Munitions and the Tactics of Trench Warfare* (New York: Continuum, 2012), 25–29.

23. A. Bortnovski, "Hand Grenades in the Russo-Japanese War" (translated from the *Voenni Shornik,* January 1910), *RUSI Journal* 54, no. 389 (1910): 918–22.

24. "Dynamite Shells," *Scientific American* 51, no. 17 (25 October 1884): 265; "The Pneumatic Dynamite Gun," *Scientific American* 55, no. 2 (10 July 1886): 16; "The New Dynamite Gun," *Scientific American* 58, no. 18 (5 May 1888): 278; and "The Dynamite Gun," *Scientific American* 58, no. 23 (9 June 1888): 361. The Pneumatic Dynamite Gun company apparently went bankrupt. See "Gun-makers in Trouble: Misfortunes of the Pneumatic Dynamite Company," *Chicago Daily Tribune,* 4 July 1891, p. 5.

25. "Explosion of a Dynamite Gun," *Scientific American* 62, no. 23 (7 June 1890): 357; and "A Dynamite Shell Bursts a Twelve Ton Gun," *Scientific American* 63, no. 12 (20 September 1890): 183.

26. "The Dynamite Cruiser Vesuvius," *Scientific American* 59, no. 18 (3 November 1888): 277; and "Another Test Ordered: The Value of the Vesuvius as an Offensive Weapon to be Determined," *The Sun* (Baltimore), 10 June 1891. In 1905, the Vesuvius was refitted to become an experimental torpedo boat and in 1915, when one of its torpedoes executed a circular path, it was damaged by its own torpedo.

27. A. D. Harvey, "The Hand Grenade in the First World War," *RUSI Journal* 138, no. 1 (1993): 44–47.

28. Kelly (2004), p. 229.

29. J. P. Cundill, *A Dictionary of Explosives* (Chatham, UK: W. & J. Mackay, 1889; 2nd ed. 1895), 110. Entry # 654, "Nitro-magnite or Dyna-magnite," Lieut. Col. Cundill was a British explosives inspector under the Explosives Act of 1875.

30. W. J. Reader, *Imperial Chemical Industries: A History,* Vol. 1: *The Forerunners, 1870–1926* (Oxford, UK: Oxford University Press, 1970), 76; and Lundström, "Alfred Nobel's Dynamite Companies," Nobelprize.org, Nobel Media AB, 2014, at https://www.nobelprize.org/alfred_nobel/biographical/articles/lundstrom/.

31. Cundill (1895), pp. 42–47.

32. Cundill (1895). Captain J. H. Thomson, also an inspector of explosives, brought this second edition up to date. See also Marcellin Berthelot, *Explosives and Their Power* (London: John Murray, 1892). The different mixtures and types of dynamite were very useful to law enforcement. In the aftermath of Fenian bombings in Britain, for example, it was often easy to determine by its specific components that a particular dynamite had come from the United States.

33. Ragnhild Lundström, *The Nobel Dynamite Companies: Multinational Enterprise of Their Time—and Ours?*, Research Report No. 12, Uppsala Papers in Economic History 1986, Department of Economic History, Uppsala University, Uppsala, Sweden, pp. 12–13. Nobel fought patent battles over many of his inventions, including nitroglycerine (in the United States against Taliaferro Preston Shaffner—Nobel won) and ballistite (in England, against Frederic Abel and James Dewar, the inventors of cordite—Nobel lost). See R. W. Reid, *Tongues of Conscience: Weapons Research and the Scientists' Dilemma* (New York: Walker, 1969), chapter 1; and Fant (1991), 80–83, 308–10.

34. A scramble ensued between Nobel's Explosives and the German company Krebs, which produced Lithofracteur. With higher production rates and lower costs, Nobel held Krebs' share of the Australian market to about 20 percent. Reader (1970), 67–89.

35. Dynamo consisted of perchlorate, nitrate, and picrate of potash, tar, and sawdust.

36. Jones Dynamite was 35 percent nitroglycerine plus kieselguhr and sulphate of lime.

37. Fant (1991), 129.

38. Barbe's father, who was an ironmaster living near Nancy in France, agreed to bankroll Nobel's projected French business. Reader (1970), 70.

39. Barbe, who was serving in the French artillery, was captured at the Battle of Sedan in September 1870. He negotiated the agreement upon his release in 1871. Reader (1970), 70, and Fant (1991), 128.

40. Notably he did not work with bankers. Reader (1970), 24.

41. Lundström (1986), 7.

42. Ibid., 6–8.

43. In a May 1875 speech before the Society of Industrial Arts in London, Nobel told the audience that his production had gone from 785 tons of dynamite in 1871, to 1350 tons in 1872, growing to 2,050 tons in 1873 and 3,120 tons in 1874. He projected that production at the Hamburg plant would be about 1,200 tons for that year, with some 400 exported. Fant (1991), 134–5. The Krummel plant was bombed at the end of World War II, ironically flattened by explosives that Nobel had helped to invent. In the 1950s it became the location of a nuclear power plant.

44. Water displaced the oil in the putty, and the nitroglycerine oil was essentially pushed out.

45. Berthelot (1892), 436–7.

46. "The Manufacture of Dynamite," *Scientific American* 68, no. 9 (4 March 1893): 132–3; and Reader (1970), 28–29 and plates 15–17.

47. Two explosions happened on 17 March 1870, killing three men and a boy. There is ambiguous historical evidence as to exactly where it was sited. See Douglas F. Job, "Indiana Job and the Nitroglycerine Factory," *Newslettter of the Bergen County Historical Society* (early winter 2008): 12–15, at bergencountyhistory.org.

48. E. Rose, "Giant Powder Company: Historical Essay," published in the Winter 2007/2008 edition of the Glen Park News, at http://www.foundsf. org/index.php?title=Giant_Powder_Company. Giant Powder began manufacturing dynamite at a factory in Rock Canyon, California, in March 1868, where the first explosion occurred, killing two people and injuring nine others. Giant relocated their plant farther from San Francisco to a location south of Golden Gate Park, but that complex was destroyed by another explosion in January 1879, killing four. The factory was then moved to Fleming Point, beginning operation in fall 1879, where the next three deadly explosions occurred.

49. Alfred Nobel to J. Downie, April 1869, Alfred Nobel's business correspondence, archives of the Nobel Institute, Stockholm; quoted Reader (1970), 23 and 35, note 13. Also cited by Pellew (1974), 181.

50. "Terrible Explosion of Nitroglycerine," *Western Daily Press*, Bristol, England, 2 July 1869; and "Frightful Explosion of Nitro-Glycerine, Near Carnarvon," *Cardiff Times*, South Glamorgan, Wales, 3 July 1869.

51. "Frightful Explosion of Nitro-Glycerine," *The Derby Mercury*, 7 July 1869, 6.

52. Cundill (1895), xxx.

53. This was not the first act regulating nitroglycerine in England. In 1866, the Carriage and Deposit of Dangerous Goods Act declared that nitroglycerine must be clearly labeled "specially dangerous," and establishing that as a category for all comparable explosives to follow. But there was no means of enforcing the act. Jill H. Pellew, "The Home Office and the Explosives Act of 1875," *Victorian Studies* 18, no. 2 (December 1974): 181; and Fant (1991), 141.

54. Scottish industrialist Charles Tenant was his local partner in Scotland, and the company was headquartered in Glasgow. The factory was renamed Nobel's Explosive Company in 1877. See also Reader (1970), 26–27.

55. Fant (1991), 145.

56. Reader (1970), 30.

57. The act applied to "gunpowder, nitro-glycerine, dynamite, gun-cotton, blasting powders, fulminate of mercury or of other metals" and other explosives, plus all elements that can cause an explosion or pyrotechnic effect. Explosives Act of 1875, Section 3, at http://www.legislation.gov.uk/ukpga/Vict/38-39/17/section/3.

58. On 2 October 1874, there was an accident on the Regent's Canal, right in the tony neighborhood of Regent's Park, also threatening the London Zoo. A barge carrying in its open hold a dangerous combination of gunpowder and petroleum spectacularly blew up, killing three people on the barge and damaging the surrounding property. The London *Times* described the damage to houses, as well as to the elephant, monkey, and giraffe houses, reporting, "The giraffes were found huddled together in terrible fear." "Gunpowder Explosion," *The Times* (London), 3 October 1874, 9.

59. Pellew (1974), 187–9.

60. Ibid., 193.

61. Dynamite was not protected by patents in Germany during the 1870s. Lundström, "Alfred Nobel's Dynamite Companies," Nobelprize.org, Nobel Media AB, 2014, at https://www.nobelprize.org/alfred_nobel/biographical/articles/lundstrom/.

62. Lundström (1986), 11. According to Lundström, Nobel delayed the merger Barbe had suggested, until the competition between the British and the German companies became very intense and merger was unavoidable.

63. Fant (1991), 135–6 and 152.
64. Brown (1998), 32–33.
65. The companies were: E. I. du Pont de Nemours and Co., the Laflin and Rand Powder Co., the Oriental Powder Mills, the American Powder Co., Miami Powder Co., and the Hazard Powder Co. William Stevens, "The Powder Trust, 1872–1912," *Quarterly Journal of Economics* 26 (May 1912): 445.
66. Stevens (1912), 447–8.
67. The 1888 deal was superseded by an even stronger one in 1897, which covered the sale of all explosives in the United States. Mira Wilkins, *Foreign Investment in the United States to 1914* (Cambridge, MA: Harvard University Press, 1989), 387–8.
68. Alfred Nobel to Thomas Johnston, June 1889, business correspondence, archives of the Nobel Institute, Stockholm, Sweden; cited by Reader (1970), 21 and 35, note 7.
69. William Stevens, "Notes and Memorandum: The Dissolution of the Powder Trust'," *Quarterly Journal of Economics* 27 (November 1912): 202–7.
70. Lundström (1986), 12.
71. Brandon Dupont, Drew Keeling, and Thomas Weiss, "Fares for Overseas Travel in the 19th and 20th Centuries," Paper prepared for the Annual Meeting of the Economic History Association, 21–23 September 2012, EHA paper 8-15-2012, *passim*, but especially Figure One, p. 32, at http://eh.net/eha/wp-content/uploads/2013/11/Weissetal.pdf. Steerage fares did not vary as much. The paper investigates changes in advertised fares published by the major transatlantic cruise lines (Cunard, etc.) over time.
72. Paul A. Gilje, *Rioting in America* (Bloomington and Indianapolis: Indiana University Press, 1996), 124 and 218, note 17.
73. Gilje (1996), 124–5.
74. George Woodcock, ed., *The Anarchist Reader* (Brighton, Sussex: Harvester Press, 1977), 43.
75. James Joll, *The Anarchists* (Cambridge, MA: Harvard University Press, 1980), 109–11; and Caroline Cahm, *Kropotkin and the Rise of Revolutionary Anarchism, 1872–1886* (Cambridge, UK: Cambridge University Press, 1989), 157–8. Cahm translates the phrase as "methods of defence and attack."
76. Richard Bach Jensen, *The Battle against Anarchist Terrorism* (2014), 17–18.
77. Joll (1980), 109–11; and Constance Bantman, "Internationalism without an International? Cross-Channel Anarchist Networks, 1880–1914," in *Revue belge de philologie et d'histoire*, tome 84, fasc. 4, 2006. Histoire medievale, moderne et contemporaine—Middeleeuwse. moderne en hedendaagse geschiedenis. p. 965, at http://www.persee.fr/doc/rbph_0035-0818_2006_num_84_4_5056.
78. The mass media tended to label any violent activist as "anarchist," which may have made the anarchists—and anarchist terrorists—seem more powerful than they actually were.

79. Jensen (2014), 2017; Jensen (2004), 127–9.
80. See, for example, "Dynamite Eloquence: Forming an Army of Irish Crusaders to Blow Up England," *New York Times*, 31 December 1883, p. 5.
81. T. Lizius, "Dynamite," *The Alarm*, 21 February 1885, p. 3; *Illinois vs. August Spies et al.* trial evidence, 29 July 1886, at http://www.chicagohistory.org/hadc/transcript/exhibits/X000-050/X0390.htm. The article was also published in Arbeiter-Zeitung. See also http://www.chicagohistoryresources.org/dramas/act1/fromTheArchive/anarchistAmmunition_f.htm.
82. Quoted by Paul Avrich, *The Haymarket Tragedy* (Princeton, NJ: Princeton University Press, 1984), 167. *Vorbote* was published in Geneva from 1866 to 1871.
83. An earlier title was *Military Science for Revolutionaries.* Johann Most, *Science of Revolutionary Warfare*. Originally published 1884; subsequently translated and published as a pamphlet by Desert Publications, Eldorado, AZ, 1978, 74 pages. Most's manual also recommends chemical and biological terrorism, including the use of prussic acid and the seeds of thorn apple weed. It was written in German but was easily understandable by many Americans at the time.
84. Most (1884), 1.
85. Ibid., 3.
86. Ibid., 11–12.
87. Czolgosz shot McKinley at point blank range. Historian Richard Bach Jensen refutes the claim that Goldman's speech inspired Czolgosz's actions. For his detailed analysis of Czolgosz's motivations see Jensen, *The Battle against Anarchist Terrorism* (2014), 240–8.
88. "Remarkable Detailed Confession by the Anarchist Czolgosz," *St. Louis Republic*, 8 September 1901; and "Emma Goldman, High Priestess of Anarchy, Whose Speeches Inspired Czolgosz to His Crime," *Chicago Daily Tribune*, 8 September 1901.
89. The Anarchist Newspaper and Other Periodical Publication Data Set (1867 to 1934), including references to 644 anarchist and publications produced between 1867 and 1934, was compiled for this book and its associated projects. Variables collected and coded include publication name and type, production location (city and country), dates of publication, language of publication, and, when available, circulation numbers and publication interval (e.g., daily, weekly, monthly, etc.). See Appendix B and forthcoming book website at audreykurthcronin.com.
90. Merriman (2009), 111–12.
91. The full quote is, "There seems to be an epidemic of fake news from the city of Lincoln, and it all comes from Mr. Bryan's 'friends'—names not given." Williams Jennings Bryan, "Fake News from Lincoln," *The Commoner Condensed*, Vol. 7 (Chicago: The Henneberry Company, 1908), 173.

92. William Jennings Bryan, *The Commoner Condensed* (Spencer County, IN: The Abbey Press, 1902), 112–13.

93. *Davenport Daily Republican*, Davenport, Iowa, 29 August 1896. See also F. W. Blackmar, "The Promises of Democracy: Have They Been Fulfilled?," *The Forum*, June 1896, p. 437.

94. Adrienne LaFrance, "How the Fake News Crisis of 1896 Explains Trump," *The Atlantic*, 19 January 2017, at https://www.theatlantic.com/technology/archive/2017/01/the-fake-news-crisis-120-years-ago/513710/.

95. "The Growth of 'Lloyd's' to Over a Million," *Lloyd's Weekly Newspaper*, 23 February 1896, p. 10. Parliament removed the Paper Duty tax on 1 October 1861.

96. Jensen (2014), 53.

97. Anthony Fellow, *American Media History* (Boston: Wadsworth, 2013), 152.

98. Merriman (2009), 89–90.

99. Mike Davis (2007), 1–3.

100. The Anarchist Bombing Incident Data Set (1867–1934) contains records of 1,291 worldwide political violence bombing incidents (not just those who strictly speaking were official "anarchists"), including date, location, and casualties. The title refers to the name of the first wave of modern terrorism. The data were extracted primarily from historical newspaper articles including the London *Times, New York Times*, and other contemporaneous newspaper sources in numerous languages from around the world, double-checked with data compiled or cited by Avrich, Jensen, Rappoport, and Whelan.
The Anti-Anarchist, Explosive Control, and Other Key Laws/Key Legal Event Data Set (1867 to 1934) contains the country, provisions, and enactment date of anti-anarchist and explosive control legislation passed and implemented during the period, along with country and dates for other key legal events such as the filing of patents, formation of explosive companies and legal trusts, arrest and persecution of key bombing suspects, and the signing of anarchist-related international agreements. Data were collected from historical newspaper articles and key secondary sources. All of the data are available on the book's associated website at audreykurthcronin.com.

101. Richard Bach Jensen gathered the dates of all the legislation. See Jensen, "The International Campaign against Anarchist Terrorism, 1880–1930s," *Terrorism and Political Violence* 21, no. 1 (2009): 91–92; and Jensen (2014), 92–93; and some new research and updates shared by personal email on 15 September 2017.

102. There were also some city ordinances, such as in New York City.

103. Mark Aldrich, "Regulating Transportation of Hazardous Substances: Railroads and Reform, 1883–1930," *The Business History Review* 76, no. 2 (summer 2002): 267–97. Shortly thereafter, in April 1908, Congress gave the Interstate Commerce Commission the power to

uphold the private regulation. As a result of the 1866 Wells Fargo building nitroglycerine explosion in San Francisco, there was a federal law prohibiting shipment of explosives in passenger vessels, but it was unenforced and largely ignored.

104. The Dynamite Factory Data Set (1867–1934) includes a total of 161 factories producing dynamite and other high explosives in over twenty countries between 1867 and 1934. Please note that the data we collected are for high explosives factories, which invariably produced dynamite (the first, most common, and easiest high explosive to produce) but sometimes other high explosives, as well. Except for in Britain (where after 1875, explosives inspectors provided an annual tally), we can find no good accessible data detailing exactly what mix of additional explosives each factory produced. As was the custom of the day, we are referring to them all as "dynamite factories." See Appendix B and book website at audreykurthcronin.com.

105. Proximity data in Europe and North America were calculated using ArcGIS for distances between bombing incidents and the closest high-explosives factory existing at the time of the bombing. In addition, a spatial regression analysis on bombing data in the United States and Western Europe showed a statistically significant relationship between bombing frequency and their proximity to high-explosive factories and anarchist publications. Details on the regression results and methodologies used can be found in the forthcoming manuscript entitled "Dynamite as a Disruptive Technology during the 'First Wave' of Modern Terrorism (1867–1934)" by Audrey Kurth Cronin, John Gudgel, and Laurie Schintler.

106. Jensen (2014), 147–8 and 270–5. This is a summary only; please see Jensen.

107. Richard Bach Jensen, "The United States, International Policing, and the War against Anarchist Terrorism, 1900–1914," *Terrorism and Political Violence* 13, no. 1 (2001): 32.

108. "Urges New Federal Law to Stop Bomb Outrages by Control of Explosives," *Washington Post*, 20 September 1920, p. 2. It is possible that the lack of explosives regulations contributed to German sabotage in the United States during World War I, particularly the 30 July 1916 explosion of the National Storage Company's munitions plant on Black Tom Island in Jersey City, New Jersey. The blast rocked Manhattan and Brooklyn, killing at least four people and injuring hundreds, also damaging the Statue of Liberty. The United States, which was still officially neutral, was producing munitions to be sold to the Allies. "Munition Explosions Cause Loss of $20,000,000; 2 Known to Be Dead, Many Missing, 35 Hurt; Harbor Raked by Shrapnel for Hours," *New York Times*, 31 July 1916, p. 1. Thanks to Hew Strachan for suggesting this connection.

109. William L. Chenery, "Red Record of Failure and of Innocent Victims: Fifty Years of Sporadic Anarchism, in the Light of the Bomb Theory of the Wall

Street Explosion—Failure of Law to Limit Sales and General Circulation of Dynamite and TNT," *New York Times*, 19 September 1920.

110. The first smokeless powders were made in 1846, when the German scientist Christian Schonbein invented guncotton (or nitrocellulose). It was notoriously unstable, however, and not used widely for firearms and artillery until improvements were made by French scientists 1884. The dates listed here are the dates when patents were first granted.

111. Nobel's French partner, Paul Barbe, wrote on 27 February 1880 that during work on the St. Gotthard tunnel, original dynamite blasted 40–50 feet per month, while blasting gelatin accomplished 70–80 feet per month. Fant (1991), 109.

112. Ibid., 231–3.

113. Ibid., 153.

114. Colin Schultz, "Blame Sloppy Journalism for the Nobel Prizes," Smithsonian. com, 9 October 2013.

115. Nobel's comment is recorded (in German) in Bertha von Suttner's personal memoirs. *Memoiren*, 270–272; quoted by Irwin Abrams, "Bertha von Suttner and the Nobel Peace Prize," *Journal of Central European* Affairs (1962): 296.

116. Sven Tägil, "Alfred Nobel's Thoughts about War and Peace," *Nobelprize.org*, Nobel Media AB 2014, accessed at http://www.nobelprize.org/alfred_nobel/ biographical/articles/tagil; and letter from Nobel to von Suttner, outlining the peace prize, 7 January 1893, at World Digital Library, https://www.wdl. org/en/item/11563/.

Chapter 5

1. Mikhail Kalashnikov, quoted by Kate Connolly, "Kalashnikov: 'I wish I'd made a lawnmower'," *The Guardian*, 29 July 2002, accessible at http://www. theguardian.com/world/2002/jul/30/russia.kateconnolly.

2. Martha Crenshaw rightly observes that plastic explosives were also very important to this wave; however, they had neither the symbolic resonance nor the global diffusion of the AK-47 for terrorist groups. Personal communication, 5 November 2018.

3. In the 2005 movie *Lord of War*, idealistic Interpol investigator Jack Valentine (played by Ethan Hawke and loosely based on American Lee S. Wolosky) tracks down Russian arms dealer Yuri Orlov (loosely based on Viktor Bout and several other Russian arms dealers). Valentine calls the AK-47 "the real weapon of mass destruction."

4. Larry Kahaner, "Weapon of Mass Destruction," *Washington Post*, 26 November 2006, at http://www.washingtonpost.com/wp-dyn/content/article/ 2006/11/24/AR2006112400788.html.

5. *The Alarm*, 18 October 1884; cited by Paul Avrich, *The Haymarket Tragedy* (Princeton, NJ: Princeton University Press, 1984), 166.

6. C. J. Chivers, *The Gun* (New York: Simon & Schuster, 2010), 12. Chivers's Pulitzer Prize–winning book is by far the best, most comprehensive, and most objective book on the Kalashnikov family of rifles. It was a crucial source throughout this chapter. Another useful book on the AK-47 is Michael Hodges, *AK-47: The Story of a Gun* (San Francisco: MacAdam/ Cage, 2007).

7. The Small Arms Survey calls the family of guns that followed "Kalashnikov-pattern" assault rifles, or AK-type firearms. "Documenting Small Arms and Light Weapons: A Basic Guide," Issue Brief Number 14 (July 2015), *Small Arms Survey*, at http://smallarmsurvey.org; and Jonathan Ferguson and N. R. Jenzen-Jones, "ARES Research Note 6—An Introduction to Basic AK Type Rifle Identification," Armament Research Services Pty. Ltd., 7 December 2014, at http://www.armamentresearch.com; and Chivers (2010), 16–17.

8. Chivers (2010), 15–16; and Joe Poyer, *The AK-47 and AK-74 Kalashnikov Rifles and Their Variations*, 4th ed. (Tustin, CA: North Cape Publications, 2013), 135–40.

9. Mikhail Kalashnikov with Elena Joly, *The Gun That Changed the World* (Malden, MA: Polity Press, 2006), xv; Christopher Carr, *Kalashnikov Culture: Small Arms Proliferation and Irregular Warfare* (Westport, CT: Praeger, 2008), 17–18. Mikhail Kalashnikov says there are between 60 and 80 million, and Carr puts the number at between 70 and 100 million. See also Chivers (2010), 12.

10. Phillip Killicoat, *Weaponomics: The Global Market for Assault Rifles*, Post-conflict Transitions Working Paper No. 10, World Bank Policy Research Working Paper 4202, April 2007; Jonathan Ferguson and N. R. Jenzen-Jones, "Armament Research Services (ARES), Research Note 6—An Introduction to Basic AK Type Rifle Identification," Armament Research Services Pty. Ltd., 7 December 2014, at www.armamentresearch.com; ARES (2015).

11. See Edward C. Ezell, *Small Arms of the World* (numerous editions) (New York: Stackpole Books); and Edward C. Ezell, *Small Arms Today: Latest Reports on the World's Weapons and Ammunition* (1986 and subsequent editions) (New York: Stackpole Books); Edward Clinton Ezell, *The AK47 Story Evolution of the Kalashnikov Weapons* (New York: Stackpole Books, 1986); Poyer (2013); Gordon L. Rottman, *The AK-47 Kalashnikov-series Assault Rifles* (Oxford: Osprey Publishing, 2011), among many others.

12. Kenneth Chase, *Firearms: A Global History to 1700* (Cambridge, UK: Cambridge University Press, 2013), p. 1.

13. Michael Howard, *War in European History* (Oxford, UK: Oxford University Press, 1976); Hew Strachan, *European Armies and the Conduct of War* (London: George Allen & Unwin, 1983).

14. *The Three Charters of the Virginia Company of London with Seven Related Documents 1606–1627* (Creative Space Independent Publishing Platform,

2014), 38. Most of the original Virginia settlers, all men and boys, had military training.

15. Philip J. Cook and Kristin A. Goss, *The Gun Debate: What Everyone Needs to Know* (Oxford, UK: Oxford University Press, 2014), 159.

16. General Washington to the Continental Congress, 2 September 1776, from the George Washington Papers at the Library of Congress, 1741–1799, accessed at http://memory.loc.gov/cgi-bin/query/r?ammem/mgw:@field(DOCID+@lit(gw060015)).

17. George Washington to Massachusetts General Court, 10 February 1776; The George Washington Papers at the Library of Congress, 1741–1799, at http://memory.loc.gov/cgi-bin/query/P?mgw:14:./temp/~ammem_kDWc. (Transcription from *The Writings of George Washington from the Original Manuscript Sources, 1745–1799*, edited by John C. Fitzpatrick). Note that in the early years of the Massachusetts Bay Colony, the state legislature was referred to as the "General Court," because in addition to making laws, it often sat as a judicial court of appeals.

18. Simon Head, "The Roots of Mass Production," chapter 2 of *The New Ruthless Economy: Work and Power in the Digital Age* (Oxford, UK: Oxford University Press, 2011), 17–37.

19. An ancient weapon in the Tower of London museum reportedly inspired him. Gunsmiths had tried to produce "repeater" or multi-shot long guns and muskets as early as the sixteenth century. Colts were not the first revolvers—just the first in mass production. Revolving chamber muskets and flintlock pistols long predated them. Brodie and Brodie (1962), 144; and Smithsonian, *Firearms: An Illustrated History* (New York: DK Publishing, 2014), 49 and 75.

20. Merritt Roe Smith, *Harpers Ferry Armory and the New Technology: The Challenge of Change* (Ithaca, NY: Cornell University Press, 1977). Some historians believe that Eli Whitney was the first developer of interchangeable parts; others dispute this. See Merritt Roe Smith, "Eli Whitney and the American System of Manufacturing," in *Technology in America: A History of Individuals and Ideas,* edited by Carroll Pursell Jr. (Cambridge, MA: MIT Press, 1982), 45–61; and Pamela Haag, *The Gunning of America: Business and the Making of American Gun Culture* (New York: Basic Books, 2016).

21. The last muzzle-loading rifle issued to the US Army was the Springfield Model 1863 Type II, developed in the middle of the Civil War. Smithsonian (2014), 103. The deadliness of this change in range is not a simple extrapolation, however, because the bullets ("Minnie balls") traveled slowly. According to Civil War historian Earl Hess, the bullets' path was a parabola-shaped arc, meaning they flew over the heads of men advancing at intermediate distances. Earl J. Hess, *The Rifled Musket in Civil War Combat: Reality and Myth* (Lawrence: University Press of Kansas, 2008).

22. Smithsonian (2014), 136; Chivers (2010), 36; Julia Keller, *Mr. Gatling's Terrible Marvel: The Gun That Changed Everything and the Misunderstood Genius Who Invented It* (New York: Penguin Books, 2008), 155. The cost per unit is an estimate, as it varies in different sources. Chivers cites Army ordnance manuals detailing a price of $612.50 + 20% apiece, whereas Keller claims the higher figure of $1,300 to $1,500 each but does not indicate a source.

23. The *concept* of a machine gun derived from centuries of experimentation beginning with the ancient Greeks, who experimented with a device to shoot arrows in rapid fire. Keller (2008), 145–50.

24. Ibid., 7.

25. Later refinements increased the rate to about 300 rounds per minute. C. Spencer, ed., *The Encyclopedia of North American Indian Wars, 1607–1890* (Oxford, UK: ABC-CLIO, 2011), 321.

26. Keller (2008), 153.

27. Ibid., 163 and 179.

28. Ibid., 175; John Ellis, *The Social History of the Machine Gun* (New York: Arno Press, 1981), 42.

29. Keller (2008), 167–8.

30. Workers at the Carnegie steel mill were required to work a twelve-hour day, six days a week. Ellis (1981), 42.

31. "Fought 13 Hours: Pinkerton's Men and Homestead Workmen, 250 against 5000. Killed, 11 Workmen, 10 Detectives—Wounded, 18 Workmen, 21 Detectives," *Boston Daily Globe,* 7 July 1892, p. 1; and "Blood Is Shed: Fierce Battles at Homestead. Pinkertons Met and Repulsed. Twenty Men Are Killed in the Fight. Many of the Others Are Badly Wounded. The Strikers Use Two Small Cannon and Hurl Dynamite at Their Foes," *Chicago Daily Tribune*, 7 July 1892, p. 1. It appears most of the strikers dead and wounded were shot. Pinkerton's men died of a variety of methods.

32. Tucker (2011), 321–2; Peter Smithurst, *The Gatling Gun* (New York: Bloomsbury, 2015), 33–38; and Chivers (2010), 92–97.

33. Donald R. Morris, *The Washing of the Spears: The Rise and Fall of the Zulu Nation* (New York: Simon and Schuster, 1965), 545–75.

34. Bruce Vandervort, *Wars of Imperial Conquest in Africa, 1830–1914* (London: UCL Press, 1998; Routledge, 2014), 49–50.

35. "On the whole, it would seem clear that the Ashantees had ceased to have the spirit to resist us, and when anything like a panic once takes possession of a semi-barbarous people it is fatal. Sir Garnet Wolseley's military difficulties, therefore, are happily reduced to a 'three days march through a hostile country, the shattered remains of a savage army, and an unfortified town.'" London *Times,* 12 February 1784, p. 9. "We are not surprised that the Ashantees were awestruck before the power of the Gatling gun. It is easy to understand that it is a weapon which is specially adapted to terrify

a barbarous or semi-civilized foe." "Foreign Items," *Army and Navy Journal* (US), 7 March 1874, p. 478.

36. Maxim was born in Maine but later became a naturalized British citizen and was knighted for his wide-ranging inventions.

37. The copy of the Maxim gun on the German side was the Maschinengewehr 08. Chivers (2010), 90.

38. Ibid., 88.

39. Ibid.

40. "New Machine Guns Ordered; Vickers Type Will Replace Those That Jammed," *New York Times*, 24 March 1916, p. 2.

41. Although significantly lighter than a machine gun, the BAR and did not become standard issue until 1938. It saw extensive service in World War II and the Korean War.

42. For a skeptical analysis of the MP-18s role, see Paul Cornish, *Machine Guns and the Great War* (South Yorkshire, UK: Pen and Sword Books, 2009), 116–17.

43. Chivers (2010), 163–4. The original name was Maschinenkarabiner 42, machine carbine rifle invented in 1942. A carbine is a smaller version of a full-sized rifle but lighter and with a shorter barrel. The rifles fired the 7.92mm Kurz round, which was an intermediate size cartridge. The Sturmgewehr was machine stamped and welded. The US military thought it was therefore cheap and inferior to the machine milled and forged American M1 Garand semiautomatic, manufactured at the famous Springfield Armory in Springfield, Massachusetts. They downplayed the significance of the German gun. Larry Kahaner, *AK-47: The Weapon That Changed the Face of War* (Hoboken, NJ: John Wiley & Sons, 2007), 20–21.

44. "The History of the Kalashnikov Gun," *Pradva*, 2 August 2003, at http://www.pravdareport.com/history/02-08-2003/3461-kalashnikov-0/. The AK-47 is not a copy of the German StG44. There are differences in their carrier mechanisms, receivers, and assembly, plus the AK has a cam system. Paul Graves-Brown, "Avtomat Kalashnikova," *Journal of Material Culture* 12, no. 3 (2007): 289–91.

45. Unfortunately, there are no open primary documents available to consult either on the AK-47 story or the development of its intermediate-sized ammunition. This is a serious drawback to research on this topic, because investigators must rely on Russian accounts (including Kalashnikov's several memoirs) that sometimes contradict each other.

46. S. G. Wheatcroft and R. W. Davies, "Population," chapter 4 of *The Economic Transformation of the Soviet Union, 1913–1945*, edited by R. W. Davies, Mark Harrison, and S. G. Wheatcroft (Cambridge, UK: Cambridge University Press, 2005), 77–79.

47. The local Communist Party delegates separated peasants into three categories: poor peasants (*bednyaki*), peasants of middling income (*serednyaki*),

and prosperous peasants (*kulaks*). A key distinguisher was how much livestock a family held.

48. Also called "de-kulakization." Kalashnikov and Joly (2006), 6–8.

49. Kalashnikov and Joly (2006), 26. Chivers points out that, of the seven male members of the Kalashnikov household in 1930, only two survived the next fifteen years unharmed. Chivers (2010), 185.

50. Prior to his conscription, Kalashnikov worked in the Turkestan-Siberian Railroad system, which gave him early on-the-job mechanical training. Edward Clinton Ezell, *Kalashnikov: The Arms and the Man* (Cobourg, Ontario: Collector Grade Publications, 2001), 61.

51. Kahaner (2006), "Weapon of Mass Destruction."

52. Interview with Jan Ismailovich Landau, historian at the Kalashnikov museum, Izhevsk, Russia, 15 July 2018; and Robert Fisk, "An Interview with Mikhail Kalashnikov," *The Independent*, London, 22 April 2001, accessed at http://www.worldpress.org/cover5.htm.

53. Kalashnikov was badly injured. He first went to hospital in Yelets, and then received an additional six-month отпуск (convalescent leave) to complete his recovery. He chose to go to Alma Ata (now Almaty), the capital of the Kazakh state (now largest city in Kazakhstan), and close to his childhood home. There he continued to tinker with gun designs. William H. Hallahan, *Misfire: The History of How America's Small Arms Have Failed Our Military* (New York: Charles Scribner's Sons, 1994), 404; and Ezell (2001), 65–66.

54. Soviet small arms designers had experimented with pistol-caliber submachine guns throughout the 1930s. The first was the PPD34, then the modified PPD34/38, the PPSh41, and others. But there was entrenched opposition to submachine guns in the Soviet army leadership. The German invasion prompted stepped-up Soviet manufacture of the PPSh41 and a successor, the PPS43, which by the end of the war was produced at a rate of 350,000 weapons per month. For an authoritative, detailed history and analysis, see Ezell (2001), 56–58.

55. Pyotr Kondratenko, "The History of the Kalashnikov Gun: Kalashnikov Is a Trademark, People Say," *Pravda*, 2 August 2003;

56. Kahaner (2007), 25.

57. Most World War I soldiers did not actually shoot at that sort of range, because lines of sight and the weather restricted their ability to see. They did shoot at that range in the South African War, where there was clearer air and they had better visibility. I am indebted to Hew Strachan for this point.

58. Chivers (2010), 161. Standard British and French rifle rounds were 78 mm long (or roughly 3 inches), the Americans' about 85 mm (3.3 inches), and the Russians' just over 77 mms (again about 3 inches).

59. This was certainly the case in the United States. Beginning in 1928, there was a bitter battle in the US Army between "large bore" and "small bore"

advocates as a result of the so-called pig board, that demonstrated the advantages of a small-bore, high-velocity gun. Opponents who wanted more penetration and reliability successfully resisted it. See Office Chief of Staff, Office Director of Weapons Systems Analysis, Pentagon, *Report of the M16 Rifle Review Panel*, 1 June 1968 (declassified 9 April 1984), at http://www. dtic.mil/dtic/tr/fulltext/u2/a953110.pdf.

60. The Germans settled on the M35 (which was 55 mm long—or about 2.2 inches) and the Kurz (which was 7.92 mm in diameter and 33 mm in length—or 1.3 inches—Kurz means "short").

61. The Kurz was about the same diameter as the German army's standard 8mm Mauser cartridge.

62. The US .30 caliber was also 7.62 mm in diameter and 33 mm in length—so like the Kurz, about 1.3 inches long.

63. John C. McManus, *The Deadly Brotherhood* (New York: Random House, 1998), 52.

64. Craig Riesch, *US M1 Carbines, Wartime Production*, 7th ed. (Tustin, CA: North Cape Publications, 2012), 7 and 11.

65. As mentioned above, the Soviets had already been designing pistol-caliber submachine guns; the invasion eliminated the Soviet leadership's opposition to the smaller ammunition.

66. The M43 was 7.62 mm in diameter and 39 mm long.

67. "RPD" from *Ruchnoy Pulemyot Detyaryova*, or Degtyaryov light machine gun.

68. *Samozaryadnyj Karabin sistemy Simonova*, or SKS. There is debate about whether the ammunition similarities are coincidental, were the result of spying on the Nazis during the war, or resulted from prewar Soviet-German collaboration (1939–1941). See Chivers (2010), 166–7; and Anthony G. Williams, "Assault Rifles and Their Ammunition," November 2014, at http://www.quarryhs.co.uk/Assault.htm.

69. Chivers (2010), 145.

70. Kalashnikov Museum, Izhevsk, 14 July 2018. Notably, Tula was directly on the Nazi invasion route to Moscow and heroically held out against the invasion.

71. Kalashnikov and Joly (2006), 55–59.

72. Ibid., 67–69; and Ezell (2001), 68–78.

73. A 2003 *Pravda* article gives Lyuty top credit for the Kalashnikov. See Pyotr Kondratenko, "The History of the Kalashnikov Gun," *Pravda*, 2 August 2003.

74. Kahaner (2007), 25. The Kalashnikov Museum and Exhibition Complex of Small Arms, founded in 2004 in Izevsk, Russia, is dedicated to Kalashnikov's personal legacy and the AK-47. en.museum-mtk.ru.

75. Chivers (2010), 152.

76. These Kalashnikov replaced the Simonov SKS carbine rifles, used during World War II and still carried for ceremonial purposes, such as by the

historic Presidential Regiment on Red Square. Despite the fact that AKs became standard issue, most Soviet soldiers did not have them until well into the 1950s.

77. Kalashnikov and Joly (2010), 80–82.

78. C. J. Chivers, "Mikhail Kalashnikov, Creator of AK-47, Dies at 94," *New York Times*, 23 December 2013.

79. Over time, Soviet engineers devoted hundreds more design hours to the rifle, addressing problems with the ejector, the hammer, and the return spring, and also reshaping the operating handle. Indeed, only months after its invention, a key problem with the receiver became clear. This is the central and most important part of the gun, containing the trigger mechanism and magazine, where the bolt moves forward and back when it fires. It was meant to be made out of stamped-metal parts, which would be easy to produce and interchangeable, but that was an uncommon manufacturing process in the Soviet Union of the late 1940s. They tried stamped steel but could not mass-produce it without a large number of rejects. So engineers crafted a solid-steel receiver, made from a four-pound metal block ground down to a pound-and-a-half, requiring more than 120 hours of skilled labor. Chivers (2010), 206–9. Starting from a huge block of steel and milling it down to size was unfortunately very wasteful, and in 1953 there was a further refinement invented by Kalashnikov's collaborator Mikhail Miller. Graves-Brown (2007), 288. The stamped receiver was reintroduced in 1959.

80. The British Lee-Enfield rifle had a longer range, greater accuracy, and more devastating wounding potential because of the larger diameter of its cartridge (.303-inch caliber).

81. Kahaner (2007), 31. The US Garand was superior to the AK-47 in terms of range, lethality, and accuracy.

82. The rifle was deliberately designed for untrained marksmen firing at a range of 200 to 300 yards. Keller Wilkinson, "AK-47: Weapon of the Century," *Military History*, September 2013, pp. 28–35.

83. Chivers (2010), 200; citing Kalashnikov, *From a Stranger's Doorstep to the Kremlin*, p. 231. See also Kalashnikov and Joly (2006), 66–68.

84. It was less than 26 inches long overall. Chivers (2010), 209.

85. Kalashnikov and Joly (2006), 85–86; Carr (2008), 20. The "M" stands for "modernized"—thus, Avtomat Kalashnikova Modernizirovannyj.

86. It weighed only 3.14 kilograms (or just under 7 pounds) while the original AK-47 model was 4.3 kilograms (or 9.5 pounds). Graves-Brown (2007), 288; Carr (2008), 19–20. The RPK Kalashnikov model light machine gun was designed at the same time as the AKM. Again, all of the rifles' parts were interchangeable: if any piece broke or malfunctioned, the user could replace it with parts from another AKM or RPK. Kalashnikov and Joly (2006), 85–86.

87. Graves-Brown (2007), 291.

88. Kalashnikov and Joly (2006), xiii; and Carr, 20–21. The AK-74 used smaller 5.45 mm caliber (or .221 inch) ammunition.

Chapter 6

1. C. J. Chivers, "How did the AK-47 Become the most Abundant Weapon on Earth?" *The Independent*, 5 November 2010, at http://www.independent. co.uk/news/world/how-did-the-ak-47-become-the-most-abundant-weapon-on-earth-2124407.html.
2. Kahaner (2006).
3. Chivers (2010), 219. The earliest public display of the AK-47 with its characteristic banana-shaped magazine was in the 1955 Soviet film *Maksim Perepelitsa*, which is a comedy about a young man who is drafted into the Soviet Army. But the film was almost unknown outside the Soviet bloc.
4. For differing perspectives of what happened, see Johanna Granville, "1956 Reconsidered: Why Hungary and Not Poland?," *Slavonic and East European Review* 80, no. 4 (October 2002): 656–87; and T. David Curp, "The Revolution Betrayed? The Poznan Revolt and the Polish Road to Nationalist Socialism," *Polish Review* 51, no. 3–4 (2006): 307–24.
5. Isotta Poggi, "The Photographic Memory and Impact of the Hungarian 1956 Uprising during the Cold War Era," *Getty Research Journal* 7 (2015): 197. The photograph was taken by Michael Rougier. See also Chivers (2010), photograph section in the center of the book, p. 5.
6. This was not the only highly influential photo. See Eszter Balazs and Phil Casoar, "An Emblematic Picture of the Hungarian 1956 Revolution: Photojournalism during the Hungarian Revolution," *Europe-Asia Studies* 58, no. 8 (2006): 1241–60. Also, *Time* magazine named a generic "Hungarian Freedom Fighter" (depicted in a drawing) as its 1956 "Man of the Year."
7. Mark Kramer, "The Soviet Union and the 1956 Crises in Hungary and Poland: Reassessments and New Findings," *Journal of Contemporary History* 33, no. 2 (April 1998): 182.
8. UN General Assembly, Report of the Special Committee on the Problem of Hungary, New York, 1957, p. 20, at http://mek.oszk.hu/01200/01274/ 01274.pdf.
9. Kramer (1998), 180, note 65.
10. "Shifrtelegramma iz Budapeshta," Cable from A. Mikoyan and M. Suslov to the CPSU Presidium, 24 October 1956 (Strictly Secret), in AVPRF, F. 059a, Op. 4, Pap. 6, D. 5, L. 2; cited by Mark Kramer (1998), 185, note 81.
11. Chivers (2010), 223–4.
12. National Security Archives, "The 1956 Hungarian Revolution: A History in Documents," edited by Malcolm Byrne, 4 November 2002, at http:// nsarchive.gwu.edu/NSAEBB/NSAEBB76/.

13. Russian military archive, TsAMO, F. 32, Op. 701291, D. 17, Ll. 33–48; cited by Mark Kramer (1998), p. 185, note 82.

14. Kramer (1998), 185.

15. This famous photograph comes from the archives of the Hungarian National Museum. See also Max Hastings, "The Most Influential Weapon of Our Time," *New York Review of Books*, 10 February 2011, at https://www.nybooks.com/articles/2011/02/10/most-influential-weapon-our-time/.

16. Kramer (1998), 188–9.

17. Ibid.

18. The number of casualties is contested, both because Soviet archival documents are unavailable and because Hungarians hid their wounded and secretly buried their dead. Thousands of Hungarians were arrested, imprisoned, and deported to Soviet camps in the aftermath of the uprising. Many simply disappeared. See Report by Soviet Deputy Interior Minister M. N. Holodkov to Interior Minister N. P. Dudorov, 15 November 1956, at http://nsarchive.gwu.edu/NSAEBB/NSAEBB76/doc8.pdf.

19. Library of Congress, Soviet Union: A Country Study, p. 1, at https://cdn.loc.gov/master/frd/frdcstdy/so/sovietunioncount00zick/sovietunioncount00zick.pdf.

20. Chivers (2010) writes that the USSR used them as a political currency to gain sway with allies (p. 203). I'm suggesting that because of the need to barter for imports, the practice went beyond that and essentially took the place of traditional currency.

21. This was documented most extensively with respect to missiles and strategic systems; however, it was also true of other Soviet weapons. The source is a debriefing of former senior Soviet officials shortly after the Cold War ended. OSD-Net Assessment, Contract #MDA903-92-C-0147, 22 September 1995; BDM Federal, Inc., *Soviet Intentions 1965–1985*, Vol. 1: *An Analytical Comparison of US-Soviet Assessments During the Cold War*; "Role of the Industrialists," 59–65, at http://nsarchive.gwu.edu/nukevault/ebb285/.

22. John Walter, *Rifles of the World*, 3rd ed. (Iola, WI: Krause Publications, 2006), 202.

23. What's more, the research that has been conducted has been done in the field, almost always on sales in illicit markets with different local economic conditions and a range of different types of Kalashnikovs. A thorough assessment of the effect of price on diffusion is therefore impossible. It is pointless to compare the cost of an AK-47 (which uses the 7.62 × 39 mm cartridge) directly to the price of an AK74 (which uses the 5.45 × 39 mm cartridge), for example: the newer rifle tends to be at least twice as expensive. More than fifty variants are available from different countries, with a range of different features and conditions that all affect price. For example, Russian-made rifles tend to capture higher prices than local models, which is why local gun dealers in global arms markets may lie about a rifle's provenance or features. Nicolas Florquin, chapter 11, "Price Watch: Arms and Ammunition

at Illicit Markets," *Small Arms Survey 2013*, 248–81, at http://www.
smallarmssurvey.org/publications/by-type/yearbook/small-arms-survey-2013.
html. According to Florquin, it is easier to gain insight from the price of
ammunition, because the cost of each cartridge varies less than the cost of
the Kalashnikov variant rifles does. But for our purposes there are no good
historical records of ammunition production and stockpiles, not least because
governments like to keep such information secret.

24. Phillip Killicoat, *Weaponomics: The Global Market for Assault Rifles*, Post-
Conflict Transitions Working Paper No. 10, World Bank Policy Research
Working Paper 4202, April 2007; and Phillip Killicoat, *Weaponomics: The
Economics of Small Arms*, CSAE WPS/2006-13, September 2006; and Philip
Killicoat, chapter 8: "What Price the Kalashnikov?," *Small Arms Survey
2007*, 256–87, at http://www.smallarmssurvey.org/fileadmin/docs/A-
Yearbook/2007/en/Small-Arms-Survey-2007-Chapter-08-summary-EN.pdf.

25. This included the return of the Soviet military base in Port Arthur, provision
of large amounts of industrial equipment, and the help of thousands of Soviet
advisers who went to China. Vladislav M. Zubok, *A Failed Empire: The Soviet
Union in the Cold War from Stalin to Gorbachev* (Chapel Hill: University of
North Carolina Press, 2009), 111.

26. Chivers (2010), 217.

27. The Czechs had their own sophisticated weapons design bureaus and
chose to stick with their own assault weapon design (the Vz.58), which
they considered lighter, more accurate and more compact than the AK-
47 or AKM. It was modified to shoot the same cartridge, however, for
interoperability within the Warsaw Pact. Chivers (2010), 217–18.

28. Chivers (2010), 215, 245–7, and 399. The names of the manufacturing
companies are Radom (Poland), Arsenal (Bulgaria), and Romtechnica
(Romania). Hungary's manufacturing company, now closed, was known as
F. E. G. Note that we are not specifically tracing the equally rapid diffusion
of ammunition. As the Kalashnikov spread globally, so did the ammunition
produced to supply it (sometimes within the same manufacturing complex;
sometimes not), beginning in the USSR, then China, Egypt, Eastern
Europe, Cuba, parts of Africa, Finland, and so on. Today a flood of cheap
Kalashnikov-compatible ammunition is globally available, with postwar
Albanian stockpiles reportedly selling for as little as 2.2 cents a round. See
"Q&A with C. J Chivers about 'The Gun'," *New York Times*, 13 October
2010. See also "Workshops and Factories: Products and Producers,"
Small Arms Survey 2003 (Geneva), p. 20. And the supply of ammunition
continues to increase. The Small Arms Survey group tracked a dramatic
increase in small-caliber ammunition globally available between 2001
and 2006 (*Small Arms Survey 2009*, 13–17). Small arms ammunition
manufacturing occurs on every populated continent, including in eleven

African countries that do not otherwise produce armaments (*Small Arms Survey 2010*, p. 22).

29. According to Guy Laron (and drawing from recently declassified Russian and Czech documents), the arms deal was connected to the February 1955 Israeli raid on Gaza, in which Czech-provided arms had proven superior to those of the Egyptian armed forces. Negotiations for the deal began in earnest in April 1955 and were concluded later that year. Guy Laron, *Cutting the Gordian Knot: The Post-WWII Egyptian Quest for Arms and the 1955 Czechoslovak Arms Deal*, Cold War International History Project Working Paper #55, February 2007, at https://www.wilsoncenter.org/sites/default/files/WP55_WebFinal.pdf.

30. Chivers (2010), 217.

31. By a 1941 law, an inventor in the USSR could obtain either a patent or a certificate of invention, a dubious distinction because both ultimately belonged to the state. If a patent were deemed of "special importance," the Council of People's Commissars of the USSR could vote to take it in exchange for a remuneration they would set, while the owner of an inventor's certificate *automatically* transferred rights to the state for a standard honorarium and sometimes other benefits. Certificates were strongly favored. During the four years 1956–1959, more than 20,000 certificates were issued to Soviet citizens, and only six patents. There were heavy fees to apply for a patent, whereas an author's certificate was free. Either way the state received a royalty-free right to produce the invention. M. Hoser, "Patents of the USSR," *Journal of the Patent Office Society* 40, no. 4 (April 1958): 241–51; and P. J. Federico, "Soviet Law on Inventions and Patents," *Journal of the Patent Office Society* 63, no. 1 (1961): 9 and 39.

32. 2 Civil Law 254 (1944); see I. Gsovski, *Soviet Civil Law* 603 (1948); quoted and cited by Bernie R. Burrus, "The Soviet Law of Inventions and Copyright," *Fordham Law Review* 30, no. 4 (1962): 709.

33. Kalashnikov commented in later years that he had "never received a single kopek" for arms sales—though he was paid a handsome salary, as well as given an apartment, a dacha (summer house), and many special privileges. He never benefited directly from royalties on the sale of his invention. Kalashnikov and Joly (2006), 81.

34. Federico (1961), 44. On the legal status of inventors certificates, see Peter B. Maggs, "Legal Problems of Patents, Industrial Designs, Technical Data, Trademarks, and Copyrights in Soviet-American Trade," *Journal of International Law and Policy* 5, no. 311 (1975): 313.

35. Peter B. Maggs and James W. Jerz, "The Significance of Soviet Accession to the Paris Convention for the Protection of Industrial Property," *Journal of the Patent Office Society* 48, no. 4 (April 1966): 242. The USSR did not sign on to the Universal Copyright Convention until 1973.

36. "Tito Visits Egypt: Nasser Hails Ties of Friendship," *Chicago Tribune*, 29 December 1955, p. 6. See also Chivers (2010), 250–1.

37. Chivers (2010), 250–1.

38. Ibid., 265.

39. Xiaoming Zhang, "Communist Powers Divided: China, the Soviet Union, and the Vietnam War," in *International Perspectives on Vietnam*, edited by Lloyd C. Gardiner and Ted Gittinger (College Station: Texas A&M Press, 2000), 82–83; cited by Record (2009), 50.

40. Mike O'Connor, "Albanian Village Finds Boom in Gun-Running," *New York Times*, 24 April 1997, at http://www.nytimes.com/1997/04/24/world/albanian-village-finds-boom-in-gun-running.html?_r=0.

41. Chivers (2010), 248–50.

42. Richard R. Hallock, *M-16 Rifle Case Study*, Prepared for the Chairman of the President's Blue Ribbon Defense Panel, 16 March 1970, p. 10; in the Col. Richard R. Hallock Papers (MC 284), Series 17, Columbus Status University Archives, Columbus, Georgia, and accessible at http://pogoarchives.org/labyrinth/09/02.pdf.

43. Associated Press, "Yisrael Galili, Weapons Inventor, 72," *New York Times*, 11 March 1995.

44. Joe Poyer, *The AK-47 and AK-74 Kalashnikov Rifles and their Variants*, 4th ed. (Tustin, CA: North Cape Publications, 2013), 83–84; N. R. Jenzen-Jones, *Chambering the Next Round: Emergent Small-calibre Cartridge Technologies*, A Working Paper of the Small Arms Survey, with support from the Federal Foreign Office of Germany (Geneva: Small Arms Survey, 2016), 17. See also Matthew Ford, *Weapon of Choice: Small Arms and the Culture of Military Innovation* (Oxford, UK: Oxford University Press, 2017), chapter 6: "Alliance Politics and NATO Standardization," 117–40.

45. Poyer, 4th ed. (2013), 81.

46. Walter (2006), 2010.

47. Small Arms Survey, "Small Arms of the Indian State," Issue Brief, no. 4, January 2014, p. 5.

48. People's Daily Online (China), "Nigeria to Mass-produce Nigerian Version of AK-47 Rifles," at http://en.people.cn/200610/02/eng20061002_308128.html.

49. For much more detailed firsthand information about diffusion of Chinese-produced Type 56 rifles and other small arms throughout the Sahel, see Savannah de Tessières, *At the Crossroads of Sahelian Conflicts: Insecurity, Terrorism, and Arms Trafficking in Niger* (Geneva: Small Arms Survey, January 2018), at http://www.smallarmssurvey.org/fileadmin/docs/U-Reports/SAS-SANA-Report-Niger.pdf.

50. Patent WO1999005467A1, dated 24 December 1997 and assigned to Izhmash, is accessible here: https://patents.google.com/patent/WO1999005467A1.

51. Reportedly China, Turkey, and Slovenia began paying royalties. Maxim Pyadushkin with Maria Haug and Anna Matveeva, *Beyond the Kalashnikov: Small Arms Production, Exports, and Stockpiles in the Russian Federation*, Occasional Paper No. 10, Small Arms Survey, Geneva, August 2003, p. 17.

52. The AK-100 series uses various calibers of ammunition.

53. Ivan Safronov Junior, "It Is Impossible to Seduce Them by Kalashnikov," *Kommersant-Dengi*, no. 45, 12–18 November 2012, p. 46.

54. "Putin Gives Green Light to Kalashnikov Merger Plan," *Jane's Defence Weekly*, 28 November 2012.

55. Headquartered in Moscow, Russia, Rostec (Ростех) is short for State Corporation for Assistance to Development, Production, and Export of Advanced Technology Industrial Product (Государственная корпорация по содействию разработке, производству, и экспорту высокотехнологичной промышленной продукции). The *Wall Street Journal* reports that the partnership invested about 12 billion rubles into the business. James Marson and Thomas Grove, "Kalashnikov Finds Success Even after US Sanctions; Russian Arms Maker More Than Doubled Revenue Last Year, as Markets Opened Up in Asia, Africa," *Wall Street Journal*, 13 July 2017.

56. "Kalashnkov, AK-47 Maker, Goes Private as Russian Government Sheds Stake," *New York Times*, 13 November 2017; Rottman (2011), 37. There was also reportedly a factory licensed in Ethiopia, but this could not be confirmed. Negotiations with India to license newer models, such as the AK-103, were ongoing at the time of writing.

57. See Kalashnikov USA website at: http://www.kalashnikov-usa.com/. RWC Group was formed in 2011. See http://www.bloomberg.com/profiles/companies/1045057D:US-rwc-group-llc.

58. The ArmaLite Rifle (AR-15, the civilian version of the M-16) was invented in the 1950s. It should be noted that the carefully constructed image of the Viet Cong as a scrappy, ill-equipped guerrilla force is overstated. In addition to AK-47s, Viet Cong fighters had submachine guns, grenades, rocket launchers, and other Chinese-supplied weapons, while the South Vietnamese armed forces made due mainly with US cast-off M-1 rifles until late in the war. Lan Cao, "The Vietnam War: Five Myths," *Washington Post*, 1 October 2017, p. B3.

59. Department of the Army Field Manual FM 23-8, "US Rifle 7.62-MM M14," Headquarters, Department of the Army, December 1959, at https://archive.org/stream/FM23-81968/FM23-81968_djvu.txt.

60. For side-by-side comparison of the two rifles, M14 and M16 (or AR-15), see R. Blake Stevens and Edward C. Ezell, *The Black Rifle: M16 Retrospective* (Cobourg, Ontario: Collector Grade Publications, 2015), 110–15.

61. Eugene Stoner, the weapon's designer, had built the weapon around a cleaner burning gunpowder (Improved Military Rifle or IMR 4475), but the US

Army switched to ball-type propellant because it was cheaper and safer to manufacture. The rifle was therefore ill-suited to the propellent used. Ezell (1984), 210–13.

62. The M-14's barrel was designed for a 7.62 mm caliber cartridge, while the M-16 fit the smaller 5.56 mm cartridge.

63. Email correspondence with Bernard Cole, 18 July 2018. Cole served as a Naval Gunfire Liaison Officer with the Third Marine Division in Vietnam from June 1967 to June 1968.

64. Catherine LeRoy, "Une colline sanglante entre dans l'histoire: Vietnam: les marines au corps a dans l'enfer de la cote 881," *Paris Match*, 13 May 1967, p. 67; cited by Ezell (1984), 208–9 and note 35; "Defense, under Fire," *Time* magazine, 9 June 1967; "The M16 Mess," *Philadelphia Inquirer*, 25 October 1967.

65. "Hearings before the Special Subcommittee on the M16 Rifle Program of the Committee on Armed Services of the House of Representatives, 90th Congress, First Session. 15 May–22 August 1967; and "Report of the Special Subcommittee on the M16 Rifle Program of the Armed Services Committee of the House of Representatives," 19 October 1967.

66. "M-16 Rifle Report," *CQ Almanac* 1967, 23rd ed. (Washington, DC: Congressional Quarterly, 1968), at http://library.cqpress.com/cqalmanac/cqal67-1313156; Richard R. Hallock, "M-16 Rifle Case Study," Prepared for the Chairman of the President's Blue Ribbon Defense Panel, 16 March 1970. Failure to use a chromium-plated barrel was surprising: previous American rifles had routinely had this coating to resist rust. Chivers (2010), 291–2. Improvements made the M16A1 the US standard by the late 1970s, then the M16A2 in the 1980s. Edward Clinton Ezell, *The Great Rifle Controversy: Search for the Ultimate Infantry Weapon from World War II through Vietnam and Beyond* (New York: Stackpole Books, 1984).

67. Chivers (2010), 309. Using enemy Kalashnikovs increased the risk of friendly-fire incidents, as the telltale clattering meant Americans mistook each other for the enemy.

68. The first two weeks overlapped with an annual vacation shutdown, but the strike still outraged many observers. Congressman Paul Findley blasted Colt and its workers. "Here we are, in the midst of a war. . . . [A]nd yet the sole producer of the single, most vitally needed weapon for combat—a weapon seriously in short supply [for] allied forces—is permitted to take a 2-week vacation." US House of Representatives, "The M-16 Rifle Procurement Scandal," *Congressional Record—House*, 17 July 1967, p. 19028.

69. Ultimately the US government arranged for two more firms to produce the rifles, General Motors Corporation and Harrington & Richardson; but the Army had to pay Colt both $4.5 million in cash for licensing rights and a 5.5 percent royalty charge for each rifle produced at non-Colt production

lines. Even then, Colt was unwilling to share practical know-how with the newer producers, especially as it began to troubleshoot the rifle. Ezell (1984), 221–8. In 1988, Colt lost its contract to produce the M16 rifle to a Belgian-owned concern, FN Manufacturing of Columbia, South Carolina, which underbid its per rifle price $477.50 (Colt) to $420 (FN). "Army Drops Colt as M16 Rifle Maker," *New York Times*, 3 October 1988, p. B4. It then produced the M4 rifle but lost its contract in 2013. Colt filed for Chapter 11 bankruptcy in 2015. Matt Jarzemsky, "Colt Defense Files for Chapter 11 Bankruptcy Protection," *Wall Street Journal*, 15 June 2015.

70. Chivers (2010), 263–336. Space constraints prevent going into greater detail here. Chivers's account of the M-16 debacle is strongly recommended.

71. Currently the M16A2. Stevens and Ezell (2015). Among other things, M-16s (and variants) generally have longer ranges and more accuracy than Kalashnikovs do.

72. N. R. Jenzen-Jones, *Global Development and Production of Self-loading Service Rifles, 1896–Present*, Working Paper #25 (Geneva: Small Arms Survey, 2017), Table 1, p. 31. See the combined figure for AR10 and AR15 rifles at http://www.smallarmssurvey.org/fileadmin/docs/F-Working-papers/SAS-WP25-Self-loading-rifles.pdf.

73. Simon Reeve, *One Day in September* (New York: Arcade Publishing, 2000), 43–47. Reeve says the supplies came from Libya. Chivers (2010), 338–9, argues that they came in from Algeria. In the movie version of *One Day in September*, the sole surviving Palestinian says the team flew to Germany from Libya. No open source evidence about the arms is available either way.

74. Even today the Kalashnikov plays an outsized role in Russian affairs. A plan for Russia and Greece to jointly manufacture Kalashnikov assault rifles was championed by a member of the Trump transition team. Griff Witte, "In Greece, Trump Aide Rose Fast, Relished Limelight," *Washington Post*, 11 December 2017, pp. A1 and A8.

75. Chivers (2010), 349–50.

76. Carr (2008), chapter 3, pp. 26–40.

77. Ibid.

78. Chivers (2010), 361–2. According to Steve Coll, the CIA secretly bought surplus Soviet weaponry from dissident Polish officers, negotiated deals for Chinese-made Kalashnikovs through the CIA station chief in Beijing, and purchased old Soviet weapons from the Egyptians. Steve Coll, *Ghost Wars* (New York: Penguin, 2004), 66.

79. Carr, passim.

80. The gun is nicknamed the "Krinkov." Because of its small size, Soviet armor and helicopter crews use it. Chivers (2010), 11 and 383. See also "The Krinks of Osama bin Laden," *Thefirearmblog.com*, at https://www.thefirearmblog.com/blog/2018/07/12/the-krinks-of-osama-bin-laden/.

81. Carr (2008), 21–22.

82. Carr (2008), 26–28. According to Carr, the Chinese were fully aware of their destination.
83. Chivers (2010), 13; and Rottman (2011), 51.
84. Amnesty International, "Taking Stock: The Arming of the Islamic State, December 2015," 11, at https://www.justice.gov/sites/default/files/pages/attachments/2016/02/11/dec._2015_taking_stock.pdf; and Conflict Armament Research, "Islamic State Weapons in Iraq and Syria," September 2014, at http://conflictarm.com/wp-content/uploads/2014/09/Dispatch_IS_Iraq_Syria_Weapons.pdf. US M-16s were also captured in northwestern Iraq.
85. Chivers (2010), 411–13.
86. Please note that details about Kalashnikov Concern and Kalashnikov USA are contradictory in open sources, not least because there is apparently an ongoing investigation about potential US sanctions violations. I have tried to present the most accurate, fair, and verifiable information available.
87. Anastasia Kravchenko, "Kalashnikov Brand in the US Is Being Registered by an American Company," *Izvestia*, 28 January 2015, p. 3; "Kalashnikov Finds Success Even after US Sanctions," *Wall Street Journal*, 13 July 2017, at https://www.wsj.com/articles/kalashnikov-finds-success-even-after-u-s-sanctions-1499943603; and Michael Smith and Stephanie Baker, "This Florida Warehouse Is Producing 'Made in America' Kalashnikovs," *Bloomburg Business Week*, 8 March 2018, at https://www.bloomberg.com/news/articles/2018-03-08/this-florida-warehouse-is-producing-made-in-america-kalashnikovs.
88. Steve Johnson, "Exclusive: Alexey Krivoruchko, the CEO of the Kalashnikov Group, Has Answered YOUR Questions," *Thefirearmblog.com*, 8 April 2015.
89. In 2017, the Russian government also reportedly sold its shares to private investors, making Kalshnikov Concern a completely private company. "Kalashnikov, AK-47 Maker, Goes Private as Russian Government Sheds Stake," *New York Times*, 13 November 2017.
90. Aaron Smith and Abigail Brooks, "Kalashnikov Cranking Up AK-47 Factory in Florida," *CNN Business*, 27 January 2016, at http://money.cnn.com/2016/01/27/news/companies/kalashnikov-florida-factory/.
91. The 11 April 2018 letter from Rep. Deutch to the US Department of the Treasury is at https://teddeutch.house.gov/news/documentsingle.aspx?DocumentID=399302.
92. Todd Prince, "SHOT Show 2019: Kalashnikov USA May Begin Exporting Products," *Las-Vegas Review-Journal*, 26 January 2019, at https://www.reviewjournal.com/business/conventions/shot-show/shot-show-2019-kalashnikov-usa-may-begin-exporting-products-video-1582310/.
93. Note that in the United States, AR-15s (the civilian version of the M-16) continue to be more popular than Kalashnikovs. The NRA estimates there are about 8 million AR-15s in the United States. The AR-15 has been used in numerous US mass shooting tragedies, including the 2012 Newtown

shooting (27 people killed), the 2017 Las Vegas shooting (58 killed) and the 2018 Parkland, Florida, shooting (17 killed).

94. Aaron Smith, "American-made Kalashnikovs Are Now for Sale," *CNN Money*, 30 June 2015, at http://money.cnn.com/2015/06/30/news/companies/kalashnikov-usa/?iid=EL. On 23 March 2018, Kalashnikov Concern in Moscow confirmed that it was not doing business with Kalashnikov USA. "Russia's Kalashnikov Says It Is Not Doing Business with Kalashnikov USA, Other US Companies," *Interfax: Russia & CIS General Newswire*, 23 March 2018.

95. The Correlates of War Project is at http://www.correlatesofwar.org.

96. Ivan Arreguin-Toft, "How the Weak Win Wars: A Theory of Asymmetric Conflict," *International Security* 26, no. 1 (summer 2001): 96–97, especially figures one and two.

97. Max Boot, *Invisible Armies: An Epic History of Guerrilla Warfare from Ancient Times to the Present* (New York: W. W. Norton, 2013), Appendix, pp. 579–89. Although they are very close, the figures in the description of the data on p. 559 do not quite match those laid out in the Appendix charts. I have relied on the latter.

98. Stathis Kalyvas and Laia Balcells, "International System and Technologies of Rebellion: How the End of the Cold War Shaped Internal Conflict," *American Political Science Review* 104, no. 3 (August 2010): 415–29; Jeffrey Record, "Why the Strong Lose," *Parameters* (winter 2005–2006): 16–31; Jeffrey Record, *Beating Goliath: Why Insurgencies Win* (Washington, DC: Potomac Books, 2009).

99. Jason Lyall and Isaiah Wilson, "Rage against the Machines: Explaining Outcomes in Counterinsurgency Wars," *International Organization* 63 (winter 2009): 67–106.

100. The full quote (translated from the Chinese) is: "Many people think it impossible for guerrillas to exist for long in the enemy's rear. Such a belief reveals lack of comprehension of the relationship that should exist between the people and the troops. The former may be likened to water and the latter to the fish who inhabit it. How may it be said that these two cannot exist together? It is only undisciplined troops who make the people their enemies and who, like the fish out of its native element, cannot live." Mao Tse-Tung, *Mao Tse-Tung on Guerrilla Warfare* (La Vergne, TN: BN Publishing, 2007), xlix–l.

101. Steven Biddle, *Military Power: Explaining Victory and Defeat in Modern Battle* (Princeton, NJ: Princeton University Press, 2004).

102. Daniel Byman, Peter Chalk, Bruce Hoffman, William Rosenau, and David Brannan, *Trends in Outside Support for Insurgent Movements* (Santa Monica, CA: RAND, 2001), 93–95; and Rohan Gunaratna, "Terrorists Threats Target Asia," *Janes Intelligence Review*, 1 July 2000, at http://fore.thomson.com.

103. The two factories in Izhevsk were known as IZHMASH and IZHMEKH. IZHMASH was an acronym that stood for Izhevsk Machine Building Plant

(or Ижевский Машиностроительный Завод). IZHMEKH stood for Izhevsky Mechanical Plant (Ижевский Механический Завод). They merged to becomeKalashnikov Concern (see also notes 86, 87, and 88). "Leading Russian Arms Makers to Unite under Kalashnikov Name," *Russia Today*, 6 November 2012, at https://www.rt.com/business/russia-kalashnikov-plant-merger-068/.

104. Henry Miller, "Kalashnikov Inventor Laments Proliferation," *Washington Post*, 11 June 2006.
105. Kalashnikov and Joly (2006), 79–83.
106. *Noviye Izvestia*, 19 February 2003, p. 1.
107. "Kalashnikov Gives Name to 'Manly' but Less Lethal Products," *The Scotsman*, 19 February 2003, at http://www.scotsman.com/news/world/kalashnikov-gives-name-to-manly-but-less-lethal-products-1-546757. There was even a limited number of AK-47-shaped bottles of vodka sold.
108. Taras Kuzio, "Ukraine: Look into Arms Exports," *Christian Science Monitor*, 12 February 2002,; at http://www.csmonitor.com/2002/0212/p09s02-woeu.html.
109. "Unsurprisingly, the Kalashnikov family of rifles has been in use with all parties involved in the conflict." According to Armament Research Services, the most common is the AK-74M, which is standard issue for the Russian Federation forces and not commonly seen elsewhere. Jonathan Ferguson and N. R. Jenzen-Jones, "Raising Red Flags: An Examination of Arms & Munitions in the Ongoing Conflict in Ukraine," Research Report No. 3, November 2014, at http://armamentresearch.com/Uploads/Research Report No. 3 - Raising Red Flags.pdf.
110. *Small Arms Survey* 2015, p. 139; and Independent Balkan News Agency, "Albania Donates Iraq 22 Million Cartridges, Shells, and Grenades, 10 Thousand Kalashnikovs to Afghanistan," 30 September 2014.
111. "In the Line of Fire: Elephant and Rhino Poaching in Africa," chapter 1 of *Small Arms Survey* 2015 (Cambridge, UK: Cambridge University Press 2015), 17–21.
112. Sometimes M-16s, Mausers, 12-gauge shotguns, homemade and imported guns and ammunition are also used. Daniel Stiles, *Elephant Meat Trade in Central Africa: Summary Report*, International Union for Conservation of Nature (IUCN) (Gland, Switzerland: IUCN, 2011), 48–50, at https://portals.iucn.org/library/sites/library/files/documents/ssc-op-045.pdf. Anti-poaching patrols also often carry Kalashnikovs.
113. The price of an AK-47 in Central African Republic is reportedly the lowest in the world. This is due to a glut of older weapons entering the country from Libya, and to the pilfering of local arms depots. See "Country Brief: Central African Republic," *The Sentry*, July 2015, p. 4; at https://thesentry.org/wp-content/uploads/2015/07/19103553/Country-Brief_CAR.pdf. The Small Arms Survey cited a minimum price of $12 in Africa, but where and when that

occurred could not be confirmed so I have used a more conservative estimate. Philip Killicoat, "What Price the Kalashnikov? The Economics of Small Arms," chapter 8, *Small Arms Survey 2007*, Table 8.1, p. 261. See also note #28 for C. J. Chivers's mention of unusually cheap Albanian ammunition.

114. Andrew E. Kramer, "Kalashnikov, Maker of AK-47, Looks to Rebrand," *New York Times*, 6 June 2016; and Kyle Mizokami, " 'Hovercycle' from the Makers of the AK-47," *Popular Mechanics*, 26 September 2017, at https://www.popularmechanics.com/military/aviation/a28397/kalashnikov-hovercycle/.

115. United Nations General Assembly, "Address by Mikhail T. Kalashnikov," as translated from the original Russian, Annex to the note verbale dated 29 June 2006 from the Permanent Mission of the Russian Federation addressed to the Secretariat of the United Nations Conference to Review Progress Made in the Implementation of the Programme of Action to Prevent, Combat, and Eradicate the Illicit Trade in Small Arms and Light Weapons in All Its Aspects, A/CONF.192/2006/RC/6, 29 June 2006, at http://www.un.org/events/smallarms2006/pdf/rc.6-e.pdf.

116. "Перед смертью Калашников написал покаянное письмо патриарху" (Before his death, Kalashnikov wrote a penitential letter to the Patriarch), *Izvestia*, 13 January 2014. The letter is accessible at https://iz.ru/news/563827.

Chapter 7

1. Scott Shane, *Objective Troy: A Terrorist, a President, and the Rise of the Drone* (New York: Random House, 2015), 159–61; and Scott Shane, "The Lessons of Anwar al-Awlaki," *New York Times Magazine*, 27 August 2015. "Call to Jihad" was part of Awlaki's "Hereafter" series.

2. Shane, *Objective Troy* (2015), 304–5.

3. US District Court, District of Massachusetts, Indictment against Dzokhar Tsarnaev, p. 3, at http://www.nytimes.com/interactive/2013/06/27/us/28tsarvaev-indictment.html.

4. Deborah Katz, "Injury Toll from Marathon Bombs Reduced to 264," *Boston Globe*, 24 April 2014, at https://www.bostonglobe.com/lifestyle/health-wellness/2013/04/23/number-injured-marathon-bombing-revised-downward/NRpaz5mmvGquP7KMA6XsIK/story.html; and Jennifer Levitz and Jon Kamp, "Struggles of Boston Amputees Mount," *Wall Street Journal*, 20 September 2013, at https://www.wsj.com/articles/struggles-of-boston-amputees-mount-1379730729.

5. "Dzhokhar Tsarnaev Was Enrolled in 'Intro to Ethics' Class during Boston Marathon Bombing," *Boston.com*, 24 March 2015, at https://www.boston.com/news/local-news/2015/03/24/dzhokhar-tsarnaev-was-enrolled-in-intro-to-ethics-class-during-boston-marathon-bombing.

6. Masha Gessen, "'This Year I Lost Too Many of My Loved Relatives,' Tsarnaev Wrote," *Washington Post*, 25 March 2015, at https://www.washingtonpost.com/news/post-nation/wp/2015/03/25/this-year-i-lost-too-many-of-my-loved-relatives-tsarnaev-wrote/?utm_term=.4ac0fa39cf9f.

7. Michele R. McPhee, *Maximum Harm: The Tsarnaev Brothers, the FBI, and the Road to the Marathon Bombing* (Lebanon, NH: University Press of New England, 2017), pp. 81–144. Immediately following the homicide, between January and July 2012, Tamerlan Tsarnaev traveled to Dagestan and Chechnya.

8. Ibid.

9. "Boston Marathon Bomber Tamerlan Tsarnaev Bought Three Pounds of Black Powder from N.H. Fireworks Store," *New York Daily News*, 24 April 2013.

10. Dzhokhar was not directly connected to a terrorist group.

11. Shane, *Objective Troy* (2015), 181. Awlaki was not the first to produce videos about jihadism. A famous predecessor was Abu Musab al-Suri, who taped hundreds of hours of lectures about jihadist strategy in the late 1990s, designed to mobilize followers remotely.

12. Shane, *Objective Troy* (2015), 159–61.

13. The tweet was from Dzhokhar Tsarnaev, using the screen name Ghuraba ("stranger" or "foreigner") and the Twitter handle @Al_firdausiA. Scott Shane, "The Enduring Influence of Anwar al-Awlaki in the Age of the Islamic State," *CTC Sentinel*, July 2016, p. 18, at https://ctc.usma.edu/the-enduring-influence-of-anwar-al-awlaki-in-the-age-of-the-islamic-state/.

14. Counter Extremism Project, *Anwar al-Awlaki's Ties to Extremists*, November 2016, at https://www.counterextremism.com/anwar-al-awlaki.

15. Charlie Savage, "Court Releases Large Parts of Memo Approving Killing of American in Yemen," *New York Times*, 23 June 2014, at https://www.nytimes.com/2014/06/24/us/justice-department-found-it-lawful-to-target-anwar-al-awlaki.html.

16. Counter Extremism Project, *Anwar al-Awlaki's Ties to Extremists*, November 2016, at https://www.counterextremism.com/anwar-al-awlaki.

17. Scott Shane, "The Enduring Influence of Anwar al-Awlaki in the Age of the Islamic State," *CTC Sentinel* (July 2016), 15. See also "By the Numbers: The Lasting Influence of Anwar al-Awlaki," fact sheet, Fordham University Law School's Center on National Security, 2015; and Peter Bergen and David Sterman, "The Man Who Inspired the Boston Bombings," http://www.cnn.com/2014/04/11/opinion/bergen-boston-bombing-awlaki-jihadists/.

18. This according to then Director of the CIA Mike Pompeo, "A Discussion on National Security with CIA Director Mike Pompeo," Center for Strategic and International Security, Washington, DC, 13 April 2017. See also Dan Frosch, "Wichita Airport Technician Charged with Terrorist Plot," *New York*

Times, 13 December 2014, at https://www.nytimes.com/2013/12/14/us/man-accused-of-airport-bombing-attempt-in-kansas.html.

19. Scott Shane points out the irony that the US government had to have a court order if it wanted to intercept Awlaki's communications but not to target and kill him with a UAV. *Objective Troy* (2015), 245.

20. "US Drone Strikes in Yemen," New America Foundation, at https://www.newamerica.org/in-depth/americas-counterterrorism-wars/us-targeted-killing-program-yemen/; and "Drone Strikes in Yemen," Bureau of Investigative Journalism, at https://www.thebureauinvestigates.com/projects/drone-war/yemen. These are the two best open sources for tracking drone strikes, but for various reasons their numbers do not align. I have used a conservative estimate here.

21. Scott Shane, *Objective Troy* (2015), 247–53.

22. In October 2011 Tamerlan Tsarnaev's name was added to the NCTC's Terrorist Identities Datamart Environment (or TIDE) list, but the spelling and the birth dates entered into the system (transliterations of Russian, provided by the Russian government) reportedly did not match the correct information. Eric Schmitt and Michael S. Schmidt, "2 US Agencies Added Boston Bomb Suspect to Watch Lists," *New York Times*, 24 April 2013, at http://www.nytimes.com/2013/04/25/us/tamerlan-tsarnaev-bomb-suspect-was-on-watch-lists.html.

23. Cliff Saran, "Apollo 11: The Computers That Put Man on the Moon," *Computer Weekly*, July 2009, at http://www.computerweekly.com/feature/Apollo-11-The-computers-that-put-man-on-the-moon.

24. Gordon E. Moore, "Cramming More Components onto Integrated Circuits," *Electronics*, 19 April 1965, pp. 114–17. Intel was founded three years later, in 1968.

25. Rachel Courtland, "Gordon Moore: The Man Whose Name Means Progress: The Visionary Engineer Reflects on 50 Years of Moore's Law," *IEEE Spectrum*, 30 March 2015, at http://spectrum.ieee.org/computing/hardware/gordon-moore-the-man-whose-name-means-progress.

26. Ibid.

27. Intel Corporation Annual report (Form 10-K), 2 December 2016, p. 2, at http://files.shareholder.com/downloads/INTC/867590276x0xS50863-16-105/50863/filing.pdf. The report claims Intel introduces "the next generation of silicon process technology every two to three years."

28. Andrew Bunnie Huang, "The Death of Moore's Law Will Spur Innovation: As Transistors Stop Shrinking, Open-source Hardware Will Have Its Day," *IEEE Spectrum*, 31 March 2015, at https://spectrum.ieee.org/semiconductors/design/the-death-of-moores-law-will-spur-innovation. Chris Mack argues that Moore's law will go in a new direction that he calls "Moore 3.0," meaning integrating new kinds of capabilities into chips. "The Multiple Lives of Moore's Law," *IEEE Spectrum*, 20

March 2015, at https://spectrum.ieee.org/semiconductors/processors/
the-multiple-lives-of-moores-law.

29. Jason Burke, "How the Changing Media Is Changing Terrorism," *The Guardian*, 25 February 2016.

30. Abbottabad Document submitted to the trial of Abid Naseer, Government Exhibit #427, 10-CR-019(S-4)(RJD); accessed at http://kronosadvisory.com/Abid.Naseer.Trial_Abbottabad.Documents_Exhibits.403.404.405.420thru433.pdf.

31. See also Office of the Director of National Intelligence, "Bin Laden's Bookshelf," at https://www.dni.gov/index.php/features/bin-laden-s-bookshelf?start=1.

32. Sean D. Hamill, "Man Accused in Pittsburgh Killings Voiced Racist Views Online," *New York Times*, 7 April 2009, at https://www.nytimes.com/2009/04/07/us/07pittsburgh.html.

33. William Yardley, "White House Shooting Suspect's Path to Extremism," *New York Times*, 20 November 2011, at https://www.nytimes.com/2011/11/21/us/oscar-ortega-white-house-shooting-suspect-struggled-with-mental-illness.html. He had seen the film a year earlier; however, it seemed to be a watershed in his gradual decline into delusion.

34. "Looking Behind the Mug-Shot Grin," *New York Times*, 15 January 2011, at https://www.nytimes.com/2011/01/16/us/16loughner.html. Note that I am not claiming that any of these three individuals would have been a peaceable citizen in the absence of online activity. Without deep analysis of each individual case, it is hard to determine the degree of preexisting personal motivation vs. online instigation.

35. Matt Pearce and Joseph Tanfani, "Virginia Gunman Hated Republicans, and 'Was Always in His Own World'," *Los Angeles Times*, 14 June 2017, at https://www.latimes.com/nation/la-na-pol-virginia-shooter-profile-20170614-story.html.

36. Another example is a video suggesting that David Hogg, a student from the Parkland, Florida, shooting who spoke out about gun control, was actually a "crisis actor." This conspiracy theory that claimed he was "bought and paid by CNN and George Soros" was the #1 Trending video on YouTube, yielding more than 200,000 views. Emanuel Maiberg, "The #1 Trending Video on You Tube Right Now Suggests That a Student From the Parkland Shooting Is a Crisis Actor," *Vice*, 21 February 2018, at https://motherboard.vice.com/en_us/article/mb5p4y/youtube-david-hogg-parkland-shooting-conspiracy-theory. On their well-established role in US history, see Mark Fenster, *Conspiracy Theories: Secrecy and Power in American Culture* (Minneapolis: University of Minnesota Press, 2008).

37. I am not arguing that Internet access is available everywhere. In 2017 the International Telecommunication Union estimated that roughly 75 percent of Africans are not connected to the Internet, followed by the Arab States

and the Asia-Pacific region, which are tied at 58 percent offline. But this is likely to change quickly. According to the mobile industry, there were roughly 226 million smartphone users in Africa in 2015, and that is predicted to grow to 720 by 2020. "Africa Hits 557M Unique Mobile Subs[cribers]; Smartphones to Dominate by 2020," *Mobile World Live*, 26 July 2016, at https://www.mobileworldlive.com/featured-content/home-banner/africa-hits-557m-unique-mobile-subs-smartphones-to-dominate-by-2020/; Imme Philbeck, *Connecting the Unconnected: Working Together to Achieve Connect 2020 Agenda Targets,* World Economic Forum 2017, p. 4, at http://broadbandcommission.org/Documents/ITU_discussion-paper_Davos2017.pdf; and Chenai Chair and Broc Rademan "Does Free Wi-Fi Improve Internet Accessibility in South Africa?," Net Politics Blog, Council on Foreign Relations, 21 February 2017, at https://www.cfr.org/blog/does-free-wi-fi-improve-InternetInternet-accessibility-south-africa.

38. See Gilbert Ramsay, " 'Terrorist' Use of the Internet: An Overblown Issue," *Middle East: Topics & Arguments* 6 (2016): 88–96.

39. Pew Research Center Global Attitudes and Trends, *Egyptians Embrace Revolt Leaders, Religious Parties and Military, As Well*, 25 April 2011, p. 2, at http://www.pewglobal.org/2011/04/25/egyptians-embrace-revolt-leaders-religious-parties-and-military-as-well/.

40. Dorothy Denning, "Activism, Hacktivism, and Cyberterrorism: The Internet as a Tool for Influencing Foreign Policy," paper presented at Internet and International Systems: Information Technology and American Foreign Policy Decision-making Workshop at Georgetown University; Dorothy Denning, "Cyberterrorism," testimony before the US House Committee on Armed Services, Special Oversight Panel on Terrorism, 107th Cong., 1st sess., 23 May 2001.

41. Sometimes the two collude. Matthew Burroughs, *The Future, Declassified: Megatrends That Will Undo the World unless We Take Action* (New York: Palgrave Macmillan, 2014), 146–7. See also Shelley, *Dirty Entanglements.*

42. Interesting recent books that address online mobilization include Zeynep Tufecki, *Twitter and Tear Gas: The Power and Fragility of Networked Protest* (New Haven, CT: Yale University Press, 2017); and David Patrikarakos, *War in 140 Characters: How Social Media Is Reshaping Conflict in the Twenty-first Century* (New York: Basic Books, 2017).

43. Sun Tzu, *The Art of War*, translated by Samuel B. Griffith (New York: Oxford University Press, 1971), 66.

44. For more on this, see Carnes Lord, "Psychological-Political Instruments," chapter 9 of *Attacking Terrorism: Elements of a Grand Strategy*, edited by Audrey Kurth Cronin and James M. Ludes (Washington, DC: Georgetown University Press, 2004).

45. Nicole Perlroth and David E. Sanger, "Cyberattacks Put Russian Fingers on the Switch at Power Plants, US Says," *New York Times*, 15 March 2018; and Dana Priest and Michael Birnbaum, "In Europe, Fake News from Russia Is Old News," *Washington Post*, 26 June 2017, pp. A1 and A11.

46. Valery Gerasimov, "The Value of Science in Prediction," *Military-Industrial Kurier {Voenno-promyshlennyi kur'er}*, 27 February 2013, translated by Rob Coalson, RFE/RL, available at "The 'Gerasimov Doctrine' and Russian Non-Linear War," https://inmoscowsshadows.wordpress.com/2014/07/06/the-gerasimov-doctrine-and-russian-non-linear-war/.

47. P. W. Singer and Emerson Brooking, "Terror on Twitter: How ISIS Is Taking War to Social Media—and Social Media Is Fighting Back," *Popular Science*, 11 December 2015.

48. Ned Parker, Isabel Coles, and Raheem Salman, "How Mosul Fell—An Iraqi General Disputes Baghdad's Story," *Reuters*, 20 October 2014, at http://www.reuters.com/article/us-mideast-crisis-gharawi-special-report-idUSKCN0I30Z820141014.

49. Suadad Al-Salhy and Tim Arango, "Sunni Militants Drive Iraqi Army Out of Mosul," *New York Times*, 10 June 2014, at https://www.nytimes.com/2014/06/11/world/middleeast/militants-in-mosul.html.

50. Ned Parker, Isabel Coles, and Raheem Salman, "How Mosul Fell—An Iraqi General Disputes Baghdad's Story," *Reuters*, 20 October 2014, at http://www.reuters.com/article/us-mideast-crisis-gharawi-special-report-idUSKCN0I30Z820141014.

51. Counter Extremism Project, at https://www.counterextremism.com/extremists/aqsa-mahmood.

52. Guy Tuysuz and Ivan Watson, "Missing UK Girls Believed to Be in Syria, Police Say," *CNN*, 24 February 2015, at http://www.cnn.com/2015/02/24/europe/turkey-uk-missing-girls/. See also Supna Zaida Peery, "Girls in CrISIS," Counter Extremism Project, 17 March 2015, at https://www.counterextremism.com/blog/girls-crisis.

53. Languages include Arabic, English, Bosnian, Kurdish, French, German, Indonesian, Pashtun, Russian, Turkish, and Uighur.

54. See, for example, "CEP Guide to Online Propagandists," Counter Extremism Project at https://www.counterextremism.com.

55. Thomas Tracy, "ISIS Has Mastered Social Media, Recruiting 'Lone Wolf' Terrorists to Target Times Square: Bratton," *New York Daily News*, 17 September 2014.

56. See Ann Larabee, *The Wrong Hands: Popular Weapons Manuals and Their Historic Challenges to a Democratic Society* (Oxford, UK: Oxford University Press, 2015).

57. Fiona Hamilton, "Manchester Killer Used YouTube to Build Bomb," London *Times*, 23 June 2017.

58. "Manchester Bomber Salman Abedi Learned to Make Explosive Device from Internet," *Asharq Al-Awsat* (English edition), 25 June 2017.

59. Zachary K. Goldman, Ellie Maruyama, Elizabeth Rosenberg, Edoardo Saravalle, and Julia Solomon-Strauss, *Terrorist Use of Virtual Currencies: Containing the Potential Threat*, Center for a New American Security, 3 May 2017, at https://www.cnas.org/publications/reports/terrorist-use-of-virtual-currencies.

60. Michael Freeman and Moyara Ruehsen, "Terrorism Financing Methods: An Overview," *Perspectives on Terrorism* 7, no. 4 (2013), at http://www.terrorismanalysts.com/pt/index.php/pot/article/view/279/html.

61. Emilie Oftedal, "The Financing of Jihadi Terrorist Cells in Europe," Norwegian Defence Research Establishment (FFI), 6 January 2015, p. 16, at https://www.ffi.no/no/Rapporter/14-02234.pdf.

62. Stephan Heibner, Peter R. Neumann, John Holland-McCowan, and Raja Basra, *Caliphate in Decline: An Estimate of Islamic State's Financial Fortunes* (London: ISCR, 2017), at http://icsr.info/wp-content/uploads/2017/02/ICSR-Report-Caliphate-in-Decline-An-Estimate-of-Islamic-States-Financial-Fortunes.pdf.

63. Peter R. Neumann, "Don't Follow the Money," *Foreign Affairs*, July/August 2017, at https://www.foreignaffairs.com/articles/2017-06-13/dont-follow-money. See also Benjamin P. Nickels and Samuel E. Cleaves, "What Money Means for Terrorism in Africa," Africa Center Policy Brief (Washington, DC: The Wilson Center, December 2018), at https://www.wilsoncenter.org/sites/default/files/policy_brief_-_what_money_means_for_terrorism_in_africa.pdf.

64. For more on the dark web, see Gabriel Weimann, "Terrorist Migration to the Dark Web," *Perspectives on Terrorism* 10, no. 2 (2016); and Gabriel Weimann, *Terrorism in Cyberspace: The Next Generation* (Washington, DC: Woodrow Wilson Center Press, 2015).

65. Rukmini Callimachi, "ISIS and the Lonely Young American," *New York Times*, 27 June 2015, at https://www.nytimes.com/2015/06/28/world/americas/isis-online-recruiting-american.html.

66. PRI International, "The Story of One Young Woman's Almost-Recruitment from Washington to the ISIS Caliphate," aired 29 June 2015, at https://www.pri.org/stories/2015-06-29/story-one-young-womans-almost-recruitment-washington-isis-caliphate.

67. Maura Conway, "Determining the Role of the Internet in Violent Extremism and Terrorism: Six Suggestions for Progressing Research," *Studies in Conflict and Terrorism* 40, no. 1 (2017): 77–98.

68. Joby Warrick, "The 'App of Choice' for Jihadists: ISIS Seizes on Internet Tool to Promote Terror," *Washington Post*, 23 December 2016, at https://www.washingtonpost.com/world/national-security/the-app-of-choice-for-jihadists-isis-seizes-on-internet-tool-to-promote-terror/2016/12/

23/a8c348c0-c861-11e6-85b5-76616a33048d_story.html?utm_
term=.5ffb778fcf6f.

69. Daxton R. "Chip" Stewart and Jeremy Littau, "Up Periscope: Mobile Streaming Video Technologies, Privacy in the Public, and the Right to Record," *Journalism & Mass Communication Quarterly* 93, no. 2 (2016): 312–31.

70. T. J. McCue, "Top 10 Video Marketing Trends and Statistics Roundup 2017," *Forbes*, 22 September 2017, at https://www.forbes.com/sites/tjmccue/2017/09/22/top-10-video-marketing-trends-and-statistics-roundup-2017/ - 1b59f4997103.

71. For much more on this topic, see Maura Conway and Joseph Dillon, *Future Trends: Live-streaming Terrorist Attacks?*, Network of Excellence for Research in Violent Online Political Extremism, *VOX-Pol* (n.d.), at http://www.voxpol.eu/download/vox-pol_publication/Live-streaming_FINAL.pdf.

72. Ibid., 2.

73. Ibid., 4; and Alexander Meleagrou-Hitchens, Shiraz Maher, and James Sheehan, *Lights, Camera, Jihad: Al Shabaab's Western Media Strategy*, International Centre for the Study of Radicalisation and Political Violence (London: King's College, 2012), at https://icsr.info/2012/11/21/icsr-report-lights-camera-jihad-al-shabaabs-western-media-strategy/ . J. M. Berger notes that this was not the first time al-Shabaab tweeted an event, having tweeted bombing attacks on Mogadishu and the attempted assassination of the Somali president. The Westgate incident appears to have been more instantaneous. "Al-Shabab Showed Gruesome Social Media Savvy during Attack," CBS News, 24 September 2014, at https://www.cbsnews.com/news/al-shabab-showed-gruesome-social-media-savvy-during-attack/.

74. David Mair, "#Westgate: A Case Study: How al-Shabaab Used Twitter during an Ongoing Attack," *Studies in Conflict & Terrorism* 40, no. 1 (2017): 24–43; and Robyn Kriel, "TV, Twitter, and Telegram: Al-Shabaab's Attempts to Influence Mass Media," *Defence Strategic Communications* 4 (spring 2018): 28–29.

75. Alan Blinder, Frances Robles, and Richard Perez-Pena, "Omar Mateen Posted to Facebook amid Orlando Attack, Lawmaker Says," *New York Times*, 16 June 2016, at https://www.nytimes.com/2016/06/17/us/orlando-shooting.html?_r=0.

76. Tim Hume, Lindsay Isaac, and Paul Cruickshank, "French Terror Attacker Threatened Euro 2016 in Facebook Video, Source Says," CNN, 14 June 2016, at http://www.cnn.com/2016/06/14/europe/french-policeman-terror-attack/.

77. Meagan Flynn, "No One Who Watched New Zealand Shooter's Video Live Reported It to Facebook, Company Says," *Washington Post*, 19 March 2019, at https://www.washingtonpost.com/nation/2019/03/19/

new-zealand-mosque-shooters-facebook-live-stream-was-viewed-thousands-times-before-being-removed/?utm_term=.9f2166aa8d47.

78. P. W. Singer and Emerson T. Brooking, *LikeWar: The Weaponization of Social Media* (New York: Eamon Dolan and Houghton Mifflin Harcourt, 2018), 246–7.

79. "Cultivate the Brand," *Wired*, April 2016.

80. See, for example, Adrian Chen, "Is Livestreaming the Future of Media, or the Future of Activism?," *New York Magazine*, 7 December 2014, at http://nymag.com/daily/intelligencer/2014/12/livestreaming-the-future-of-media-or-activism.html; and Lexi Pandell and Kia Kayyali, "Hey Activists: You Need to Think Twice before Livestreaming Protests," *Wired*, 19 March 2017, at https://www.wired.com/2017/03/hey-activists-need-think-twice-livestreaming-protests/.

81. Journalists Marie Colvin and Rémi Ochlik were also killed at the same time, when the media center was deliberately targeted. The Syrian Army had cut phone lines into the city and were reportedly bombing locations emitting satellite phone signals. Naomi Hunt, "Foreign Journalists and Syrian Videographer Killed in Homs Shelling; Activists reportedly fear targeted attacks," *International Press Institute*, 22 February 2012, at https://ipi.media/two-foreign-journalists-and-syrian-videographer-killed-in-homs-shelling/; and Anne Barnard, "Syrian Forces Aimed to Kill Journalists, US Court Is Told," *New York Times*, 9 April 2018, at https://www.nytimes.com/2018/04/09/world/middleeast/syria-marie-colvin-death.html.

82. "How Livestreaming Is Transforming Activism around the World," *Wired*, 16 November 2016, at https://www.wired.com/2016/11/livestreaming-transforming-activism/.

83. Jason Burke, "How the Changing Media Is Changing Terrorism," *The Guardian*, 25 February 2016, at https://www.theguardian.com/world/2016/feb/25/how-changing-media-changing-terrorism. Al Jazeera did not broadcast the material.

84. Burke, "How the Changing Media Is Changing Terrorism." The video even automatically played in Twitter and Facebook user feeds. Samuel Gibbs, "Facebook and Twitter Users Complain over Virginia Shooting Videos Autoplay," *The Guardian*, 27 August 2015, at https://www.theguardian.com/technology/2015/aug/27/facebook-twitter-users-complain-virginia-shooting-videos-autoplay.

85. Burke, "How the Changing Media Is Changing Terrorism."

86. Adrian Chen, "The Agency," *New York Times Magazine*, 2 June 2015, at https://www.nytimes.com/2015/06/07/magazine/the-agency.html.

87. Ibid.

88. Dara Kerr, "Romney Campaign Suspiciously Gets 116K Twitter Followers in One Day," *CNET*, 6 August 2012, at https://www.cnet.com/news/mitt-romney-suspiciously-gets-116k-twitter-followers-in-one-day/.

89. David Adeleke, "How Fake Twitter Followers Are Generated," *Techcabal*, 11 November 2015, at https://techcabal.com/.

90. "Twitter 'Actively Suspends' Accounts That Share Images of James Foley Beheading," *Fox8*, 20 August 2014, at http://myfox8.com/2014/08/20/twitter-actively-suspend-accounts-that-share-images-of-james-foley-beheading/.

91. Brookings, *The Twitter Census*; and Rick Gladstone and Vindu Goel, "ISIS Is Adept on Twitter, Study Finds," *New York Times*, 5 March 2015. For reference, this is out of some 288 million Twitter accounts.

92. Ibid.

93. Singer and Brooking (2015).

94. See, for example, Scott Spence, "Why You Should Have Your Own Twitter Bot, and How to Build One in Less Than 30 Minutes," *Medium*, 27 January 2017, at https://medium.freecodecamp.org/easily-set-up-your-own-twitter-bot-4aeed5e61f7f.

95. Adam Frank, "Computational Propaganda: Bots, Targeting, and he Future," *NPR*, 9 February 2018, at https://www.npr.org/sections/13.7/2018/02/09/584514805/computational-propaganda-yeah-that-s-a-thing-now.

96. Hackers are usually on the lazy side or lack resources, so instead of trying to break encryption, they typically try to get to the message before it is encrypted. In the case of the average customer, an email message is sent to the ISP mail server in clear text (unencrypted) and then it may be encrypted. So the hacker will try a man-in-the-middle (MITM) attack, to grab the message when it is between the individual workstation and the ISP mail server, as it's traveling in the "clear."

97. Some readers may wonder how companies that make smartphones (or apps installed on them) don't have a key or a backdoor to access what they've made. The answer relates to how the end-to-end encryption process works. When you send a message, your computer mathematically generates both a public and a private key. The private key is secret and stays on your device, and the public key is shared with the other device. When the message arrives, the key at the other end is the only thing that can "unlock" them. This is the basic concept. Andy Greenberg, "Hacker Lexicon: What Is End-to-End Encyption?," *Wired*, 25 November 2014, at https://www.wired.com/2014/11/hacker-lexicon-end-to-end-encryption/. Note, however, that messages stored on a device are only protected if the device itself is secured.

98. Singer and Brooking, "Terror on Twitter."

99. Felicia Woron, "DIY Terror on Telegram," *Counter Extremism Project*, 13 July 2013; at https://www.counterextremism.com/blog/diy-terror-telegram. These sites have been appearing since 2016. They appear for several days, Telegram takes them down, and they are reuploaded to new channels.

100. Frank Gardner, "How Do Terrorists Communicate?," *BBC*, 2 November 2013, at http://www.bbc.com/news/world-24784756.

101. Paul Sarconi, "Now's Probably the Time to Consider One of These Burner Phones," *Wired*, 3 February 2017, at https://www.wired.com/2017/02/7-great-burner-phones/.

102. Craig Timberg, Elizabeth Dwoskin, and Ellen Nakashima, "WikiLeaks: The CIA Is Using Popular TVs, Smartphones, and Cars to Spy on Their Owners," *Washington Post*, 7 March 2017.

103. FBI Director Comey described two different encryption challenges: "data in motion and data at rest." Data in motion is information that is encrypted but relates to operatives communicating with each other when they are potentially about to commit a terrorist attack. Data at rest is stored information relating to a crime or attack that has already happened, but sometimes law enforcement cannot read the perpetrator's phone because it is encrypted.

104. Bianca Bosker, "The Binge Breaker," *The Atlantic*, November 2016, at https://www.theatlantic.com/magazine/archive/2016/11/the-binge-breaker/501122/.

105. Diana I. Tami and Jason P. Mitchell, "Disclosing Information about the Self Is Intrinsically Rewarding," *Proceedings of the National Academy of Science of the USA* 109, no. 21 (22 May 2012): 8038–43, at http://www.pnas.org/content/109/21/8038.abstract; and Bill Davidow, "Exploiting the Neuroscience of InternetInternet Addiction," *The Atlantic*, 18 July 2012, at https://www.theatlantic.com/health/archive/2012/07/exploiting-the-neuroscience-of-InternetInternet-addiction/259820/.

106. Maggie Neil, *PRI International*, 19 May 2017; "On Facebook, Nigerian Victims of Sex Trafficking Present Their Life as Far More Glamorous Than It Is," at https://www.pri.org/stories/2017-05-19/facebook-nigerian-victims-sex-trafficking-often-present-their-life-far-more.

107. B. J. Fogg, *Persuasive Technology: Using Computers to Change What We Think and Do* (New York: Morgan Kaufmann, 2003); and "A Behavior Model for Persuasive Design," Persuasive Technology Lab, Claremont, California, 2009, at http://www.bjfogg.com/fbm_files/page4_1.pdf.

108. "Millennials Check Their Phones More Than 157 Times per Day," *Social Media Week*, 31 May 2016, at https://socialmediaweek.org/newyork/2016/05/31/millennials-check-phones-157-times-per-day/.

109. P. W. Singer and Emerson Brooking, "Terror on Twitter."

Chapter 8

1. See also, Lynn E. Davis, Michael J. McNerney, James Chow, Thomas Hamilton, Sarah Harting, and Daniel Byman, *Armed and Dangerous? UAVs and US Security* (Santa Monica, CA: RAND, 2014), 2, at https://www.rand.org/nsrd/projects/armed-drones.html, . This report points out that GPS was not a huge factor in the evolution of large armed UAVs. Autopilot has been

around for a century and landing aircraft since 1947. GPS just means aircraft can locate the landing field without other assistance.

2. "The Razor: UVA's 3D-printed U.A.V.," 29 August 2014, at https://www.youtube.com/watch?v=FwRD7UBGecg.

3. Paul Marks, "3D Printing: The World's First Printed Plane," *New Scientist*, 27 July 2011, at https://www.newscientist.com/article/dn20737-3d-printing-the-worlds-first-printed-plane/; and Luke Dormehl, "U.K. Royal Navy Puts 3D-Printed Drone through Its Paces in the Antarctic," Digital Trends, 6 January 2017, at https://www.digitaltrends.com/cool-tech/royal-navy-drone-antarctic/.

4. Matthew Stock, "3D Printed Drone Launched from Warship," *Reuters*, 19 August 2015, at http://www.reuters.com/article/us-drone-warship-idUSKCN0QO19F20150819. See also "I Was Wrong about 3D Printed Drones," 8 December 2016, at https://www.youtube.com/watch?v=dACQxI5o4Mk.

5. Mike Senese, "Want a Flying Drone? These Students 3D-Printed Their Own," *Wired*, 28 November 2012, at https://www.wired.com/2012/11/3d-printed-autonomous-airplane/.

6. Ibid.

7. "Gentlemen, Start Your Robots: Smartphones as Guidance Systems," MITRE Corporation Project Stories, December 2012, at https://www.mitre.org/publications/project-stories/gentlemen-start-your-robots-smartphones-as-guidance-systems.

8. "Open, Commercial Technologies Lead to Cost-Effective Reconnaissance Solutions," MITRE Corporation Project Stories, December 2013, at https://www.mitre.org/publications/project-stories/open-commercial-technologies-lead-to-cost-effective-reconnaissance.

9. The craft is built by AeroVironment and first appeared in 1999 as the FQM-151 but developed into its current form in 2002, when Special Operations forces began to use it. It won the US Army's small UAV contest in 2005 and went into full production in 2006, and was subsequently adopted by other US military Services, as well as a large number of international militaries. US Air Force, RQB-11 Raven Fact Sheet, March 2017, at http://www.af.mil/About-Us/Fact-Sheets/Display/Article/104533/rq-11b-raven/.

10. Improvements to the capabilities of the Raven include transition to a digital link, a jam-proof GPS receiver, and better software. David Hambling, *Swarm Troopers: How Small Drones Will Conquer the World* (self-published, 2015), 67–71.

11. A few larger, state-sponsored groups like Hezbollah get drones from Iran, as we will discuss below.

12. For a thorough overview of good drone uses see Austin Choi-Fitzpatrick, "Drones for Good: Technological Innovations, Social Movements, and the State," *Journal of International Affairs* 68, no. 1 (fall/winter 2014): 19–36.

13. Also known as Unmanned Aerial Systems (UASs) and Remotely Piloted Vehicles (RPVs). Some call them "uninhabited" (not "unmanned") aerial vehicles, because they are "manned" by pilots on the ground. For more see John Villasenor, "What Is a Drone, Anyway?," *Scientific American* (blog), 12 April 2012, at https://blogs.scientificamerican.com/guest-blog/what-is-a-drone-anyway/.

14. The use of the term "drone" is imprecise and not appreciated by both the militaries who use them and the companies who make them; nonetheless, it is the widely accepted common term. Robin Shoaps and Sarah Stanley, "'Don't Say Drone': Hits and Misses in a Rhetorical Project of Naming," *Rhetorics of Names and Naming*, edited by Star Medzerian Vanguri (London: Routledge, 2016), 102–17.

15. For a general overview of the range of unmanned US systems, see Department of Defense, *Unmanned Systems Integrated Roadmap, FY 2013–2038*, at http://archive.defense.gov/pubs/DOD-USRM-2013.pdf. The best and most wide-ranging book on this topic is Peter Singer's *Wired for War* (London: Penguin, 2009).

16. The meaning of the word "drone" is evolving. The *Oxford English Dictionary* confines it to aerial vehicles: "A pilotless aircraft of missile directed by remote control." This definition is anachronistic, because autonomy increasingly characterizes them and they are no longer "remote controlled": http://www.oed.com.proxyau.wrlc.org/view/Entry/57852?rskey=4PufZn&result=1&isAdvanced=false - eid. The Merriam-Webster dictionary defines a drone as "an unmanned aircraft or ship guided by remote control or onboard computers": https://www.merriam-webster.com/dictionary/drone. This more forward-looking definition will be used here.

17. Matthew Rosenberg and John Markoff, "The Pentagon's 'Terminator Conundrum': Robots That Could Kill on Their Own," *New York Times*, 26 October 2016, at https://www.nytimes.com/2016/10/26/us/pentagon-artificial-intelligence-terminator.html?mcubz=3. Elsa Kania observed in correspondence that the word may actually be a reference to centaur chess, which Garry Kasparov invented.

18. These huge trucks are called Route Clearance Interrogation Systems or RCISs.

19. There is a similar Avatar project in the Air Force. "The Loyal Wingman" pairs unmanned F-15s or F/A-18s, flown unmanned for the first time, with newer jets like the F-22 Raptor or the F-35 Joint Strike Fighter. Dan Lamothe, "Veil of Secrecy Lifted on Pentagon Office Planning 'Avatar' Fighters and Drone Swarms," *Washington Post*, 8 March 2016.

20. On the growing global competition for driverless cars, see Danielle Muoio, "Ranked: The 18 Companies Most Likely to Get Self-Driving Cars on the Road First," *Business Insider*, September 27, 2017, http://www.businessinsider.com/

the-companies-most-likely-to-get-driverless-cars-on-the-road-first-2017-4/
#16-uber-3.

21. QinetiQ North America, at https://www.qinetiq-na.com.
22. "Dragon Runner Reconnaissance Robot, USA," at http://www.army-technology.com/projects/dragonrunnerrobots/.
23. Dan Gettinger, "Underwater Drones (Updated)," 28 October 2016, Center for the Study of the Drone at Bard College, at http://dronecenter.bard.edu/underwater-drones-updated/.
24. Unmanned Surface Vehicles are known as USVs, and unmanned undersea vehicles are called UUVs.
25. In October 2016 a Chinese remotely operated deep-sea drone also searched in vain for Malaysia Airlines Flight 370 (MH370), followed by the US firm Ocean Infinity, which searched unsuccessfully between January and May 2018 with eight autonomous submarines. *Safety Investigation Report: Malaysia Airlines Boeing B777-200ER (9M-MRO)*, issued 2 July 2018 by the Safety Investigation Team for MH370, Ministry of Transportation, Government of Malaysia, at http://mh370.mot.gov.my/MH370SafetyInvestigationReport.pdf.
26. Prices for small manned submarines range from under $200,000 for a shallow water craft, to $30 million+, depending on features.
27. Chris Dixon, "Do Humans Have a Future in Deep Sea Exploration?," *New York Times*, 14 September 2015, at https://www.nytimes.com/2015/09/15/science/piloted-deep-sea-research-is-bottoming-out.html?mcubz=3&_r=0.
28. See Robert Sparrow and George Lucas, "When Robots Rule the Waves?," *Naval War College Review* 69, no. 4 (autumn 2016): 49–78.
29. Remote Control Project (2016), 5–6.
30. An excellent resource for the full range of international unmanned aerial systems is the Center for the Study of the Drone at Bard College, at http://dronecenter.bard.edu/the-drone-database/.
31. The information in this chart was compiled from the following sources: Center for the Study of the Drone at Bard College, at http://dronecenter.bard.edu/the-drone-database/; Center for a New American Security, The Drone Database; at http://drones.cnas.org/drones/; and Department of Defense, *Unmanned Systems Integrated Roadmap, FY 2013–2038* (see note 15, above), among other DoD sites, as well as from websites of private corporations that manufacture military UAVs. UAVs have also been classified as Micro, Mini, Tactical, Medium Altitude, and High Altitude (often also called "Strategic"). See, for example, Collier C. Crouch, *Integration of Mini-UAVs at the Tactical Operations Level: Implications of Operations, Implementation, and Information Sharing*, Master's Thesis, June 2005, Naval Postgraduate School, Monterey California, pp. 12–14. The label "Strategic" is not used here because it prejudges the UAV's purpose and

effectiveness. In other words, Small UAVs can be just as "strategic" as a High Altitude UAVs, depending on their effects.

32. Examples include: *Eye in the Sky* (movie, 2016); *Grounded* (play, 2013); *Good Kill* (movie, 2015); *Unblinking Eye* (play, 2015); and a book by Matt Martin (former drone pilot), *Predator: The Remote Control War over Iraq and Afghanistan: A Pilot's Story* (New York: Zenith Press, 2010). Many other drone books have since followed.

33. Eric Schmitt, "17 Fighters Reportedly Killed as US Ends Lull in Libya Airstrikes," *New York Times*, 24 September 2017, at https://www.nytimes.com/2017/09/24/us/politics/libya-military-strike-isis.html. For databases on US use in targeted killings, see Bureau of Investigative Journalism, "Drone Wars: The Full Data," at https://www.thebureauinvestigates.com/stories/2017-01-01/drone-wars-the-full-data; and New America Foundation, "Drone Strikes: Pakistan," at https://www.newamerica.org/in-depth/americas-counterterrorism-wars/pakistan/. C:\at https\::www.newamerica.org:in-depth:americas-counterterrorism-wars:pakistan\.

34. For example, Rosa Brooks and John Abizaid, *Recommendations and Report of the Task Force on US Drone Policy*, Stimson Center, Washington, DC, June 2014; Gregoire Chamayou, *Drone Theory*, trans. Janet Lloyd (Paris: LaFabrique Editions, 2013; English version London: Penguin, 2015); Avery Plaw, Matthew S. Fricker, and Carlos R. Colon, *The Drone Debate: A Primer on the US Use of Unmanned Aircraft outside Conventional Battlefields* (New York: Rowman & Littlefield, 2015); Sarah Kreps, *Drones: What Everyone Needs to Know* (Oxford, UK: Oxford University Press, 2016); John Kaag and Sarah Kreps, *Drone Warfare* (Oxford, UK: Polity Press, 2014); Hugh Gusterson, *Drone: Remote Control Warfare* (Cambridge, MA: MIT Press, 2016); Brian Glyn Williams, *Predator* (Dulles, VA: Potomac Books, 2013).

35. Excellent resources on the uses of medium- and high-altitude armed drones include: David Cortright, Rachel Fairhurst, and Kristen Wall, *Drones and the Future of Armed Conflict: Ethical, Legal, and Strategic Implications* (Chicago: University of Chicago Press, 2015); Kaag and Kreps (2014); and Kreps (2016).

36. H. R. Everett, *Unmanned Systems of World Wars I and II* (Cambridge, MA: MIT Press, 2015), p. 247.

37. F. Stansbury Haydon, *Military Ballooning during the Early Civil War* (Baltimore: Johns Hopkins Press, 1941), introduction by Tom D. Crouch, "The Making of a Classic," (2000), pp. 17–18 and note 84. Manned balloons originate in the French Revolutionary wars, where the French Aerostatic Corps used them for reconnaissance over the battlefield. "Reconnaissance by Balloon," Imperial War Museum, London, at https://www.iwm.org.uk/learning/resources/how-has-war-in-the-air-changed-over-time; Sarah Kreps, "What Is the Historical Genealogy of Armed Drones?," *Drones: What Everyone Needs to Know* (Oxford, UK: Oxford University Press, 2016), 9–13; Chris

Cole, "Rise of the Reapers: A Brief History of Drones," *Drone Wars UK*, 10 June 2014, at https://dronewars.net/2014/10/06/rise-of-the-reapers-a-brief-history-of-drones/; Gusterson (2016), 9–28.

38. Thomas P. Ehrhard, *Air Force UAVs: The Secret History*, Mitchell Institute Study, July 2010, pp. 38–9.

39. The best source on the very interesting history of Ryan Aeronautical's drones is William Wagner's *Lightning Bugs and Other Reconnaissance Drones: The Can-do Story of Ryan's Unmanned "Spy Planes"* (Fallbrook, CA: Armed Forces Journal International, 1983).

40. US Office of the Secretary of Defense, *Unmanned Aircraft Systems Roadmap, 2005–2030*, at https://fas.org/irp/program/collect/uav_roadmap2005.pdf, p. 1. Note the US Air Force uses a different classification system.

41. Some US military sources refer to the Predator and Reaper drones as "medium-altitude," and Global Hawk as "high-altitude." The terminology is inconsistent. We have chosen to use the five categories employed in the US government's Unmanned Systems Integrated Roadmap. See https://archive. defense.gov/pubs/DOD-USRM-2013.pdf , which puts both in the "high-altitude" category, Group #5.

42. Steve Coll, *Ghost Wars: The Secret History of the CIA, Afghanistan, and bin Laden, from the Soviet Invasion to September 10, 2001* (New York: Penguin Press, 2004), 141–4; citing personal interview by Steve Coll with Duane Clarridge, 28 December 2001; Chris Woods, "The Story of America's Very First Drone Strike," *The Atlantic*, 30 May 2015, at https://www. theatlantic.com/international/archive/2015/05/america-first-drone-strike-afghanistan/394463/; Christopher J. Fuller, *See It/Shoot It: The Secret History of the CIA's Lethal Drone Program* (New Haven, CT: Yale University Press, 2017), 104–5. According to Duane Clarridge, who in 1986 helped found the CIA's Counterterrorist Center, they were initially intended as a way to gather intelligence on Hezbollah holding US hostages in Lebanon, but they were not authorized for lethal missions. Clarridge persuaded William Casey to establish the CTC in 1986 and was put in charge. It is now called the Counterterrorism Center. "Duane R. Clarridge, Brash Spy Who Fought Terror Networks, Dies at 83," *New York Times*, 10 April 2016, at https://www.nytimes.com/2016/04/11/us/duane-r-clarridge-brash-spy-who-fought-terror-dies-at-83.html.

43. Steve Coll, "The Unblinking Stare," *New Yorker*, 24 November 2014, at http://www.newyorker.com/magazine/2014/11/24/unblinking-stare.

44. John A. Tirpak, "The RPA Boom," *Air Force Magazine*, August 2010, at http://www.airforcemag.com/MagazineArchive/Documents/2010/August 2010/0810RPA.pdf.

45. The next generation should have longer ranges. General Atomics has built a prototype jet-powered drone, the Avenger (formerly Predator C) that can range 1,800 from the nearest base. Allen McDuffee, "New Jet-Powered

Drone Can Kill 1,800 Miles from Home Base," *Wired*, 21 February 2014, at https://www.wired.com/2014/02/avenger/; Davis et al. (2014), 2; and Jeffrey M. Sullivan, "Evolution or Revolution? The Rise of UAVs," *IEEE Technology and Society Magazine*, February 2006, pp. 43–9.

46. They use gyro-stabilized high-power telescopes, which are also used in Hollywood action films. Davis et al. (2014), 3.

47. Scott Shane, "The Moral Case for Drones," *New York Times*, 14 July 2012; Human Rights Watch, *Precisely Wrong: Gaza Civilians Killed by Israeli Drone-Launched Missiles*, 30 June 2009, at https://www.hrw.org/report/2009/06/30/precisely-wrong/gaza-civilians-killed-israeli-drone-launched-missiles. Recently, US drones have also been equipped with a new modified Hellfire missile (R9X) that carries an inert warhead that kills targeted individuals with no explosion, reducing the potential for unintended casualties. Gordon Lubold and Walter P. Strobel, *Wall Street Journal*, 9 May 2019, at https://www.wsj.com/articles/secret-u-s-missile-aims-to-kill-only-terrorists-not-nearby-civilians-11557403411.

48. "Who Has What: Countries with Drones Used in Combat," World of Drones Project, the New America Foundation, at https://www.newamerica.org/in-depth/world-of-drones/2-who-has-what-countries-drones-used-combat/.

49. General Atomics has built a prototype jet-powered drone, the Avenger (formerly Predator C) that can range 1,800 from the nearest base. Allen McDuffee, "New Jet-Powered Drone Can Kill 1,800 Miles from Home Base," *Wired*, 21 February 2014, at https://www.wired.com/2014/02/avenger/.

50. For example, the Northrop Grumman X-47B is a high-technology, stealthy UAV. And China has designed a solar-powered drone (CH-T4) that can potentially stay aloft for years. Jeffrey Lin and P. W. Winger, "China Just Flew a 130-foot, Solar-powered Drone Designed to Stay in the Air for Months," *Popular Science*, 6 June 2017, at http://www.popsci.com/china-solar-powered-drone.

51. Davis et al. (2014).

52. US Government Accountability Office, *Unmanned Aircraft Systems: Global Hawk Cost Increase Understated in Nunn-McCurdy Report*, GAO-06-222R, 15 December 2005, at https://www.gao.gov/assets/100/93905.pdf. Alexander Cohen, "The Drone That Wouldn't Die: How a Defense Contractor Bested the Pentagon," *The Atlantic*, 16 July 2013, at https://www.theatlantic.com/politics/archive/2013/07/the-drone-that-wouldnt-die-how-a-defense-contractor-bested-the-pentagon/277807/.

53. US Department of Defense, "Performance of the Defense Acquisition System, 2015 Annual Report," 16 September 2015, p. 43, at https://admin.govexec.com/media/gbc/docs/pdfs_edit/performance-of-defense-acquisition-system-2015.pdf; and Dan Gettinger, "Drone Spending: The MQ-9 Reaper," Drone

Center at Bard College, 12 October 2015, at http://dronecenter.bard.edu/
drone-spending-the-mq-9-reaper/.

54. The interesting thing is which countries will *not* have armed drones. See also
Patrick Tucker, "Every Country Will Have Armed Drones Within 10 Years,"
Defense One, 6 May 2014, at http://www.defenseone.com/technology/2014/05/
every-country-will-have-armed-drones-within-ten-years/83878/.

55. China has long been interested in UAVs but has recently been expanding
its capabilities faster than any other country in the world. The PRC began
building drones in the 1950s, when it reverse-engineered Soviet drones. In
the 1960s, the Chinese acquired several US Firebees that were lost over China
and North Vietnam and reverse-engineered them as well. In the 1990s,
they bought 100 Harpy armed drones from Israel—though in subsequent
years the Pentagon pressured the Israelis not to upgrade them. Since then,
the top source for technological improvement appears to be cyberespionage,
which explains why several of the latest Chinese drones bear an uncanny
resemblance to American systems. Today China has a full range of UAVs,
from small, short-range drones to long-range armed and reconnaissance
drones, and is well poised to be a top global source of high-altitude UAVs
with its Predator-clone Wing Loong (or Pterodactyl). First exported in
2011, the Wing Loong lacks the over-horizon strike capability, loiter time
and payload of the US Reaper, but it can carry two laser-guided missiles and
match Predator in endurance and flight range, at a much lower price.
China is willing to sell its drones to countries when the United States isn't, notably
to Ethiopia, Zambia, Myanmar, Turkmenistan, Algeria, Jordan, the UAE,
and Iraq, so far. Many find that large armed Chinese drones, while not as
impressive as the American ones, are good enough for what they want to do
with them. Edward Wong, "Hacking US Secrets, China Pushes for Drones,"
New York Times, 20 September 2013, at http://www.nytimes.com/2013/
09/21/world/asia/hacking-us-secrets-china-pushes-for-drones.html; Liang
Xu, "Will China Start Selling the 'AK-47' of Drones?," *The Cipher Brief*, 15
October 2017, at https://www.thecipherbrief.com/will-china-start-selling-ak-
47-drones; Sharon Weinberger, "China Has Already Won the Drone Wars,"
Foreign Policy, 10 May 2018; and Michael C. Horowitz, Sarah E. Kreps, and
Matthew Fuhrmann, "Separating Fact from Fiction in the Debate over Drone
Proliferation," *International Security* 41, no. 2 (fall 2016): 11–13.

56. Matthew Fuhrmann and Michael C. Horowitz, "Droning On: Explaining
the Proliferation of Unmanned Aerial Vehicles," *International Organization* 71
(Spring 2017): 414–15. Hezbollah has shown interest in this capability. They
are a state-like organization with an army and thus the exception that proves
the rule.

57. Arthur Holland Michel, "Iran's Many Drones," Center for the Study of
the Drone at Bard College, 25 November 2013, at http://dronecenter.
bard.edu/irans-drones/. Iran's program has been helped along by American

drones they've stolen or shot down. In 2011, the Iranians captured a US Sentinel stealth surveillance drone and began to reverse-engineer it. This was followed in 2012 by the shoot-down of a smaller US ScanEagle reconnaissance drone. In 2013, Iran unveiled a Reaper-size large drone called the Fotros (or "Fallen Angel"). Associated Press, "Iran Says It Has Gleaned Data from US Spy Drone," *San Francisco Chronicle*, 23 April 2012, at http://www.sfgate.com/world/article/Iran-says-it-has-gleaned-data-from-U-S-spy-drone-3501847.php; "Iranian TV Shows Off 'Captured US ScanEagle Drone,'" BBC News, 4 December 2012, at http://www.bbc.com/news/world-middle-east-20591336. Defense Minister Brigadier General Hossein Dehghan claimed it had a range of 1,250 miles, and could fly up to an altitude of 25,000 feet, carry missiles, and remain aloft for 16–30 hours. Kelsey D. Atherton, "Iran's Fotros Drone: Iran's Latest Unmanned Flying Camera with Bombs," *Popular Science*, 18 November 2013, at http://www.popsci.com/article/technology/iran-unveils-new-reaper-sized-drone; and Arthur Holland Michel, "Iran's Many Drones," Center for the Study of the Drone, 25 November 2013, at http://dronecenter.bard.edu/irans-drones/.

58. Benny Avni, "US Widens Iran Sanctions As Drone Is Reported in Darfur," *New York Sun*, 11 September 2008, at http://www.nysun.com/foreign/us-widens-iran-sanctions-as-drone-is-reported/85655/; and W. J. Hennigan, "US Forces Shoot Down Iranian Drone over Syria As Fighting Escalates," *Los Angeles Times*, 20 June 2017, at http://www.latimes.com/politics/washington/la-na-essential-washington-updates-us-forces-shoot-down-iranian-drone-over-1497972506-htmlstory.html.

59. Dion Nissenbaum and Warren P. Strobel, "Mideast Insurgents Enter the Age of Drone Warfare," *The Wall Street Journal*, 2 May 2019, at https://www.wsj.com/articles/mideast-insurgents-enter-the-age-of-drone-warfare-11556814441.

60. This is known as the Ansariya ambush.

61. Yaakov Katz, "Israel Air Force Initiates Drive to Encrypt All of Its UAVs," *Jane's Defence Weekly*, 26 November 2010; cited by Don Rassler, "Remotely Piloted Innovation: Terrorism, Drones, and Supportive Technology," Combating Terrorism Center at West Point, October 2016, p. 27, at https://ctc.usma.edu/posts/remotely-piloted-innovation-terrorism-drones-and-supportive-technology.

62. Gili Cohen, "IDF Strands Parts of Intercepted Hezbollah Drone at Negev Crash Site," *Haaretz*, 15 October 2012, at http://www.haaretz.com/israel-news/idf-strands-parts-of-intercepted-hezbollah-drone-at-negev-crash-site.premium-1.470191.

63. Milton Hoenig, "Hezbollah and the Use of Drones as a Weapon of Terrorism," *Public Interest Report* 67, no. 2 (spring 2014), Federation of

American Scientists; at https://fas.org/pir-pubs/hezbollah-use-drones-weapon-terrorism/.

64. Yaakov Lappin, "Israel Shoots Down UAV," *IHS Jane's Defence Weekly*, 19 September 2017, at http://www.janes.com/article/74201/israel-shoots-down-uav. Hezbollah also claimed to have fired an armed missile from a UAV over Syria in September 2014, but this has not been firmly established. Rassler (2016), 28.

65. Dan Gettinger and Arthur Holland Michel, "A Brief History of Hamas and Hezbollah's Drones," Center for the Study of the Drone at Bard College, 14 July 2014, at http://dronecenter.bard.edu/hezbollah-hamas-drones/.

66. "Israel Shoots Down Hamas Drone from Gaza Strip: Military," *Reuters*, 23 February 2017, at http://www.reuters.com/article/us-israel-palestinians-uav/israel-shoots-down-hamas-drone-from-gaza-strip-military-idUSKBN1621TL; "Hamas Blames Israel for Assassination of Drone Chief," *Al Jazeera*, 18 December 2016, at http://www.aljazeera.com/news/2016/12/hamas-blames-israel-assassination-drone-chief-161217174004082.html. His name was Muhammad al-Zawari.

67. Rafael Advanced Defense Systems, Ltd., Iron Beam Mobile High-Energy Weapon System, at http://www.rafael.co.il/5688-763-en/Marketing.aspx.

68. The Houthis have also successfully used Iranian-style water-borne improvised explosive devices to attack Saudi naval vessels. On 30 January 2017, a Houthi drone boat attacked the Saudi frigate Al-Madinah near the Yemeni port of Al Hudaydah. Conflict Armament Research, *Anatomy of a 'Drone Boat': A Water-borne Improvised Explosive Device (WBIED) Constructed in Yemen*, December 2017, at http://www.conflictarm.com/publications/.

69. Nick Waters, "Houthis Use Armed Drone to Target Yemeni Army Top Brass," *Bellingcat*, 10 January 2019, at https://www.bellingcat.com/news/mena/2019/01/10/houthis-use-armed-drone-to-target-yemeni-army-top-brass/; Aaron Stein, "Low-tec, High Reward: The Houthi Drone Attack," Foreign Policy Research Institute, 11 January 2019, at https://www.fpri.org/article/2019/01/low-tech-high-reward-the-houthi-drone-attack/; and Dion Nissenbaum and Warrn P. Strobel, "Mideast Insurgents Enter the Age of Drone Warfare," *Wall Street Journal*, 2 May 2019, at https://www.wsj.com/articles/mideast-insurgents-enter-the-age-of-drone-warfare-11556814441; and Conflict Armament Research, *Iranian Technology Transfers to Yemen: 'Kamikaze' Drones Used by Houthi Forces to Attack Missile Defense Systems*, March 2017, at http://www.conflictarm.com/publications/; and Jon Gambrell, "Bomb-laden Drones of Yemen Rebels Threaten Arabian Peninsula," *Associated Press*, 16 May 2019.

70. "Hamas Claims to Capture Israeli UAV," *Times of Israel*, 12 August 2015, cited by Rassler (2016), p. 31.

71. "US MQ-9 Reaper Drone Shot Down in Yemen: CentCom," *DefenseTech*, Military.com, 2 October 2017, at https://www.military.com/defensetech/2017/10/02/mq-9-reaper-drone-shot-yemen-centcom.

72. *Hostile Drones: The Hostile Use of Drones by Non-state Actors against British Targets*, Remote Control Project, London, January 2016, at http://remotecontrolproject.org/wp-content/uploads/2016/01/Hostile-use-of-drones-report_open-briefing.pdf.

73. "Teal Group Predicts Worldwide Civil Drone Production Will Soar $73.5 Billion over the Next Decade," 19 June 2017, at http://www.tealgroup.com/index.php/about-teal-group-corporation/press-releases/136-teal-group-predicts-worldwide-civil-drone-production-will-soar-73-5-billion-over-the-next-decade.

74. "How DJI Has Crushed the Consumer Drone Industry, and the Rivals That Could Still Take Flight," *Market Watch*, 17 February 2017, at http://www.marketwatch.com/story/how-dji-has-crushed-the-consumer-drone-industry-and-the-rivals-that-could-still-take-flight-2017-02-17.

75. Robert Spousta and Steve Chan, "Hold the Drones: Fostering the Development of Big Data Paradigms through Regulatory Frameworks," *Journal of Communication and Computing* 12 (2015): 143; and Jack Nicas and Colum Murphy, "Who Builds the World's Most Popular Drones?," *The Wall Street Journal*, 10 November 2014, at https://www.wsj.com/articles/who-builds-the-worlds-most-popular-drones-1415645659.

76. "How DJI Has Crushed the Consumer Drone Industry, and the Rivals That Could Still Take Flight," *Market Watch*, 17 February 2017, at http://www.marketwatch.com/story/how-dji-has-crushed-the-consumer-drone-industry-and-the-rivals-that-could-still-take-flight-2017-02-17.

77. DJI's Spark is controlled by hand gestures. "The New Spark is DJI's Smallest, Cheapest Drone Yet," *MarketWatch*, 24 May 2017, at https://www.marketwatch.com/story/the-new-spark-is-djis-smallest-cheapest-drone-yet-2017-05-24.

78. Jeremy Hsu, "DJI Phantom Drone Gets Flight Control Upgrades, Powerful Cameras," *IEEE Spectrum*, 13 April 2015, at https://spectrum.ieee.org/automaton/robotics/drones/dji-phantom-3-camera-drones.

79. Ben Coxworth, "DJI Phantom Can Now Perform Autonomous Flight," *New Atlas*, 2 July 2014, at https://newatlas.com/dji-phantom-ground-control-autonomous-flight/32787/.

80. Faine Greenwood, "The US Military Shouldn't Use Commercial Drones," *Future Tense*, 16 August 2017, at http://www.slate.com/articles/technology/future_tense/2017/08/the_u_s_military_shouldn_t_use_commercial_drones.html.

81. Anna Ahronheim, "IDF to Continue Using Drones That US Army Deemed Unsafe," *Jerusalem Post*, 6 August 2017, at http://www.jpost.com/Israel-News/US-Army-order-troops-to-stop-using-Chinese-made-DJI-drones-501741.

82. Not everyone thinks this is a real vulnerability. A 2016 study of the DJI S-1000 drone done by the National Oceanic and Atmospheric Administration (a US government organization that collects data on the weather) found no evidence of data transfer. DJI protested the US Army's decision and emphasized that they made drones strictly for peaceful purposes. Ben Popper, "A Government Study Found DJI Drone, Banned by US Army, Kept Data Safe," *The Verge*, 7 August 2017, at https://www.theverge.com/2017/8/7/16106810/dji-drone-banned-government-study-data-safety.

83. Ben Watson, "The US Army Just Ordered Soldiers to Stop Using Drones from China's DJI," *Defense One*, 4 August 2017, at http://www.defenseone.com/technology/2017/08/us-army-just-ordered-soldiers-stop-using-drones-chinas-dji/139999/.

84. Ibid.

85. Michael S. Schmidt and Michael D. Shear, "A Drone, Too Small for Radar to Detect, Rattles the White House," *New York Times*, 26 January 2015, at https://www.nytimes.com/2015/01/27/us/white-house-drone.html?mcubz=3.

86. Kevin Poulsen, "Why the US Government Is Terrified of Hobbyist Drones," *Wired*, 5 February 2015, at https://www.wired.com/2015/02/white-house-drone/.

87. Kazuaki Nagata, "Drone-makers Say Demand Will Take Off in Japan," *Japan Times*, 20 May 2015, at http://www.japantimes.co.jp/news/2015/05/20/business/tech/drone-makers-say-demand-will-take-japan/ -.WYtmD62ZPJw.

88. "7 Drones That Can Lift Heavy Weights," *Drones Globe*, 2 October 2017, at http://www.dronesglobe.com/guide/heavy-lift-drones/.

89. Remote Control Project (2016), 4.

90. Robert R. Rodwell, "Technology in the Streets of Ulster," *New Scientist*, 6 April 1972, pp. 16–17; and Ann Larabee, "Technology and Terrorism," chapter 29 of *The Routledge History of Terrorism* (London: Routledge, 2015), 449.

91. Richard Danzig, Marc Sageman, Terrance Leighton, Lloyd Hough, Hidemi Yuki, Rui Kotani, and Zachary M. Hosford, *Aum Shinrikyo: Insights into How Terrorists Develop Biological and Chemical Weapons*, 2nd ed., Center for a New American Security, December 2012, at https://s3.amazonaws.com/files.cnas.org/documents/CNAS_AumShinrikyo_SecondEdition_English.pdf?mtime=20160906080510.

92. Rassler has done excellent research compiling all known recent terrorist use of UAVs. Only overall trends and highlights are recapped here. See Rassler (2016). Other excellent sources include Nicholas Grossman, *Drones and Terrorism: Asymmetric Warfare and the Threat to Global Security* (London: I. B. Tauris, 2018); Steven Stalinsky and R. Sosnow, *A Decade of Jihadi Organizations' Use of Drones—from Early Experiments by Hizballah, and al-Qaeda to Emerging National Security Crisis for the West AAs ISIS Launches First Attack Drones*, The Middle East Media Research Institute (MEMRI), 21 February

2017; Larry Friese with N. R. Jenzen-Jones and Michael Smallwood, *Emerging Unmanned Threats: The Use of Commercially Available UAVs by Armed Non-State Actors*, Armament Research Services (ARES), Special Report No. 2 (2016), at http://armamentresearch.com/special-report-no-2-emerging-unmanned-threats/; and Robert J. Bunker, *Terrorist and Insurgent Unmanned Aerial Vehicles: Use, Potentials, and Military Implications*, Strategic Studies Institute, US Army War College, August 2015.

93. Rassler (2016), 32.

94. Scott Shane and Ben Hubbard, "ISIS Displaying a Deft Command of Varied Media," *New York Times*, 30 August 2014, at https://www.nytimes.com/2014/08/31/world/middleeast/isis-displaying-a-deft-command-of-varied-media.html?mcubz=0.

95. Ruth Sherlock, "Islamic State Release Drone Video of Kobane," *The Telegraph*, 12 December 2014, at http://www.telegraph.co.uk/news/worldnews/islamic-state/11287145/Islamic-State-release-drone-video-of-Kobane.html.

96. Don Rassler, *The Islamic State and Drones: Supply, Scale, and Future Threats*, Combating Terrorism Center at West Point, United States Military Academy, July 2018, at https://ctc.usma.edu/app/uploads/2018/07/Islamic-State-and-Drones-Release-Version.pdf.

97. The small explosive was reportedly disguised as a battery. Michael S. Schmidt and Eric Schmitt, "Pentagon Confronts a New Threat from ISIS: Exploding Drones," *New York Times*, 11 October 2016, at https://www.nytimes.com/2016/10/12/world/middleeast/iraq-drones-isis.html?mcubz=3.

98. Nathalie Guibert, "Irak: Paris confirme qu'un drone piégé a blessé deux membres des forces spéciales françaises à Erbil," *Le Monde*, 11 October 2016, at http://www.lemonde.fr/proche-orient/article/2016/10/11/irak-deux-commandos-francais-gravement-blesses-a-erbil-par-un-drone-piege_5011751_3218.html.

99. Rassler (2018), 4.

100. David Hambling, "Iraqi Officials Find Strange Collection of Makeshift ISIS Drones," *Popular Mechanics*, 29 November 2016, at http://www.popularmechanics.com/military/weapons/a24056/iraqi-officials-isis-drones-mosul/.

101. Schmitt (2017); Tom Westcott, "Death from Above: IS Drones Strike Terror in 'Safe' Areas of Mosul," *Middle East Eye*, 22 February 2017; both cited by Rassler (2018), 4–5.

102. Rassler (2018), 2–3.

103. David Hambling, "ISIS Is Reportedly Packing Drones with Explosives Now," *Popular Mechanics*, 16 December 2015, at http://www.popularmechanics.com/military/weapons/a18577/isis-packing-drones-with-explosives/. For more in-depth information and photographs of the drones used by ISIS and other nonstate actors in Syria and Iraq, see Dan Gettinger, *Drones Operating in Syria*

and *Iraq* (New York: Center for the Study of the Drone, 13 December 2016), at http://dronecenter.bard.edu/drones-operating-in-syria-and-iraq/; and Ben Watson, "The Drones of ISIS," *Defense One*, 12 January 2017, at https://www. defenseone.com/technology/2017/01/drones-isis/134542/.

104. Mike Mawhinney, "Islamic State Using Hobby Drones with Deadly Effect," *Sky*, 4 April 2017; cited by Rassler (2018), 3.

105. Benedetta Argentieri, "Private Surveillance Drones Take Flight over Iraq: The Kurds Had Help from an American Entrepreneur," *War is Boring*, 28 April 2015, at https://warisboring.com/private-surveillance-drones-take-flight-over-iraq/; The Third Block Group, at http://thirdblockgroup.com.

106. Remote Control Project (2016), 12.

107. This happened in Svatovo (29 October 2015), Zaporozhye (17 February 2017), Balakliya (23 March 2017), and Vinnytsya (27 September 2017). Iulia Mendel, *New York Times*, 27 September 2017, at https://www.nytimes.com/2017/09/27/world/europe/ukraine-ammunition-depot-explosion.html?_r=0; and David Hambling, "Russian Drones Attack with Grenade Weapons," *Scout*, 18 September 2017, at https://scout.com/military/warrior/Article/Small-Russian-Drones-Do-Massive-Damage-WIth-Grenade-Weapons-103103172.

108. For much more on this, see Larry Friese, N. R. Jenzen-Jones, and Michael Smallwood, *Emerging Unmanned Threats: The Use of Commercially Available UAVs by Armed Non-state Actors*, Armament Research Services, Special Report No. 2, Perth, Australia, February 2016, http://armamentresearch.com:special-report-no-2-emerging-unmanned-threats.

109. Emiko Jozuka, "A Crowdfunded UAV Is Helping Ukraine's Bootstrapped Army: The People's Project Are a Volunteer Group Crowdfunding Everything from Tanks, Uniforms, and Drones for Ukraine's Army," *Vice*, 18 June 2015, at https://motherboard.vice.com/en_us/article/8qxj5g/a-crowdfunded-uav-is-helping-ukraines-bootstrapped-army.

110. People's Project.com: Ukraine's military and civil crowdfunding, at https://www.peoplesproject.com/en/first-peoples-uav-complex/.

111. Davis et al. (2014), 3 and 14; and J. Noel Williams, "Killing Sanctuary: The Coming Era of Small, Smart, Pervasive Lethality," *War on the Rocks*, 8 September 2017, at https://warontherocks.com/2017/09/killing-sanctuary-the-coming-era-of-small-smart-pervasive-lethality/.

112. Bill Gates, "A Robot in Every Home," *Scientific American*, February 1 2008, at https://www.scientificamerican.com/article/a-robot-in-every-home-2008-02/.

113. For a great explanation, watch "Search for the Super Battery," *Nova*, Season 44, Episode 3, aired 1 February 2017, at https://www.pbs.org/video/nova-search-super-battery/.

114. Robert Rapier, "A Battery That Could Change the World," *Forbes*, 20 May 2018, at https://www.forbes.com/sites/rrapier/2018/05/20/a-battery-that-could-change-the-world/ - 541a6f3c4cf2; and "Search for the Super Battery,"

NOVA, Season 44, Episode 3, aired 1 February 2017, at https://www.pbs.org/video/nova-search-super-battery/.

115. There is also potential for nanoexplosives to increase the power in very light payloads.

116. Michael Kaplan, "How Vegas Security Drives Surveillance Technology Everywhere," *Popular Mechanics*, 1 January 2010, at https://www.popularmechanics.com/technology/security/how-to/a5226/4341499/; and Brad Chacos and Maximum Tech, "7 Casino Technologies They Don't Want You to Know About," *Gizmodo*, 10 August 2011, at https://www.gizmodo.com.au/2011/08/7-casino-technologies-they-dont-want-you-to-know-about/.

117. Robert Triggs, "Facial Recognition Technology Explained," *Android Authority*, 11 September 2017, at http://www.androidauthority.com/facial-recognition-technology-explained-800421/.

118. "All Glassholes Are Revolutionaries," *Wired* 26, no. 5 (May 2018): 29.

119. Patrick Tucker, "Tomorrow's Intelligent Malware Will Attack When It Sees Your Face," *Defense One*, 14 August 2018, at https://www.defenseone.com/technology/2018/08/tomorrows-intelligent-malware-will-attack-when-it-sees-your-face/150550/.

120. "We'll Share Our Emotional State as Willingly as We Share Our Photos," *Wired* 26, no. 6 (June 2018): 55.

121. Matthew Rosenberg and John Markoff, "The Pentagon's 'Terminator Conundrum': Robots That Could Kill on Their Own," *New York Times*, 25 October 2016, at https://www.nytimes.com/2016/10/26/us/pentagon-artificial-intelligence-terminator.html?mcubz=3.

122. Timothy B. Lee, "Watch the Pirate Party Fly a Drone in Front of Germany's Chancellor," *Washington Post*, 18 September 2013, at https://www.washingtonpost.com/news/the-switch/wp/2013/09/18/watch-the-pirate-party-fly-a-drone-in-front-of-germanys-chancellor/?utm_term=.1a6c3787ec45; and Friederike Heine, "Merkel Buzzed by Mini-Drone at Campaign Event," *Der Spiegel*, 16 September 2013, at http://www.spiegel.de/international/germany/merkel-campaign-event-visited-by-mini-drone-a-922495.html.

123. "Venezuela President Maduro Survives 'Drone Assassination Attempt,'" *BBC News*, 5 August 2018, at https://www.bbc.com/news/world-latin-america-45073385.

124. Some describe them as "loitering munitions." Davis et al. (2014), 4.

125. Peter Finn, "Mass. Man Accused of Plotting to Hit Pentagon and Capitol with Drone Aircraft," *Washington Post*, 28 September 2011, at https://www.washingtonpost.com/national/national-security/mass-man-accused-of-plotting-to-hit-pentagon-and-capitol-with-drone-aircraft/2011/09/28/gIQAWdpk5K_story.html?utm_term=.e937a57c11fd; "FBI: Drone-like Toy Planes in Bomb Plot," *Connecticut Post*, 7 April 2014, at http://www.ctpost.com/local/article/FBI-Drone-like-toy-planes-in-bomb-plot-5383658.

php. In 2015, a $400 remote-controlled quadcopter landed on White House grounds.

126. Peter Finn, "Mass. Man Accused of Plotting to Hit Pentagon and Capitol with Drone Aircraft," *Washington Post*, 28 September 2011.

127. Schmitt (2017).

128. For example, the start-up DroneDeploy emphasizes these capabilities. See Gretchen West, "The Sky's the Limit—If the FAA Will Get Out of the War," *Foreign Affairs* (May/June 2015), at https://www.foreignaffairs.com/articles/2015-05-01/drone.

129. Michael S. Schmidt and Michael D. Shear, "A Drone Too Small for Radar to Detect, Rattles the White House," *New York Times*, 26 January 2015, at https://www.nytimes.com/2015/01/27/us/white-house-drone.html.

130. Others have come to the same conclusion. See Greg Allen and Taniel Chan, *Artificial Intelligence and National Security*, a Study on Behalf of Intelligence Advanced Research Projects Activity (IARPA), Belfer Center, Harvard Kennedy School, Cambridge, MA, July 2017, p. 22.

131. Alex Silverman and Tom Kaminski, "Drone Hits Army Helicopter Flying over Staten Island," *CBS New York*, 22 September 2017, at http://newyork.cbslocal.com/2017/09/22/drone-hits-army-helicopter/.

132. Travis M. Andrews, "A Commercial Airplane Collided with a Drone in Canada, a First in North America," *Washington Post*, 16 October 2017, at https://www.washingtonpost.com/news/morning-mix/wp/2017/10/16/a-commercial-airplane-collided-with-a-drone-in-canada-a-first-in-north-america/?utm_term=.afc370868ea9.

133. Aerovel Felxrotor, at http://aerovel.com/wp-content/cache/wp-rocket/aerovelco.com/flexrotor/index.html_gzip; and T. X. Hammes, "The Democratization of Airpower: The Insurgent and the Drone," *War on the Rocks*, 17 October 2016, at https://warontherocks.com/2016/10/the-democratization-of-airpower-the-insurgent-and-the-drone/.

134. Michael Zhang, "This Guy Flew His Camera Drone onto, inside, and under a Moving Train," at https://petapixel.com/2017/09/22/guy-flew-camera-drone-onto-inside-moving-train/.

135. Union Pacific Policy for Photograph and Video Recording, at https://www.up.com/aboutup/community/safety/photo-video_policy/index.htm.

136. David Hambling, "Drone Maps Mines to Explore Unsafe Caverns and Seek Out Minerals," *New Scientist*, 11 April 2017, at https://www.newscientist.com/article/2127123-drone-maps-mines-to-explore-unsafe-caverns-and-seek-out-minerals/.

137. Hammes (2016).

138. The US Marines have what they call "maker labs" at three military facilities, and also a mobile lab that moves from base to base, training marines in 3D-printing, soldering, using sensor, and other skills. The goal is to deploy something like a forward-based trailer to create custom-made drones or other

equipment to use in combat. Randy Rieland, "Giving Marines the Tools to Build Drones on the Battlefield," *Smithsonian.com*, 19 May 2017.

139. Eugen Bogdan Petcu, "3D Bio-Printing: An Introduction to a New Approach for Cancer Patients at the Interface of Art and Medicine," *Leonardo* 50, no. 2 (2017): 195–6.

140. "A Brief History of 3D Printing," T. Rowe Price Connections, at https://individual.troweprice.com/staticFiles/Retail/Shared/PDFs/3D_Printing_Infographic_FINAL.pdf..

141. "A Tissue of Truths: Regenerative Medicine," *Economist* 422, no. 9025 (28 January 2017): 74; and Richard A. D'Aveni, "Five Myths about 3D Printing," *Washington Post*, 10 August 2018.

142. Polly Mosendz, "You Can Now Get a 3D Printer for Under $200," *The Atlantic*, 28 May 2014, at https://www.theatlantic.com/technology/archive/2014/05/you-can-now-get-a-3d-printer-for-under-200/371742/.

143. Tim Worstall, "The Liberator 3D Printed Gun Successfully Smuggled through International Transport Security," *Forbes*, 12 May 2013, at https://www.forbes.com/sites/timworstall/2013/05/12/the-liberator-3d-printed-gun-successfully-smuggled-through-international-transport-security/.

144. Lazer Berman, "Journalists Print Gun, Point It at Netanyahu," *Times of Israel*, 4 July 2013, at https://www.timesofisrael.com:journalists-print-gun-bring-it-to-netanyahu-speech. Five months later, a man in Kawasaki, Japan, was arrested and sentenced to two years in prison for 3D printing his own revolver. The Japanese gun was a six-shot revolver named the ZigZag, more sophisticated than the Liberator. Yoshimoto Imura posted videos of the revolver firing, along with the words, "Freedom of armaments to all people!! A gun makes power equal!!" James Vincent, "Japanese Man Jailed for Two Years for Creating 3D Printed Guns," *The Independent* (UK), 21 October 2014, at https://www.independent.co.uk/life-style/gadgets-and-tech/japanese-man-jailed-for-two-years-for-creating-3d-printed-guns-9807765.html; and Kelsey D. Atherton, "Man in Japan Arrested for 3D Printing Revolvers," *Popular Science*, 21 October 2014, at https://www.popsci.com/article/technology/man-japan-arrested-3-d-printing-revolvers.

145. Even though the State Department forced Defense Distributed to take the files down, they were available on a wide range of third party websites. David Kimball-Stanley, "3D-Printed Guns Hit the Courts," *Lawfare*, 30 November 2016, at https://www.lawfareblog.com/3d-printed-guns-hit-courts.

146. Cody Wilson and his Defense Distributed group in 2015 tried to evade both the State Department's legal injunction and the technological limitations of their equipment by offering to sell and ship a computer-controlled milling machine they called the "Ghost Gunner," sold for $1500. The machine allows anyone to start with a metal block of aluminum and create the lower receiver, the key firing part of an AR-15 assault rifle (and also the most heavily regulated element). The computer instructions were distributed

on a USB thumb drive. Demand for Ghost Gunner milling machines was overwhelming—the initial run of 500 ran out in three days. Andrew Greenberg, "AR-15: Secret Weapon," *Wired* 23, no. 8 (August 2015): 76.

147. Tiffany Hsu and Alan Feuer, "'Downloadable Gun' Clears a Legal Obstacle, and Activists Are Alarmed," *New York Times*, 13 July 2018. Defense Distributed did post the 3D files for printing AR-15 assault rifles for a few days, and some 1,000 downloaded them. Deanna Paul, Meagan Flynn, and Katie Zezima, "Federal Judge Blocks Posting of Blueprints for 3D Printed Guns Hours before They Were to Be Published," *Washington Post*, 31 July 2018, at https://www.washingtonpost.com/news/morning-mix/wp/2018/07/31/in-last-minute-lawsuit-states-say-3-d-printable-guns-pose-national-security-threat/?utm_term=.8fc20426c0f0. A judge extended the restraining order in August 2018. Then the case took a bizarre twist when Cody Wilson, who describes himself as a crypto-anarchist, was charged with sexually assaulting a child in Texas and apparently fled the country. "3-D Printed Gun Promoter, Cody Wilson, Is Charged With Secual Assault of Child," *New York Times*, 19 September 2018, at https://www.nytimes.com/2018/09/19/business/cody-wilson-3d-guns-sexual-assault.html. That is the last status I can find.

148. *Estimating Civilian-Owned Firearms*, Research Notes no. 9, *Small Arms Survey*, Geneva, September 2011, at http://www.smallarmssurvey.org/fileadmin/docs/H-Research_Notes/SAS-Research-Note-9.pdf.

149. Jenzen-Jones, 49 and 52–53.

150. This section merely touches upon a complex topic. For a thorough analysis of 3D printed firearms see Benjamin King and Glenn McDonald, eds., *Behind the Curve: New Technologies, New Control Challenges* (Small Arms Survey, Graduate Institute of International Development Studies, February 2015), especially N. R. Jenzen-Jones, chapter 3, "Small Arms and Additive Manufacturing: An Assessment of 3D-printed Firearms, Components, and Accessories," 43–74, at http://www.smallarmssurvey.org/about-us/highlights/highlights-2015/highlight-op32.html.

151. Jenzen-Jones, 49.

152. "'MultiFab' 3D-Prints a Record 10 Materials at Once, No Assembly Required," MIT Computer Science and Artificial Intelligence Laboratory, 27 August 2015, at http://www.csail.mit.edu/multifab_multimaterial_3D_printer.

153. Chris Welch, "World's First 3D-Printed Metal Gun Successfully Fires over 50 Rounds," *The Verge*, 7 November 2013, at https://www.theverge.com/2013/11/7/5077718/worlds-first-3d-printed-metal-gun-fires-over-50-rounds; and Megan Garber, "Don't Freak Out, but the World's First 3D-Printed Metal Gun Totally Works," *The Atlantic*, 7 November 2013, at https://www.theatlantic.com/technology/archive/2013/11/dont-freak-out-but-the-first-3d-printed-metal-gun-totally-works/281266/.

154. Jen Judson, "Marine Corps Looks to Printing to Make Spare Parts Downrange," *Defense News*, 11 September 2017, at https://www.defensenews.com/smr/equipping-the-warfighter/2017/09/11/marine-corps-looks-to-3-d-printing-to-make-spare-parts-downrange/.

155. The idea came out of the Marine Corps Logistics Innovation Challenge, a program designed to crowd source ideas about 3D printing and wearable devices. Each Raven costs more than $30,000, and a simple wing replacement is about $8,000. The Scout, which uses an open-source flight controller, open-source software for waypoint navigation, and can carry a camera (or other payload), costs less than $1,000 and can be easily produced in the field. It's not as advanced as a Raven (its range, speed, and persistence are all lower), but it would be available much faster. Kelsey E. Atherton, "The Marine Corps Wants to 3D Print Cheaper Drones," *Popular Science*, 13 September 2017, at https://www.popsci.com/marine-corps-3d-printed-drones; Sandra Erwin, "Marines Take 3D Printed Drones from the Lab to the Field," *Defense Systems*, 8 May 2017, at https://defensesystems.com/articles/2017/05/08/marinecorpprint.aspx.

156. "Johns Hopkins Team Makes Hobby Drones Crash to Expose Design Flaws," Johns Hopkins University press release, 8 June 2016, at http://releases.jhu.edu/2016/06/08/johns-hopkins-team-makes-hobby-drones-crash-to-expose-design-flaws/.

157. Derek Hawkins, "A US 'Ally' Fired a $3 Million Patriot Missile at a $200 Drone. Spoiler: The Missile Won," *Washington Post*, 17 March 2017, at https://www.washingtonpost.com/news/morning-mix/wp/2017/03/17/a-u-s-ally-fired-a-3-million-patriot-missile-at-a-200-drone-spoiler-the-missile-won/?utm_term=.0e3d7445897e.

158. Schmitt (2017).

159. Jeremy Binnie, "Russians Reveal Details of UAV Swarm Attacks on Syrian Bases," *IHS Jane's Defense Weekly*, 12 January 2018.

160. Because the small UAVs were not networked to each other, technically they may not have been "swarming."

Chapter 9

1. Benjamin Bidder, "Der Mann, der den dritten Weltkrieg verhinderte [The Man Who Prevented the Third World War," *Der Spiegel*, 21 April 2010, at http://www.spiegel.de/einestages/vergessener-held-a-948852.html.

2. David Hoffman, "'I Had a Funny Feeling in My Gut,'" *Washington Post*, 10 February 1999, p. A19.

3. Paul Scharre and Michael C. Horowitz, *An Introduction to Autonomy in Weapon Systems*, Working Paper, Center for a New American Security, February 2015; Michael C. Horowitz, "Why Words Matter: The Real-World Consequences of Defining Autonomous Weapons Systems," *Temple International and*

Comparative Law Journal 30, no. 1 (spring 2016); Vincent Bloulanin and Maaike Vergruggen, *Mapping the Development of Autonomy in Weapon Systems,* Stockholm International Peace Research Institute (SIPRI), Solna, Sweden, November 2017.

4. This way of explaining the differences in autonomy using a "sense, decide, act" loop is drawn directly from Paul Scharre's *Army of None* (New York: W. W. Norton, 2018), 28–34.

5. The 2011 Department of Defense Roadmap used four slightly different but similar categories: 1) human-operated, 2) human-delegated, 3) human-supervised, and 4) fully autonomous. See *Unmanned Systems Integrated Roadmap FY2011–2036*, p. 43, accessible at https://fas.org/irp/program/collect/usroadmap2011.pdf.

6. For a good explanation of automatic vs. automated, see Scharre (2018), 26–34.

7. United Nations Institute for Disarmament Research, *The Weaponization of Increasingly Autonomous Technologies: Concerns, Characteristics, and Definitional Approaches*, UNIDIR Resources (2017): 9, at http://www.unidir.org/programmes/emerging-security-issues/the-weaponization-of-increasingly-autonomous-technologies-phase-iii.

8. Greg Allen and Taniel Chan, *Artificial Intelligence and National Security*, July 2017, p. 13.

9. DoD Directive 3000.9, *Autonomy in Weapon Systems*, Ashton B. Carter, Department of Defense, 21 November 2012, p. 14, at http://www.esd.whs.mil/Portals/54/Documents/DD/issuances/dodd/300009p.pdf.

10. Scharre and Horowitz (2015), 9; and Scharre (2018), 29.

11. The high cost of precision-guided munitions is one reason why they have not proliferated nearly as quickly as US advocates of the Revolution in Military Affairs (RMA) predicted in 1991 that they would. Barry Watts, "Why Has the Diffusion of Reconnaissance Strike Been So Slow," in *The Evolution of Precision Strike*, Center for Strategic and Budgetary Assessments, 6 August 2013, pp. 11–12, at http://csbaonline.org/research/publications/the-evolution-of-precision-strike/publication; and Barry D. Watts, "Precision Strike: An Evolution," *The National Interest*, 2 November 2013, at http://nationalinterest.org/commentary/precision-strike-evolution-9347?page=show.

12. Watts, "Why Has the Diffusion of Reconnaissance Strike Been So Slow"; and Watts, "Precision Strike: An Evolution."

13. DoD Directive 3000.9, *Autonomy in Weapon Systems*.

14. John K. Hawley, *Patriot Wars: Automation and the Patriot Air and Missile Defense System*, Center for a New American Security, 25 January 2017, at https://www.cnas.org/publications/reports/patriot-wars.

15. Scharre and Horowitz (2015), 12 and Appendix B. See also Elinor Sloan, "Robotics at War," *Survival* 57, no. 5 (November 2015): 107–20.

16. This is just a representative sample. For more, see Scharre and Horowitz (2015).

17. According to Paul Scharre, there was an incident in Nagorno-Karabakh where a Harpy II struck a bus full of people, but the manufacturer was unwilling to supply further facts on the technical specifications of the weapon: https://futureoflife.org/2018/07/31/podcast-six-experts-explain-the-killer-robots-debate/.

18. Dan Gettinger and Arthur Holland Michel, "Loitering Munitions in Focus," Center for the Study of the Drone at Bard College, at http://dronecenter. bard.edu/files/2017/02/CSD-Loitering-Munitions.pdf; Ami Rojkes Dombe, "China Unveils a Harpy-Type Loitering Munition," *Israel Defense*, 3 January 2017, at https://www.israeldefense.co.il/en/node/28716; and Paul Scharre, "Meet the New Robot Army," *Wall Street Journal*, 11 April 2018. Some have purchased the closely related Harop (or Harpy II) Israeli-built loitering munition.

19. See Michael C. Horowitz, "Public Opinion and the Politics of the Killer Robot Debate," *Research and Politics* (January–March 2016): 2.

20. Kyle Mizokami, "US Army Tanks to Get Active Protection Systems by 2020," *Popular Mechanics*, 10 October 2017, at https://www. popularmechanics.com/military/weapons/news/a28576/us-army-tanks-to-get-active-protection-systems-by-2020/.

21. This phrase was coined by Richard Moyes, managing director of the United Kingdom's nonprofit organization Article 36, which works "to prevent the unintended, unnecessary or unacceptable harm cause by certain weapons." See http://www.article36.org/about/. The question of what the word "meaningful" means is a point of sharp contention among AI experts.

22. Allen and Chan, *Artificial Intelligence and National Security*, July 2017 .

23. Hayley Evans, "Lethal Autonomous Weapons Systems at the First and Second UN GGE Meetings," *Lawfare*, 9 April 2018, at https://www.lawfareblog. com/lethal-autonomous-weapons-systems-first-and-second-un-gge-meetings. See also Future of Life Institute, at https://futureoflife.org. Most experts emphasize the key goal of retaining "meaningful human control" in AI.

24. Cade Metz, "Microsoft Teaches Autonomous Gliders to Make Decisions on the Fly," *New York Times*, 16 August 2017.

25. Ibid.

26. An excellent, concise introduction to these concept is *The Weaponization of Increasingly Autonomous Technologies: Artificial Intelligence*, United Nations Institute for Disarmament Research, 2018.

27. Tom Keeley, "A Revolution in Military Affairs versus 'Evolution': When Machines Are Smart Enough!," *Small Wars Journal*, 8 January 2016, at http://smallwarsjournal.com/jrnl/art/a-revolution-in-military-affairs-versus-"evolution"-when-machines-are-smart-enough.

28. Michael C. Horowitz, "Who'll Want Artificially Intelligent Weapons? ISIS, Democracies, or Autocracies?," *Bulletin of the Atomic Scientists*, 29 July 2016, at https://thebulletin.org/who'll-want-artificially-intelligent-weapons-isis-democracies-or-autocracies9692.

29. Allen and Chan (2017), 24; See also Pedro Domingos, *The Master Algorithm: How the Quest for the Ultimate Learning Machine Will Remake Our World* (New York: Basic Books, 2015), 281.

30. This is a loose retelling of Elon Musk's scenario, laid out at the National Governors Association 2017 Summer Meeting. See note 32 below.

31. In January 2015, more than 150 AI experts, programmers, scientists, and ethicists signed a public letter "Research Priorities for Beneficial Artificial Intelligence: An Open Letter," that called for deep research on the broad public impact of artificial intelligence. Future of Life Institute, at https://futureoflife.org/ai-open-letter. See also *Losing Humanity: The Case against Killer Robots*, Human Rights Watch and the International Human Rights and the International Human Rights Clinic (IHRC) at Harvard Law School, November 2012, at https://www.hrw.org/report/2012/11/19/losing-humanity/case-against-killer-robots.

32. Elon Musk, Remarks at the National Governors Association 2017 Summer Meeting, 15 July 2017, at https://www.c-span.org/video/?c4676772/elon-musk-national-governors-association-2017-summer-meeting&start=3220.

33. "Putin: Leader in Artificial Intelligence Will Rule World," *CNBC*, 4 September 2017, at https://www.cnbc.com/2017/09/04/putin-leader-in-artificial-intelligence-will-rule-world.html.

34. Ibid.

35. David Silver, Juian Schrittwieser, Karen Simonyan, Ioannis Antonoglou, Aja Huang, Arthus Guez, Thomas Hubert, Lucas Baker, Matthew Lai, Adrian Bolton, Ytian Chen, Timothy Lillicrap, Fan Hui, Laurent Sifre, George van dan Driessche, Thors Graepel, and Demis Hassabis, "Mastering the Game of Go without Human Knowledge," *Nature* 550 (19 October 2017): 354–9.

36. James Vincent, "DeepMind's Go-playing AI Doesn't Need Help to Beat Us Anymore," *The Verge*, 18 October 2017, at https://www.theverge.com/2017/10/18/16495548/deepmind-ai-go-alphago-zero-self-taught.

37. Satinder Singh, "News and Views: Learning to Play Go from Scratch," *Nature* 550 (19 October 2017): 337.

38. Carl Miller, "God Is in the Machine: Carl Miller on the Terrifying, Hidden Reality of Ridiculously Complicated Algorithms," *Times Literary Supplement*, 21 August 2018, at https://www.the-tls.co.uk/articles/public/ridiculously-complicated-algorithms/.

39. "Stephen Hawking Says AI Could Be 'Worst Event in the History of Our Civilization,' " *CNBC*, 6 November 2017, at https://www.cnbc.com/2017/11/06/stephen-hawking-ai-could-be-worst-event-in-civilization.html.

40. Noah Shachtman, "Iraq Militants Brag: We've Got Robotic Weapons, Too," *Wired*, 4 October 2011, at https://www.wired.com/2011/10/militants-got-robots/.

41. Kelsey D. Atherton, "ISIS Shows Off a Driverless Carbomb: Not So Much Tesla, More Explosive," *Popular Science*, 6 January 2016, at https://www.popsci.com/isis-shows-off-driverless-carbomb.

42. The ideology of a group may make a difference, however. Those who profess that they "are not afraid of death" may prefer to be martyrs. I thank Martha Crenshaw for this point.

43. Also non-state actors won't have requirements for testing, verification, and assurance that may impede the deployment of these systems by states.

44. Allen and Chan (2017), 22.

45. Robert Wall, "Armies Race to Deploy Drone, Self-Driving Tech on the Battlefield," *Wall Street Journal*, 29 October 2017, at https://www.wsj.com/articles/armies-race-to-deploy-drone-self-driving-tech-on-the-battlefield-1509274803?mg=prod/accounts-wsj.

46. John D. Sutter, "How 9/11 Inspired a New Era of Robotics," *CNN*, 7 September 2011, at http://www.cnn.com/2011/TECH/innovation/09/07/911.robots.disaster.response/index.html.

47. Kalyan M. Kemburi, "Robotisation of Militaries: Organisational, Policy and Operational Issues," CO16134, RSIS Publications, Rajaratnam School of International Studies, 2 June 2016, at https://www.rsis.edu.sg/rsis-publication/rsis/co16134-robotisation-of-militaries-organisational-policy-and-operational-issues/ -.WimWGbbMxEI.

48. Kevin Sullivan, Tom Jackman, and Brian Fung, "Dallas Police Used a Robot to Kill. What Does That Mean for the Future of Police Robots?" *Washington Post*, 21 July 2016, at https://www.washingtonpost.com/national/dallas-police-used-a-robot-to-kill-what-does-that-mean-for-the-future-of-police-robots/2016/07/20/32ee114e-4a84-11e6-bdb9-701687974517_story.html?utm_term=.0090e5268937; and Alina Selyukn, "Bomb Robots: What Makes Killing in Dallas Different and What Happens Next," *NPR*, 8 July 2016, at http://www.npr.org/sections/thetwo-way/2016/07/08/485262777/for-the-first-time-police-used-a-bomb-robot-to-kill.

49. It was a Northrop Grumman Remotec Andros Mark V-A1, bought for $151,000 in 2008. Kevin Sullivan et al., *Washington Post*, 21 July 2016.

50. Ibid.

51. Peter W. Singer, "Police Used a Robot to Kill—The Key Questions," *CNN.com*, 10 July 2016, at http://www.cnn.com/2016/07/09/opinions/dallas-robot-questions-singer/index.html.

52. Cameron F. Kerry and Jack Karsten, "Gauging Investment in Self-driving Cars," *Brookings Institution*, 16 October 2017, at https://www.brookings.edu/research/gauging-investment-in-self-driving-cars/.

53. Stew Magnuson, "The Military Wants Self-driving Trucks to Deliver Supplies in Combat," *National Defense Magazine*, 24 February 2017.

54. Simson Garfinkel, "Hackers Are the Real Obstacle for Self-Driving Vehicles," *MIT Technology Review*, 22 August 2017. at https://www.technologyreview.com/s/608618/hackers-are-the-real-obstacle-for-self-driving-vehicles/.

55. Kevin Eykholt, Ivan Evtimov, Earlence Fernandes, Bo Li, Amir Rahmati, Chaowei Xiao, Atul Prakash, Tadayoshi Kohno, and Dawn Song, "Robust Physical-World Attacks on Deep Learning Models," *CVPR* 5, no. 10 (April 2018), at https://arxiv.org/abs/1707.08945v5.

56. *The Malicious Use of Artificial Intelligence: Forecasting, Prevention, and Mitigation*, Future of Humanity Institute, University of Oxford, et al., February 2018, p. 20, at https://maliciousaireport.com.

57. The Nice attack was 14 July 2016 (86 killed). The Berlin Christmas market was attacked on 19 December 2016 (12 killed). London attacks included Westminster Bridge (22 March 2017, 4 killed), London Bridge (3 June 2017, 22 killed), and Finsbury Park (19 June 2017, 1 killed). The Stockholm attack happened on 7 April 2017 (4 killed). Barcelona was attacked on 17 August 2017 (13 killed). New York City was 31 October 2017 (8 killed). This is not a complete listing of all vehicular attacks.

58. Bruce Schneier, "Security and Privacy in a Hyper-connected World," *Data & Society Research Institute*, 14 December 2016, at https://www.youtube.com/watch?v=cJMG34UzIyk&feature=youtu.be.

59. Jack Karsten and Darrell M. West, "Fixing Vulnerabilities in Networked Devices, from Pacemakers to Driverless Cars," *Brookings Institution*, 18 August 2017, at https://www.brookings.edu/blog/techtank/2017/08/18/fixing-vulnerabilities-in-networked-devices-from-pacemakers-to-driverless-cars/.

60. Joshua A. T. Fairfield, "The 'Internet of Things" Is Sending Us Back to the Middle Ages," *The Conversation*, 5 September 2017, at https://theconversation.com/the-internet-of-things-is-sending-us-back-to-the-middle-ages-81435; and Joshua Fairfield, *Owned: Property, Privacy, and the New Digital Serfdom* (Cambridge, UK: Cambridge University Press, 2017).

61. iRobot Roomba's security and privacy sharing rules prohibit them sharing data with third parties directly. See https://homesupport.irobot.com/app/answers/detail/a_id/964/~/irobot-roomba-privacy-and-data-sharing.

62. Janet Burns, "We-Vibe Settles for $3.7 M in 'Spying Vibrator' Data Suit," *Forbes*, 15 March 2017, at https://www.forbes.com/sites/janetwburns/2017/03/15/we-vibe-settles-for-3-7m-in-spying-vibrator-data-lawsuit/ - 7bbec2e66021.

63. Dan Goodin, "Update Gone Wrong Leaves 500 Smart Locks Inoperable," *Artstechnica.com*, at https://arstechnica.com/information-technology/2017/08/500-smart-locks-arent-so-smart-anymore-thanks-to-botched-update/.

64. Lee Rainie, "The Internet of Things Connectivity Binge: What Are the Implications?," *Pew Research Center*, 6 June 2017; Whitney Curtis, "The Jeep Hackers Are Back to Prove Car Hacking Can Get Much Worse," *Wired*, 1 August 2016, at https://www.wired.com/2016/08/jeep-hackers-return-high-speed-steering-acceleration-hacks/; Ben Wofford, "How to Hack an Election in 7 Minutes," *Politico.com*, 5 August 2016, at https://www.politico.com/magazine/story/2016/08/2016-elections-russia-hack-how-to-hack-an-election-in-seven-minutes-214144; Andy Greenberg, "Crash Override: The Malware That Took Down a Power Grid," *Wired*, 12 June 2017, at https://www.wired.com/story/crash-override-malware/; Eyal Ronen, Adi Shamir, Achi-Or Weingarten, and Colin O'Flynn, "IoT Goes Nuclear: Creating a ZibBee Chain Reaction," 2017 IEEE Symposium on Security and Privacy, 26 June 2017, at https://ieeexplore.ieee.org/document/7958578/.

65. Edward J. Markey, *Tracking and Hacking: Security and Privacy Gaps Put American Drivers at Risk*, February 2015, at https://www.markey.senate.gov/imo/media/doc/2015-02-06_MarkeyReport-Tracking_Hacking_CarSecurity 2.pdf.

66. Elizabeth Weise, "Johnson & Johnson Warns of Insulin Pump Hack Risk," *USA Today*, 4 October 2016, at https://www.usatoday.com/story/tech/news/2016/10/04/johnson-johnson-warns-insulin-pump-hack-risk-animas/91542522/.

67. Schneier (2016).

68. Bruce Schneier, "Could Your Plane Be Hacked?," *CNN*, 16 April 2015, at https://www.schneier.com/essays/archives/2015/04/could_your_plane_be_.html.

69. There have been limited attempts to predict terrorist activity in urban environments, including through the Internet of Things. See Raul Sormani, John Soldatos, Spyros Vassilaras, Georgios Kioumourtzis, George Leventakis, Ilaria Giordani, and Francesco Tisato, "A Serious Game Empowering the Prediction of Potential Terrorist Actions," *Journal of Policing, Intelligence, and Counter Terrorism* 11, no. 1 (2016): 30–48.

70. Scharre (2018), 17.

71. On swarming, see John Arquilla and David Ronfeldt, *Swarming and the Future of Conflict*, RAND, National Defense Research Institute, 2000, at https://www.rand.org/pubs/documented_briefings/DB311.html; Sean Edwards, *Swarming on the Battlefield; Past, Present, and Future*, RAND, National Defense Research Institute, 2000, at https://www.rand.org/pubs/monograph_reports/MR1100.html; and David Hambling (2015).

72. Kyle Mizokami, "US Navy Wants Smart Guided Cannon to Fend Off Iran's Boat Swarms," *Popular Mechanics*, 20 January 2017, at https://www.popularmechanics.com/military/weapons/a24838/mad-fires-cannon-rounds/.

73. Companies like Raytheon are developing smart cannon rounds that can change course en route, to simultaneously attack multiple targets. The relative mobility of the small targets remains a challenge. Ibid.

74. Emily Feng and Charles Clover, "Drone Swarms vs. Conventional Arms: China's Military Debate," *Financial Times*, 24 August 2017, at https://www.ft.com/content/302fc14a-66ef-11e7-8526-7b38dcaef614.n.

75. Scharre (2018), 20. Scharre has a great, in-depth explanation of swarming in chapter 1.

76. Scharre (2018), 12.

77. Jeffrey Lin and P. W. Singer, "China Is Making 1,000-UAV Drone Swarms Now," *Popular Science*, 8 January 2018.

78. "Intel Breaks Guinness World Records Title for Drone Light Shows in Celebration of 50th Anniversary," 18 July 2018, at https://newsroom.intel.com/news/intel-breaks-guinness-world-records-title-drone-light-shows-celebration-50th-anniversary/.

79. Ibid.

80. Dan Lamothe, "Veil of Secrecy Lifted on Pentagon Office Planning 'Avatar' Fighters and Drone Swarms," *Washington Post*, 8 March 2016, at https://www.washingtonpost.com/news/checkpoint/wp/2016/03/08/inside-the-secretive-pentagon-office-planning-skyborg-fighters-and-drone-swarms/?utm_term=.aa7a10c828a8.

81. "Watch Perdix, the Secretive Pentagon Program Dropping Tiny Drones from Jets," *Washington Post*, 8 March 2016, at https://www.washingtonpost.com/news/checkpoint/wp/2016/03/08/watch-perdix-the-secretive-pentagon-program-dropping-tiny-drones-from-jets/?utm_term=.1f38297697af.

82. Kyle Mizokami, "The Pentagon's Autonomous Swarming Drones Are the Most Unsettling Thing You'll See Today," *Popular Mechanics*, 9 January 2017, at http://www.popularmechanics.com/military/aviation/a24675/pentagon-autonomous-swarming-drones/; and Dan Lamonthe, "Watch Perdix, the Secretive Pentagon Program Dropping Tiny Drones from Jets," *Washington Post*, 8 March 2016.

83. Hambling (2015),. 193–4; and Scharre (2018), 22.

84. Andrew Tate, "China Launches Record-breaking UAV Swarm," *IHS Jane's Defence Weekly*, 22 June 2017, at http://www.janes.com/article/71624/china-launches-record-breaking-uav-swarm -.WUulhfyfErg.twitter.

85. Bryan Harris, "South Korea to Create 'Drone-bot Combat Unit' to Swarm North," *Financial Times*, 6 December 2017, at https://www.ft.com/content/6878ba90-da1a-11e7-a039-c64b1c09b482.

86. David Hambling, "Boeing 'Base Station' Concept Would Autonomously Refuel Military Drones: Overcoming the Biggest Obstacle for Small Drones," *Popular Science*, 21 November 2016, at https://www.popsci.com/boeing-has-patented-drone-battle-station.

87. This according to a report by the Conflict Armament Research group, *Frontline Perspective: Iranian Technology Transfers to Yemen: "Kamikaze" Drones Used by Houthi Forces to Attack Coalition Missile Defense Systems*, March 2017, at http://www.conflictarm.com/perspectives/iranian-technology-transfers-to-yemen/; Thomas Biggons-Neff, "Houthi Forces Appear to Be Using Iranian-made Saudi Air Defenses in Yemen, Report Says," *Washington Post*, 22 March 2017; and David Hambling, "Swarms of Cheap Drones Are Attacking Missile Defences in Yemen," *New Scientist*, 8 March 2018.

88. David Hambling, "If Drone Swarms Are the Future, China May Be Winning," *Popular Mechanics*, at http://www.popularmechanics.com/military/research/a24494/chinese-drones-swarms/.

89. Paul Scharre, "Why You Shouldn't Fear 'Slaughterbots,'" *IEEE Spectrum*, 22 December 2017, at https://spectrum.ieee.org/automaton/robotics/military-robots/why-you-shouldnt-fear-slaughterbots.

90. Nicholas Weaver, "'Slaughterbots' and Other (Anticipated) Autonomous Weapons Problems," *Lawfare*, 28 November 2017, at https://www.lawfareblog.com/slaughterbots-and-other-anticipated-autonomous-weapons-problems.

91. For further discussion of these threats, see *Malicious Use of Artificial Intelligence: Forecasting, Prevention, and Mitigation*, February 2018, at https://img1.wsimg.com/blobby/go/3d82daa4-97fe-4096-9c6b-376b92c619de/downloads/1c6q2kc4v_50335.pdf.

92. Allen and Chan (2017), 31–32.

93. The Chinese government's July 2017 Artificial Intelligence plan is accessible at http://www.gov.cn/zhengce/content/2017-07/20/content_5211996.htm. See also, Elsa B. Kania, "Artificial Intelligence and Chinese Power: Beijing's Push for a Smart Military—and How to Respond," *Foreign Affairs.com*, 5 December 2017; at https://www.foreignaffairs.com/articles/china/2017-12-05/artificial-intelligence-and-chinese-power?cid=nlc-fa_fatoday-20171205.

94. Dennis S. Hoadley and Nathan J. Lucas, *Artificial Intelligence and National Security*, CRS Report for Congress # R45178, 26 April 2018, p. 5; Cecilia Kang and Alan Rappeport, "The New US-China Rivalry: A Technology Race," *New York Times*, 6 March 2018; at https://www.nytimes.com/2018/03/06/business/us-china-trade-technology-deals.html; John D. McKinnon, "Trump Administration Vows to Maintain US Edge in AI Technology," *Wall Street Journal*, 10 May 2018; at https://www.wsj.com/articles/trump-administration-vows-to-maintain-u-s-edge-in-ai-technology-1525972043.

95. Allen and Chan (2017), 46.

96. Economists Daron Acemoglu and Pascual Restrepo were able to demonstrate a clear quantitative effect on unemployment. Increasing by one robot per thousand workers lowered the employment to population rate by about 0.18–0.34 percentage points, and lowered wages by

0.25–0.5%. See Daron Acemoglu and Pascual Restrepo, "Robots and Jobs: Evidence from US Labor Markets," *NBER Working Paper Series* No. 23285, March 2017, JEL No. J23, J24, National Bureau of Economic Research, Cambridge, MA.

97. OECD (2015), *In It Together: Why Less Inequality Benefits All*, OECD Publishing, Paris, p. 15, at http://dx.doi.org/10.1787/9789264235120-en.

98. Carl Benedikt Frey and Michael A. Osborne, *The Future of Employment: How Susceptible Are Jobs to Computerisation?*, Oxford Martin School, University of Oxford, 17 September 2013, at https://www.oxfordmartin.ox.ac.uk/downloads/academic/The_Future_of_Employment.pdf.

99. Brad Hershbein and Lisa B. Kahn, "Do Recessions Accelerate Routine-Biased Technological Change? Evidence from Vacancy Postings," National Bureau of Economic Research, NBER Working Paper No. 22762, Cambridge, MA, September 2017, at http://www.nber.org/papers/w22762; and Derek Thompson, "When Will Robots Take All the Jobs?," *The Atlantic*, 31 October 2016, at https://www.theatlantic.com/business/archive/2016/10/the-robot-paradox/505973/.

Conclusion

1. For two impassioned presentations about the benefits of popular experimentation with biotechnology, see Ellen Jorgensen, "Biohacking: You Can Do It, Too," TEDGlobal 2012, at https://www.ted.com/talks/ellen_jorgensen_biohacking_you_can_do_it_too?language=en; and Cathal Garvey, "Bringing Biotechnology into the Home," TEDx Talk, 22 October 2013, at https://www.youtube.com/watch?v=g_ZswrLFSdo.

2. "Easy to Purchase Dynamite: No Law Requiring Would-be Purchasers to Show That Explosive Is to Be Used for Legitimate Purposes, but Smallest Amount Sold Costs $10," *New York Times*, 31 May 1903, p. 35.

3. Paul Scharre, "Why You Shouldn't Fear 'Slaughterbots'," *IEEE Spectrum* (Automaton blog), 22 December 2017, at https://spectrum.ieee.org/automaton/robotics/military-robots/why-you-shouldnt-fear-slaughterbots.

4. National Academies of Sciences, Engineering and Medicine, *Counter-Unmanned Aircraft System Capability for Battalion-and-Below Operations: Abbreviated Version of a Restricted Report*, The National Academics of Sciences, Engineering, and Medicine (Washington, DC: The National Academies Press, 2018), at https://doi.org/10.17226/24747.

5. US Senate, Committee on Armed Services, Nomination-Dunford, 26 September 2017, transcript, p. 74, at https/::www.armed-services.senate.gov:imo:media:doc:17-80_09-26-17.pdf.

6. Robert Beckhusen, "Chinese Robo-Boats Swarm the South China Sea: Beijing Is Well Situated to Exploit Unmanned Machines," *The National Interest*, 30 August 2018, at https://nationalinterest.org/blog/buzz/chinese-robo-boats-swarm-south-china-sea-30052.

7. The Board calls for tailored cyber deterrence against them, an approach that may work well against states but has yet to demonstrate any effect on the range of malicious non-state actors targeting the US, Europe, and other allies. Office of the Under Secretary of Defense for Acquisition, Technology and Logistics, US Department of Defense, *Report of the Defense Science Board Task Force on Cyber Deterrence*, February 2017.

8. Verentsev et al., *Information-Psychological War Operations: A Short Encyclopedia and Reference Guide* (Moscow: Hotline-Telecom, 2011); quoted by Peter Pomerantsev, "Inside the Kremlin's Hall of Mirrors," *The Guardian*, 9 April 2015, at https://www.theguardian.com/news/2015/apr/09/kremlin-hall-of-mirrors-military-information-psychology.

9. Fred Charles Iklé, *Annihilation from Within: The Ultimate Threat to Nations* (New York: Columbia University Press, 2006).

10. "Tracking and Hacking: Security and Privacy Gas Put American Drivers at Risk," report written by the staff of Senator Edward J. Markey, February 2015, at https://www.markey.senate.gov/imo/media/doc/2015-02-06_MarkeyReport-Tracking_Hacking_CarSecurity%202.pdf.

11. Jeff Plungis, "Who Owns the Data Your Car Collects?," *Consumer Reports*, 2 May 2018, at https://www.consumerreports.org/automotive-technology/who-owns-the-data-your-car-collects/.

12. Those regulations that had emerged were overturned by the Trump administration in April 2017.

13. Todd Spangler, "Facebook Loses $120 Billion in Market Value, as Stock Slides on Fears Growth Is Hitting a Wall," *Variety*, 26 July 2018, at https://variety.com/2018/digital/news/facebook-market-cap-120-billion-user-growth-slow-1202886792/. The EU's GDPR came into force on 25 May 2018.

14. Pew Research Center, *Teens, Social Media, and Technology 2018*, 31 May 2018, at http://www.pewinternet.org/2018/05/31/teens-social-media-technology-2018/.

15. Mark Zuckerberg, "The Internet Needs New Rules. Let's Start in These Four Areas," *The Washington Post*, 30 March 2019.

16. "Russia's Neighbors Respond to Putin's 'Hybrid War,'" *Foreign Policy*, 12 October 2017, at https://foreignpolicy.com/2017/10/12/russias-neighbors-respond-to-putins-hybrid-warlatvia-estonia-lithuania-finland/.

Epilogue

1. U.S. Department of Justice, U.S. Attorney's Office, District of Minnesota, "St. Paul Man Sentenced to Prison, $12 Million in Restitution For Minneapolis Police Third Precinct Arson," Press Release, 5 May 2021; at https://www.justice.gov/usao-mn/pr/st-paul-man-sentenced-prison-12-million-restitution-minneapolis-police-third-precinct

2. Numbers cited are accurate through April 2021. See "Terrorism in America after 9/11," New America Foundation, https://www.newamerica.org/in-depth/terrorism-in-america/why-do-they-commit-terrorist-acts/ accessed 26 June 2021; Seth G. Jones, Catrina Doxsee, and Nicholas Harrington, "The Escalating Terrorism Problem in the United States," Center for Strategic and International Studies, June 2020; at https://csis-website-prod.s3.amazonaws.com/s3fs-public/publication/200612_Jones_DomesticTerrorism_v6.pdf; and Robert O'Harrow Jr., Andrew Ba Tran, and Derek Hawkins, "The Rise of Domestic Extremism in America: Data Shows a Surge in Homegrown Incidents not see in a Quarter-Century," *The Washington Post*, 12 April 2021; at https://www.washingtonpost.com/investigations/interactive/2021/domestic-terrorism-data/.

3. Diego A. Martin, Jacob N. Shapiro, Julia G. Ilhardt, "Trends in Online Influence Efforts," Version 2.0, 5 August 2020, pp. 14-15; working paper accessible at https://scholar.princeton.edu/sites/default/files/jns/files/trends_in_online_influence_efforts_v2.0_aug_5_2020.pdf

4. Michael Howard, "The Forgotten Dimensions of Strategy," *Foreign Affairs* (Summer 1979), pp. 975-986.

5. See Akhil Reed Amar, *The Words that Made Us: America's Constitutional Conversation*, 1760–1840 (New York: Basic Books, 2021).

Appendix B

1. R. Lundstrom, "Alfred Nobel Dynamite Companies," at https://www.nobelprize.org/alfred_nobel/biographical/articles/lundstrom/.

2. Kenne Fant, *Alfred Nobel: A Biography* (New York: Arcade Publishing, 1991), 135.

3. Bureau of Explosives, "Report of the Chief Inspector: The Bureau for the Safe Transportation of Explosives and Other Dangerous Articles," New York, February 1908; Appendix 13 List of Manufacturers of Explosive (as of 31 December 1901), pp. 98–107.

4. "E. I. du Pont de Nemours Power Company's High Explosives Price List No. 1, Effective December 1, 1907," issued November 29, 1907, accessed at the Hagley Museum and Library in Wilmington, Delaware.

BOOKS CITED

Acton, Edward, V. Iu Cherniaev, and William G. Rosenberg, eds. *Critical Companion to the Russian Revolution, 1914–1921*. Bloomington: Indiana University Press, 1997.

Adamic, Louis. *Dynamite: The Story of Class Violence in America*. 2nd ed. New York: Chelsea House Publishing, 1958.

Adamsky, Dima. *The Culture of Military Innovation: The Impact of Cultural Factors on the Revolution in Military Affairs of Russia, the US, and Israel*. Stanford, CA: Stanford University Press, 2010.

Adey, Peter, Mark Whitehead, and Alison Williams. *From Above: War, Violence, and Verticality*. London: Hurst, 2013.

Alter, Adam. *Irresistible: The Rise of Addictive Technology and the Business of Keeping Us Hooked*. New York: Penguin Press, 2017.

Andrade, Tonio. *The Gunpowder Age: China, Military Innovation, and the Rise of the West in World History*. Princeton, NJ: Princeton University Press, 2016.

Archetti, Cristina. *Understanding Terrorism in the Age of Global Media: A Communication Approach*. New York: Palgrave Macmillan, 2013.

Armacost, Michael. *The Politics of Weapons Innovation: The Thor-Jupiter Controversy*. New York: Columbia University Press, 1969.

Art, Robert J., and Patrick M. Cronin, eds. *The United States and Coercive Diplomacy*. Washington, DC: US Institute of Peace, 2003.

Avant, Deborah. *Political Institutions and Military Change: Lessons from Peripheral Wars*. Ithaca, NY: Cornell University Press, 1994.

Avant, Deborah D. *The Market for Force: The Consequences of Privatizing Security*. Cambridge, UK: Cambridge University Press, 2005.

Avrich, Paul. *The Haymarket Tragedy*. Princeton, NJ: Princeton University Press, 1984.

Avrich, Paul. *The Russian Anarchists*. Chica, CA: AK Press, 2005.

Avrich, Paul. *Sacco and Venzetti: The Anarchist Background*. Princeton, NJ: Princeton University Press, 1991.

Bacevich, Andrew. *The Pentomic Era: The US Army between Korea and Vietnam*. Washington, DC: National Defense University Press, 1986.

Baer, James A. *Anarchist Immigrants in Spain and Argentina*. Champaign: University of Illinois Press, 2015.

Baldwin, Ralph B. *The Deadly Fuze: Secret Weapon of World War II*. San Rafael, CA: Presidio Press, 1980.

Barnhardt, Richard K., Stephen B. Hottman, Douglas M. Marshall, and Eric Shappee. *Introduction to Unmanned Aircraft Systems*. Boca Raton, FL: CRC Press, 2012.

Bartlett, W. B. *The Assassins: The Story of Medieval Islam's Secret Sect*. London: Sutton Publishing, 2001.

Belfiore, Michael. *The Department of Mad Scientists: How DARPA Is Remaking Our World, from the Internet to Artificial Limbs*. New York: HarperCollins, 2009.

Bell, J. (John) Bowyer. *The Secret Army: The IRA 1916–1979*. Cambridge, MA: MIT Press, 1980.

Bergen, Peter L., and David Rothenberg, eds. *Drone Wars: Transforming Conflict, Law, and Policy*. Cambridge, UK: Cambridge University Press, 2015.

Bergerson, Frederic. *The Army Gets an Air Force*. Baltimore: Johns Hopkins University Press, 1980.

Berthelot, Marcellin. *Explosives and Their Power*. London: John Murray, 1892.

Betz, David. *Carnage & Connectivity*. Oxford, UK: Oxford University Press, 2015.

Bickel, Keith. *Mars Learning: The Marine Corps Development of Small Wars Doctrine, 1915–1940*. Boulder, CO: Westview Press, 2000.

Biddle, Stephen. *Military Power: Explaining Victory and Defeat in Modern Battle*. Princeton, NJ: Princeton University Press, 2014.

Bijker, Wiebe E., Thomas P. Hughes, and Trevor Pinch, eds. *The Social Construction of Technological Systems: New Directions in the Sociology and History of Technology*. Cambridge, MA: MIT Press, 1987; 2012.

Black, Jeremy. *War and Technology*. Bloomington: Indiana University Press, 2013.

Black, Jeremy. *War and the World: Military Power and the Fate of Continents, 1450–2000*. New Haven, CT: Yale University Press, 1998.

Bloom, James. *The Jewish Revolts against Rome, A.D. 66–135*. London: McFarland, 2010.

Boot, Max. *Invisible Armies: An Epic History of Guerrilla Warfare from Ancient Times to the Present*. New York: W. W. Norton, 2013.

Bridge, Maureen, and John Pegg. *Call to Arms: A History of Military Communications from the Crimean War to the Present Day*. Devon, UK: Focus Publishing, 2001.

Brodie, Bernard, and Fawn Brodie. *From Crossbow to H-Bomb: The Evolution of the Weapons and Tactics of Warfare*. Bloomington: Indiana University Press, 1973.

Brodie, Bernard, ed. *The Absolute Weapon: Atomic Power and World Order.* New York: Harcourt, Brace, 1946.

Brown, G. I. *The Big Bang: A History of Explosives.* Stroud, UK: Sutton Publishing, 1998.

Brown, Stephen R. *A Most Damnable Invention: Dynamite, Nitrates, and the Making of the Modern World.* New York: St. Martin's Press, 2005.

Bryan, William Jennings. *The Commoner Condensed.* Spencer County, IN: The Abbey Press, 1902.

Brynjolfsson, Eric, and Andrew McAfee. *The Second Machine Age: Work, Progress, and Prosperity in a Time of Brilliant Technologies.* New York: W. W. Norton, 2014.

Burrough, Bryan. *Days of Rage: America's Radical Underground, the FBI, and the Forgotten Age of Revolutionary Violence.* New York: Penguin Press, 2015.

Burrows, Mathew. *The Future, Declassified: Megatrends That Will Undo the World unless We Take Action.* New York: Palgrave Macmillan, 2014.

Butcher, Tim. *The Trigger: Hunting the Assassin Who Brought the World to War.* New York: Grove Press, 2014.

Cahm, Caroline. *Kropotkin and the Rise of Revolutionary Anarchism, 1872–1886.* Cambridge, UK: Cambridge University Press, 1989.

Calaprice, Alice. "On Peace, War, the Bomb, and the Military." In *The New Quotable Einstein.* Princeton, NJ: Princeton University Press, 2005.

Callwell, C. E. *Small Wars.* New York: Presidio Press, 1991.

Carr, Christopher. *Kalashnikov Culture: Small Arms Proliferation and Irregular Warfare.* Westport, CT: Praeger Security International, 2008.

Carr, E. H. *Michael Bakunin.* London: Macmillan, 1937.

Castle, Ian. *The First Blitz: Bombing London in the First World War.* Oxford, UK: Osprey Publishing, 2015.

Chaliand, Gerard, and Arnaud Blin, eds. *The History of Terrorism: From Antiquity to Al Qaeda.* Los Angeles: University of California Press, 2007.

Chamayou, Grégoire. *Drone Theory.* London: Penguin Random House, 2015.

Chameau, Jean-Lou, William F. Ballhaus, and Herbert S. Lin, eds. *Emerging and Readily Available Technologies and National Security—A Framework for Addressing Ethical, Legal, and Societal Issues.* Prepublication copy. Washington, DC: National Academic Press, 2014.

Chase, Kenneth. *Firearms: A Global History to 1700.* Cambridge, UK: Cambridge University Press, 2013.

Cheung, Tai Ming, Thomas G. Mahnken, and Andrew L. Ross. "Frameworks for Analyzing Chinese Defense and Military Innovation." In *Forging China's Military Might: A New Framework for Assessing Innovation,* edited by Tai Ming Cheung. Baltimore: Johns Hopkins University Press, 2014.

Chivers, C. J. *The Gun.* New York: Simon & Schuster, 2010.

Christensen, Clayton. *The Innovator's Dilemma.* New York: HarperBusiness, 2011.

Choucri, Nazli. *Cyberpolitics in International Relations.* Cambridge, MA: MIT Press, 2012.

Cipolla, Caro. *Guns, Sails, and Empires: Technology Innovation and the Early Phases of European Expansion, 1400–1700*. Manhattan, KS: Sunflower University Press, 1985.

Clark, Christopher. *The Sleepwalkers: How Europe Went to War in 1914*. New York: HarperCollins, 2013.

Clarke, David, ed. *Technology and Terrorism*. New Brunswick, NJ: Transaction Publishers, 2004.

Clarke, Richard A., and Robert K. Knake. *CyberWar: The Next Threat to National Security and What to Do about It*. New York: HarperCollins, 2010.

Clausewitz, Carl von. *On War*, edited and translated by Michael Howard and Peter Paret. Princeton, NJ: Princeton University Press, 1976.

Coker, Christopher. *Warrior Geeks: How 21st Century Technology Is Changing the Way We Fight and Think about War*. Oxford, UK: Oxford University Press, 2013.

Coll, Steve. *Ghost Wars*. New York: Penguin, 2004.

Connable, Ben, and Martin Libicki. *How Insurgencies End*. Santa Monica, CA: RAND Corporation, 2010.

Cook, Philip J., and Kristin A. Goss, *The Gun Debate: What Everyone Needs to Know*. Oxford, UK: Oxford University Press, 2014.

Cooper, Paul, and Stanley R. Kurowski. *Introduction to the Technology of Explosives*. New York: Wiley-VCH, 1996.

Cornish, Paul. *Machine Guns and the Great War*. South Yorkshire, UK: Pen & Sword Books, 2009.

Cortada, James W. *The Digital Flood: The Diffusion of Information Technology across the US, Europe, and Asia*. Oxford, UK: Oxford University Press, 2012.

Cote, Owen, *The Politics of Innovative Military Doctrine: The US Navy and Fleet Ballistic Missiles*. Cambridge, MA: PhD dissertation, MIT 1998.

Cragin, Kim, Peter Chalk, Sara A. Daly, and Brian A. Jackson. *Sharing the Dragon's Teeth: Terrorist Groups and the Exchange of New Technologies*. Santa Monica, CA: RAND Corporation, 2007.

Crawford, Neta C. *Accountability for Killing*. Oxford, UK: Oxford University Press, 2013.

Crenshaw, Martha. *Explaining Terrorism: Causes, Processes, and Consequences*. London: Routledge, 2011.

Croft, Lee B. *Nikolai Ivanovich Kibalchich: Terrorist Rocket Pioneer*. Tempe, AZ: Institute for Issues in the History of Science, 2006.

Cronin, Audrey Kurth. *How Terrorism Ends: Understanding the Decline and Demise of Terrorist Campaigns*. Princeton, NJ: Princeton University Press, 2009.

Cronin, Audrey Kurth, and James Ludes, eds. *Attacking Terrorism: Elements of a Grand Strategy*. Washington, DC: Georgetown University Press, 2004.

Cundill, J. P. (and John Ponsonby), *A Dictionary of Explosives*. Chatham, UK: W. & J. Mackay, 1889; 2nd ed. 1895.

David, Paul A. *Technical Choice Innovation and Economic Growth: Essays on American and British Experience in the Nineteenth Century*. Cambridge, UK: Cambridge University Press, 1975.

David, R., ed. *The Routledge History of Terrorism*. London and
New York: Routledge, 2015.

Davies, R. W., Mark Harrison, and S. G. Wheatcroft, eds. *The Economic
Transformation of the Soviet Union, 1913–1945*. Cambridge, UK: Cambridge
University Press, 2005.

Davis, Mike. *Buda's Wagon: A Brief History of the Car Bomb*. New York: Verso, 2007.

Davis, Vincent, *The Politics of Innovation: Patterns in Navy Cases*. Denver,
CO: University of Denver Press, 1967.

Dedijer, Vladimir. *The Road to Sarajevo*. London: MacGibbon & Kee, 1967.

DeVries, Kelly. *Medieval Military Technology*. Peterborough, Ontario: Broadview
Press, 1994.

Diamond, Jared. *Guns, Germs, and Steel*. New York: Norton Press, 1999.

Dinnis, Heather Harrison. *Cyber Warfare and the Laws of War*. Cambridge,
UK: Cambridge University Press, 2012.

Dolnik, Adam. *Understanding Terrorist Innovation: Technology, Tactics, and Global
Trends*. London and New York: Routledge, 2007.

Domingos, Pedro. *The Master Algorithm: How the Quest for the Ultimate Learning
Machine Will Remake Our World*. New York: Basic Books, 2015.

Domm, Robert. *Michigan Yesterday and Today*. Minneapolis: Voyageur Press.

Doron, Assa, and Robin Jeffrey. *The Great Indian Phone Book: How the Cheap Cell
Phone Changes Business, Politics, and Daily Life*. Cambridge, MA: Harvard
University Press, 2013.

Dupuy, Trevor N. *The Evolution of Weapons and Warfare*. New York: Da Capo
Press, 1984.

Dutton, William S. *One Thousand Years of Explosives: From Wildfire to H-bomb*.
New York: Rinehart and Winston, 1960.

Einstein, Arthur W., Jr. *"Ask the Man Who Owns One": An Illustrated History of
Packard Advertising*. Jefferson, NC: MacFarland, 2016.

Ellis, John. *The Social History of the Machine Gun*. New York: Arno Press, 1981.

Engelbrecht, Helmuth Carol, and F. C. Hanighen. *Merchants of Death*.
New York: Garden City Publishing, 1937.

English, Richard. *Irish Freedom: The History of Nationalism in Ireland*.
London: Macmillan, 2006.

Everett, H. R. *Unmanned Systems of World Wars I and II*. Cambridge,
UK: Cambridge University Press, 2015.

Eyal, Nir, with Ryan Hoover. *Hooked: How to Build Habit-Forming Products*.
New York: Penguin, 2014.

Ezell, Edward Clinton. *The AK47 Story: Evolution of the Kalashnikov Weapons*.
New York: Stackpole Books, 1986.

Ezell, Edward Clinton. *The Great Rifle Controversy: Search for the Ultimate Infantry Weapon
from World War II through Vietnam and Beyond*. New York: Stackpole Books, 1984.

Ezell, Edward Clinton. *Kalashnikov: The Arms and the Man*. Cobourg,
Ontario: Collector Grade Publications, 2001.

Fant, Kenne. *Alfred Nobel: A Biography*. New York: Arcade Publishing, 1991.

Farrell, Theo, Frans Osinga, and James A. Russell. *Military Adaptation in Afghanistan*. Stanford, CA: Stanford University Press, 2013.

Farrell, Theo, and Terry Terriff. *The Sources of Military Change: Culture, Politics, Technology*. London: Lynn Rienner, 2002

Feldman, Burton. *The Nobel Prize: A History of Genius, Controversy, and Prestige*. New York: Arcade, 2000.

Figner, Vera. *Memoirs of a Revolutionist*. DeKalb, IL: Northern Illinois Press, 1991.

Finnegan, Terrence J. *Shooting the Front: Allied Aerial Photographic Interpretation on the Western Front—World War I*. Washington, DC: National Defense Intelligence College, 2006.

Fischer, Hannah. *United States Military Casualty Statistics: Operation Iraqi Freedom and Operation Enduring Freedom, CRS Report for Congress, #RS22452, 9 September 2008*.

Footman, David. *Red Prelude*. London: Barrie & Rockliff, the Cresset Press, 1944.

Ford, Mathew. *Weapon of Choice: Small Arms and the Culture of Military Innovation*. Oxford, UK: Oxford University Press, 2017.

Forest, James. *Teaching Terror: Strategic and Tactical Learning in the Terrorist World*. New York: Rowman & Littlefield, 2006.

Fraser, Antonia. *Faith and Treason*. New York: Random House, 1996.

Freedman, Lawrence. *The Evolution of Nuclear Strategy*. 3rd ed. New York: Palgrave Macmillan, 2003.

Freedman, Lawrence. *The Future of War: A History*. New York: Public Affairs, 2017.

Freedman, Lawrence. *The Revolution in Strategic Affairs*. Adelphi Paper 318. International Institute for Strategic Studies. Oxford, UK: Oxford University Press, 1998.

Fuller, J. F. C. *The Influence of Armament on History from the Dawn of Classical Warfare to the End of the Second World War*. New York: Da Capo Press, 1998.

Gage, Beverly. *The Day Wall Street Exploded: A Story of America in Its First Age of Terror*. Oxford, UK: Oxford University Press, 2009.

Galula, David, *Counterinsurgency Warfare: Theory and Practice*. New York: Praeger, 2006.

Gambetta, Diego, and Stephen Hertog. *Engineers of Jihad: The Curious Connection between Violent Extremism and Education*. Princeton, NJ: Princeton University Press, 2016.

Gambetta, Diego, ed. *Making Sense of Suicide Missions*. Oxford, UK: Oxford University Press, 2005.

Gardiner, Lloyd C., and Ted Gittinger, eds. *International Perspectives on Vietnam*. College Station: Texas A&M Press, 2000.

Geifman, Anna. *Thou Shalt Kill: Revolutionary Terrorism in Russia, 1894–1917*. Princeton, NJ: Princeton University Press, 1993.

Gilje, Paul A. *Rioting in America*. Bloomington and Indianapolis: Indiana University Press, 1996.

Giustozzi, Antonio. *Koran, Kalashnikov, and Laptop: The Neo-Taliban Insurgency in Afghanistan*. London: Hurst, 2007.

Godwin, William. *An Enquiry Concerning Political Justice*. Oxford World's Classics, with an introduction and notes by Mark Philp. Oxford, UK: Oxford University Press, 2013.

Goldman, Emily O., and Leslie C. Eliason, eds. *The Diffusion of Military Technology and Ideas*. Stanford, CA: Stanford University Press, 2003.

Goldstone, Jack. *Revolution and Rebellion in the Early Modern World*. Berkeley: University of California Press, 1991.

Goldstone, Jack. *Why Europe? The Rise of the West in World History, 1500–1850*. New York: McGraw-Hill, 2018.

Goodman, Marc. *Future Crimes: Inside the Digital Underground and the Battle for Our Connected World*. New York: Anchor Books, 2015.

Graham, Bob, and Jim Talent. *World at Risk: The Report of the Commission for the Prevention of Weapons of Mass Destruction Proliferation and Terrorism*. New York: Vintage Books, 2008.

Gray, Colin. *Modern Strategy*. Oxford, UK: Oxford University Press, 2000.

Greene, Owen, and Nicholas Marsh, eds. *Small Arms, Crime, and Conflict: Global Governance and the Threat of Armed Conflict*. London: Routledge, 2012.

Gref, Lynn G. *The Rise and Fall of American Technology*. New York: Algora Publishing, 2010.

Grossman, Nicholas. *Drones and Terrorism: Asymmetric Warfare and the Threat to Global Security*. London: I. B. Tauris, 2018.

Gundmundsson, Bruce. *Stormtroop Tactics: Innovation in the German Army, 1914– 1918*. Westport, CT: Praeger, 1995.

Haag, Pamela. *The Gunning of America: Business and the Making of American Gun Culture*. New York: Basic Books, 2016.

Hacking, Ian. *The Emergence of Probability: A Philosophical Study of Early Ideas about Probability, Induction, and Statistical Inference*. 2nd ed. Cambridge, UK: Cambridge University Press, 2006.

Hacking, Ian. *The Taming of Chance*. Cambridge, UK: Cambridge University Press, 1990.

Hall, Bert S. *Weapons and Warfare in Renaissance Europe: Gunpowder, Technology, and Tactics*. Baltimore and London: Johns Hopkins University Press, 1997.

Hallahan, William H. *Misfire: The History of How America's Small Arms Have Failed Our Military*. New York: Charles Scribner's Sons, 1994.

Halasz, Nicholas. *Nobel; A Biography of Alfred Nobel*. New York: Orion Press, 1959.

Hambling, David. *Swarm Troopers: How Small Drones Will Conquer the World*. Self-Published, 2015.

Hambling, David. *Weapons Grade: How Modern Warfare Gave Birth to Our High-Tech World*. New York: Carroll & Graf, 2005.

Harris, Shane. @War: The Rise of the Military-Internet Complex. New York: Houghton Mifflin Harcourt, 2014.

Harrison, Richard M., and Trey Herr. Cyber Insecurity: Navigating the Perils of the Next Information Age. London: Rowman & Littlefield, 2016.

Hartley, Scott. The Fuzzy and the Techie: Why the Liberal Arts Will Rule the Digital World. New York: Houghton Mifflin, 2017.

Haug, Karl Erik, and Ole Jørgen Maaø, eds. Conceptualizing Modern War. New York: Columbia University Press, 2011.

Haworth, W. Blair. The Bradley and How It Got That Way: Technology, Institutions, and the Problem of Mechanized Infantry in the United States Army. Westport, CT: Greenwood Press, 1999.

Head, Simon. The New Ruthless Economy: Work and Power in the Digital Age. Oxford, UK: Oxford University Press, 2011.

Headrick, Daniel R. Power over Peoples: Technology, Environment, and Western Imperialism, 1400 to the Present. Princeton, NJ: Princeton University Press, 2010.

Healey, Jason. A Fierce Domain: Conflict in Cyberspace, 1986 to 2002. Washington, DC: Cyber Conflict Studies Association in Partnership with the Atlantic Council, 2013.

Herrera, Geoffrey L. Technology and International Transformation: The Railroad, the Atom Bomb, and the Politics of Technological Change. Albany: State University of New York Press, 2016.

Herta, Ernestine Pauli, Alfred Nobel: Dynamite King—Architect of Peace. New York: L. B. Fischer, 1942.

Hess, Earl J. The Rifled Musket in Civil War Combat: Reality and Myth. Lawrence: University Press of Kansas, 2008.

Heuser, Beatrice. The Evolution of Strategy: Thinking War from Antiquity to the Present. Cambridge, UK: Cambridge University Press, 2010.

Hodges, Michael. AK47: The Story of a Gun. San Francisco: MacAdam Cage, 2007.

Hoffman, Bruce. Inside Terrorism. New York: Columbia University Press, 2017.

Horowitz, Michael C. The Diffusion of Military Technology: Causes and Consequences for International Affairs. Princeton, NJ: Princeton University Press, 2010.

Howard, Michael. War in European History. Oxford, UK: Oxford University Press, 1976.

Howard, Michael, and John F. Guilmartin. Two Historians in Technology and War. Carlisle, PA: Strategic Studies Institute, 1994.

Howe, Timothy, and Lee L. Brice, eds. Brill's Companion to Insurgency and Terrorism in the Ancient Mediterranean. Leiden, Netherlands: Koninklijke Brill, 2016.

Hulse, Michael. The Sorrows of Young Werther. London: Penguin Classics, 1989.

Huntley, W. W., and F. M. Robinson. Catalogue of Standard List-Price of Material Used by Railroads 1900. Richmond, VA: I. N. Jones, 1900.

Inkster, Nigel. China's Cyber Power. Abingdon, UK: Routledge, 2016.

Isaacson, Walter. The Innovators: How a Group of Hackers, Geniuses, and Geeks Created the Digital Revolution. New York: Simon & Schuster, 2014.

Jackson, Brian A., et al. *Breaching the Fortress Wall: Understanding Terrorist Efforts to Overcome Defensive Technologies*. Santa Monica, CA: RAND Corporation, 2007.

Jacobsen, Anne M. *The Pentagon's Brain*. New York: Little, Brown, 2015.

Jensen, Benjamin. *Forging the Sword: Doctrinal Change in the US Army*. Stanford, CA: Stanford University Press, 2016.

Jensen, Richard Bach. *The Battle against Anarchist Terrorism: An International History, 1878–1934*. Cambridge, UK: Cambridge University Press, 2014.

Jentz, Thomas. *Dreaded Threat: The 8.8 cm Flak 18/36/37 in the Anti-Tank Role*. Boyds, MD: Panzer Tracts, 2001.

Johnson, David E. *Fast Tanks and Heavy Bombers: Innovation in the US Army, 1917–1945*. Ithaca, NY: Cornell University Press, 1998.

Joll, James. *The Anarchists*. 2nd ed. Cambridge, MA: Harvard University Press, 1980.

Kaag, John, and Sarah Kreps. *Drone Warfare*. Cambridge, UK: Polity Press, 2014.

Kahaner, Larry. *AK-47: The Weapon That Changed the World*. Hoboken, NJ: John Wiley & Sons, 2007.

Kalashnikov, Mikhail, with Elena Joly. *The Gun That Changed the World*. Malden, MA: Polity Press, 2006.

Kaplan, Fred. *The Wizards of Armageddon*. New York: Touchstone Books, 1983.

Kauffman, James, Vlad P. Glaveanu, and John Baer, eds. *The Cambridge Handbook of Creativity across Domains*. Cambridge, UK: Cambridge University Press, 2017.

Keegan, John. *History of Modern Warfare*. New York: Vintage Books, 1993.

Keen, Andrew. *The Cult of the Amateur: How Today's Internet Is Killing Our Culture*. New York: Doubleday, 2007.

Keen, Andrew. *The Internet Is Not the Answer*. London: Atlantic Books, 2015.

Keller, Julia. *Mr. Gatling's Terrible Marvel: The Gun That Changed Everything and the Misunderstood Genius Who Invented It*. New York: Penguin Books, 2008.

Kelly, Jack. *Gunpowder, Alchemy, Bombards, and Pyrotechnics*. New York: Basic Books, 2004.

Kenna, Shane. *War in the Shadows: The Irish-American Fenians Who Bombed Victorian Britain*. Newbridge, Ireland: Irish Academic Press, 2014.

Kerksis, Sydney, and Thomas S. Dickey. *Field Artillery of the American Civil War*. Atlanta: Phoenix Press, 1968.

Kier, Elizabeth. *Imagining War: French and British Military Doctrine between the Wars*. Princeton, NJ: Princeton University Press, 1997.

Kierman, Frank, and John K. Fairbank eds. *Chinese Ways in Warfare*. Cambridge, MA: Harvard University Press, 1974.

Kilcullen, David. *The Accidental Guerrilla: Fighting Small Wars in the Midst of a Big One*. Oxford, UK: Oxford University Press, 2009.

Kilcullen, David. *Out of the Mountains: The Coming Age of the Urban Guerrilla*. Oxford, UK: Oxford University Press, 2013.

Killicoat, Phillip. *Weaponomics: The Economics of Small Arms*. CSAE WPS/2006-13, Department of Economics, University of Oxford, September 2006.

Killicoat, Phillip. *Weaponomics: The Global Market for Assault Rifles*. World Bank Policy Research Working Paper 4202, April 2004.

Kitson, Frank. *Low Intensity Operations: Subversion, Insurgency, and Peacekeeping*. London: Faber and Faber, 2011.

Knox, MacGregor, and Williamson Murray. *The Dynamics of Military Revolution, 1300–2050*. Cambridge, UK: Cambridge University Press, 2001.

Kreps, Sara E. *Drones: What Everyone Needs to Know*. Oxford, UK: Oxford University Press, 2016.

Lefebure, V. *Riddle of the Rhine: Chemical Strategy in Peace and War*. London: W. Collins Sons, 1921.

LaFree, Gary, Laura Dugan, and Erin Miller. *Putting Terrorism in Context: Lessons from the Global Terrorism Database*. London: Routledge, 2015.

Laqueur, Walter. *The Age of Terrorism*. New York: Little, Brown, 1987.

Larabee, Ann. *The Wrong Hands: Popular Weapons Manuals and Their Historic Challenges to a Democratic Society*. Oxford, UK: Oxford University Press, 2015.

Larsson, Ulf. *Alfred Nobel: Networks of Innovation*. Stockholm, Sweden: Archives of the Nobel Museum, 2008.

Le Caron, Henri. *Twenty-five Years in the Secret Service*. Heinemann, London, 1892.

Lee, Wayne. *Waging War: Conflict, Culture, and Innovation in World History*. New York: Oxford University Press, 2015.

Levine, Linda. *The Labor Market during the Depression and the Current Recession*. Report #R40655, US Congressional Research Service, 19 June 2009.

Levy, Joel. *Fifty Weapons That Changed the Course of History*. New York: Firefly Books, 2014.

Lieber, Keir A. *War and the Engineers: The Primary of Politics over Technology*. Ithaca, NY: Cornell University Press, 2005.

Lifton, Robert. *Destroying the World to Save It*. New York: Henry Holt, 1999.

Long, David E. *The Anatomy of Terrorism*. New York: Free Press, 1990.

Lupfer, Timothy. *The Dynamics of Doctrine: The Change in German Tactical Doctrine during the First World War*. Leavenworth, KS: US Army Combat Studies Institute, 1981.

Lynn, John. *Battle: A History of Combat and Culture*. Oxford, UK: Westview Press, 2003.

Macmillan, Margaret. *Paris 1919: Six Months That Changed the World*. New York: Random House, 2003,

Mahnken, Thomas. *Technology and the American Way of War since 1945*. New York: Columbia University Press, 2008.

Malkasian, Carter *War Comes to Garmser: Thirty Years of Conflict on the Afghan Frontier.* Oxford, UK: Oxford University Press, 2013.

Manucy, Albert. *Artillery through the Ages*. National Park Service Interpretive History Series No. 3. Washington, DC: US Government Printing Office, 1949.

McCormick, Charles H. *Hopeless Cases: The Hunt for the Red Scare Terrorist Bombers*. Lanham, MD: University Press of America, 2005.

McCullough, David. *The Wright Brothers*. New York: Simon & Schuster, 2015.

McFate, Sean, *The New Rules of War: Victory in the Age of Durable Disorder*. New York: William Morrow, 2019.

McIvor, Anthony D, ed. *Rethinking the Principles of War*. Annapolis, MD: Naval Institute Press, 2005.

McKenna, Joseph. *The Irish-American Dynamite Campaign: A History 1881–1896*. Jefferson, NC: McFarland, 2012.

McManus, John C. *The Deadly Brotherhood*. New York: Random House, 1998.

McNeil, William H. *The Pursuit of Power: Technology, Armed Force, and Society since A.D. 1000*. Chicago: University of Chicago Press, 1984.

McPherson, James J. *Battle Cry of Freedom: The Civil War Era*. Oxford, UK: Oxford University Press, 2003.

Mahnken, Thomas G. *Technology and the American Way of War since 1945*. New York: Columbia University Press, 2008.

Martin, Matt J., with Charles W. Sasser. *Predator: The Remote-Control Air War over Iraq and Afghanistan: A Pilot's Story*. Minneapolis: Zenith Press, 2010.

Merriman, John. *The Dynamite Club: How a Bombing in Fin-de-Siècle Paris Ignited the Age of Modern Terror*. Boston: Houghton Mifflin, 2009.

Moghadam, Assaf. *The Globalization of Martyrdom*. Baltimore: Johns Hopkins University Press, 2008.

Mokyr, Joel. *The Lever of Riches: Technological Creativity and Economic Progress*. Oxford, UK: Oxford University Press, 1990.

Morris, Donald R. *The Washing of the Spears: The Rise and Fall of the Zulu Nation*. New York: Simon and Schuster, 1965.

Morris, Ian. *Why the West Rules—for Now*. New York: Farrar, Straus, and Giroux, 2010.

Moss, Walter G. *Alexander II and His Times: A Narrative History of Russia in the Age of Alexander II, Tolstoy, and Dostoevsky*. London: Anthem Press, 2002.

Moten, Matthew, ed. *Between War and Peace: How America Ends Its Wars*. New York: Free Press, 2009.

Müller, Simone M. *Wiring the World: The Social and Cultural Creation of Global Telegraph Networks*. New York: Columbia University Press, 2016.

Murray, Robert K. *Red Scare: A Study in National Hysteria, 1919–1920*. Minneapolis: University of Minnesota Press, 1955.

Murray, Williamson. *Military Adaptation in War: With Fear of Change*. Cambridge, UK: Cambridge University Press, 2011.

Murray, Williamson, and Alan R. Millett. *Military Innovation in the Interwar Period*. Cambridge, UK: Cambridge University Press, 1996.

Nacos, Brigitte L. *Mass-Mediated Terrorism: Mainstream and Digital Media in Terrorism and Counterterrorism*. Lanham, MD: Rowman and Littlefield, 2016.

Nagl, John. *Learning to Eat Soup with a Knife*. Chicago: University of Chicago Press, 2002.

Nielsen, Suzanne. *An Army Transformed: The US Army's Post-Vietnam Recovery and the Dynamics of Change in Military Organizations*. Carlisle, PA: US Army War College, 2010.

Nixon, Richard. *RN: The Memoirs of Richard Nixon*. New York: Warner
Books, 1978.

Nursey-Bray, Paul, ed. *Anarchist Thinkers and Thought: An Annotated Bibliography*.
Westport, CT: Greenwood Press, 1992.

O'Connell, Robert. *Of Arms and Men: A History of War, Weapons, and Aggression*.
Oxford, UK: Oxford University Press, 1989.

O'Hanlon, Michael. *Technological Change and the Future of Warfare*. Washington,
DC: Brookings Institution Press, 2000.

Oppenheimer, A. R. *IRA: The Bombs and the Bullets: A History of Deadly Ingenuity*.
Dublin, Ireland, and Portland, OR: Irish American Press, 2009.

Owens, William, and Edward Offley, *Lifting the Fog of War*. Baltimore: Johns
Hopkins University Press, 2000.

Pape, Robert A. *Bombing to Win: Air Power and Coercion in War*. Ithaca, NY: Cornell
University Press, 1996.

Pape, Robert A. *Dying to Win: The Strategic Logic of Suicide Terrorism*.
New York: Random House, 2005.

Pape, Robert A., and James K. Feldman. *Cutting the Fuse: The Explosion of Global
Suicide Terrorism and How to Stop It*. Chicago and London: University of Chicago
Press, 2010.

Parker, Geoffrey *The Military Revolution: Military Innovation and the Rise of the
West, 1500–1800*. Cambridge, UK: Cambridge University Press, 1988; 2nd
ed. 1996.

Parrott, David. *The Business of War: Military Enterprise and Military Revolution in
Early Modern Europe*. Cambridge, UK: Cambridge University Press, 2012.

Partington, J. R. *A History of Greek Fire and Gunpowder*. Baltimore: Johns Hopkins
University Press, 1960.

Patrikarakow, David. *War in 140 Characters: How Social Media Is Reshaping Conflict
in the Twenty-first Century*. New York: Basic Books, 2017.

Pennock, J. Roland, and John W. Chapman. *Anarchism*. New York: New York
University Press, 1978.

Pierce, Terry C. *Warfighting and Disruptive Technologies: Disguising Innovation*.
London: Frank Cass, 2004.

Pipes, Richard. *Russia under the Old Regime*. London: Penguin, 1974, 1995.

Pocock, Chris. *50 Years of the U-2: The Complete Illustrated History of the Dragon Lady*.
Atglen, PA: Schiffler, 2005.

Pomper, Phillip. *Lenin's Brother: The Origins of the October Revolution*. New York: W.
W. Norton, 2009.

Porter, Bernard. *The Origins of the Vigilant State: The London Metropolitan Police Special
Branch before the First World War*. Woodbridge, UK: The Boydell Press, 1987.

Posen, Barry. *The Sources of Military Doctrine: France, Britain, and Germany between the
World Wars*. Ithaca, NY: Cornell University Press, 1984.

Poyer, Joe. *The AK-47 and AK-74 Kalashnikov Rifles and Their Variants*. 4th ed.
Tustin, CA: North Cape Publications, 2013.

Pursell, Carroll, Jr., ed. *Technology in America: A History of Individuals and Ideas.* Cambridge, MA: MIT Press, 1982.

Radzinsky, Edvard. *Alexander II: The Last Great Tsar.* Translated by Antonia W. Bouis. London: Free Press, 2005.

Rae, James DeShaw. *Analyzing the Drone Debates: Targeted Killing, Remote Warfare, and Military Technology.* New York: Palgrave Macmillan, 2014.

Ramsey, Gilbert. *Jihadi Culture on the World Wide Web.* New York: Bloomsbury Academic, 2013.

Ranstorp, Magnus, ed. *Mapping Terrorism Research: State of the Art, Gaps, and Future Direction.* London: Routledge, 2006.

Ranstorp, Magnus, and Magnus Normark. *Understanding Terrorism Innovation and Learning: Al Qaeda and Beyond.* London: Routledge, 2015.

Rasmussen, Maria J., and Hafez, Mohammed M. *Terrorist Innovations in Weapons of Mass Effect: Preconditions, Causes, and Predictive Indicators.* Workshop Report, Defense Threat Reduction Agency, 2010.

Rasmussen, Mikkel Vedby. *The Risk Society at War: Terror, Technology, and Strategy in the Twenty-first Century.* Cambridge, UK: Cambridge University Press, 2006.

Ratcliff, R. A. *Delusions of Intelligence: Enigma, Ultra, and the End of Secure Ciphers.* Cambridge, UK: Cambridge University Press, 2006.

Raudzens, George. "War-Winning Weapons: The Measurement of Technological Determinism in Military History." *Journal of Military History* 54, no. 4 (1990): 403–433.

Reader, Ian. *Religious Violence in Contemporary Japan: The Case of Aum Shinrikyo.* Honolulu: University of Hawaii Press, 2000.

Reader, W. J. *Imperial Chemical Industries: A History.* Vol. 1: *The Forerunners, 1870–1926.* Oxford, UK: Oxford University Press, 1970.

Reeve, Simon. *One Day in September: The Full Story of the 1972 Munich Olympics Massacre and the Israeli Revenge Operation "Wrath of God."* New York: Arcade Publishing, 2000.

Reid, R. W. *Tongues of Conscience: Weapons Research and the Scientists' Dilemma.* New York: Walker, 1969.

Rideal, Charles. *Stories from Scotland Yard.* London: George Routledge and Sons, 1890.

Riesch, Craig. *US M1 Carbines, Wartime Production.* 7th ed. Tustin, CA: North Cape Publications, 2012.

Robinson, James Harvey, and Charles Beard, eds. *Readings in Modern European History.* Vol. 2. Boston: Ginn, 1980.

Rodriquez, Michael. *R. E. Olds and Industrial Lansing.* Chicago: Arcadia, 2004.

Rogers, Everett M. *Diffusion of Innovations.* 5th ed. New York: Free Press, 2003.

Rogger, Hans. *Russia in the Age of Modernisation and Revolution, 1881–1917.* London: Longman, 1983.

Roland, Alex. *War and Technology: A Very Short Introduction.* Oxford, UK: Oxford University Press, 2016.

Ronfeldt, David, and William Sater. *The Mindsets of High-Technology Terrorists: Future Implications from an Historical Analog.* RAND #N-1610-SL. Sandia Laboratories, March 1981.

Ropp, Theodore. *The Development of a Modern Navy: French Naval Policy 1871–1904.* Annapolis, MD: Naval Institute Press, 1987.

Rosen, Stephen Peter. *Winning the Next War: Innovation and the Modern Military.* Ithaca, NY: Cornell University Press, 1991.

Rosenfeld, Jean E., ed. *Terrorism, Identity, and Legitimacy: The Four Waves Theory and Political Violence.* London: Routledge, 2011.

Rottman, Gordon L. *The AK-47 Kalashnikov-Series Assault Rifles.* Oxford, UK: Osprey, 2011.

Sageman, Marc. *Turning to Political Violence: The Emergence of Terrorism.* Philadelphia: University of Pennsylvania Press, 2017.

Sapolsky, Harvey. *Polaris System Development: Bureaucratic and Programmatic Success in Government.* Cambridge, MA: Harvard University Press, 1972.

Saunders, Anthony. *Reinventing Warfare 1914–18: Novel Munitions and the Tactics of Trench Warfare.* New York: Continuum, 2012.

Saunders, David. *Russia in the Age of Reaction and Reform, 1801–1881.* London: Longman, 1982.

Schaack, Michael. *Anarchy and Anarchists: A History of the Red Terror and the Social Revolution in America and Europe.* New York and Philadelphia: W. A. Houghton, 1889.

Schaerf, Carlo, Brian Holden Reid, and David Carlton, eds. *New Technologies and the Arms Race.* New York: St. Martin's Press, 1989.

Scharre, Paul. *Army of None: Autonomous Weapons and the Future of War.* New York: W. W. Norton, 2018.

Schneier, Bruce. *Carry On: Sound Advice from Schneier on Security.* New York: John Wiley & Sons, 2014.

Schumpeter, J. A. *Business Cycles.* New York, 1939.

Schumpeter, Joseph. *Capitalism, Socialism, and Democracy.* New York: Harper & Brothers, 1942.

Schwab, Klaus. *The Fourth Industrial Revolution.* Geneva: World Economic Forum, 2016.

Segal, Adam. *The Hacked World Order: How Nations Fight, Trade, Maneuver, and Manipulate in the Digital Age.* New York: Public Affairs, 2016.

Seth, Ronald. *The Russian Terrorists.* London: Barrie and Rockliff, 1966.

Seton-Watson, Hugh. *The Decline of Imperial Russia, 1855–1914.* London: Methuen, 1952.

Shane, Scott. *Objective Troy: A Terrorist, a President, and the Rise of the Drone.* New York: Random House, 2015.

Shapiro, Jacob N. *The Terrorist's Dilemma: Managing Violent Covert Organizations.* Princeton, NJ: Princeton University Press, 2013.

Sharpe, James. *Remember, Remember: A Cultural History of Guy Fawkes Day*. Cambridge, MA: Harvard University Press, 2005.

Shilin, Val, and Charlie Cutshaw. *Legends and Reality of the AK: A Behind-the-Scenes Look at the History, Design, and Impact of the Kalashnikov Family of Weapons*. Boulder, CO: Paladin Press, 2000.

Short, K. R. M. *The Dynamite War: Irish-American Bombers in Victorian Britain*. Atlantic Highlands, NJ: Humanities Press, 1979.

Showalter, Dennis. *Railroads and Rifles: Soldiers, Technology, and the Unification of Germany*. Hamden, CT: Archon Books, 1975.

Simon, Jeffrey D. *Lone Wolf Terrorism*. New York: Prometheus Books, 2013.

Simpson, Emile. *War from the Ground Up: 21st Century Combat as Politics*. Oxford, UK: Oxford University Press, 2012.

Singer, Jane. *The Confederate Dirty War: Arsons, Bombings, Assassinations, and Plots for Chemical and Germ Attacks on the Union*. London: McFarland, 2005.

Singer, P. W. *Corporate Warriors: The Rise of the Privatized Military Industry*. Ithaca, NY: Cornell University Press, 2003.

Singer, P. W., and Emerson T. Brooking, *LikeWar: The Weaponization of Social Media*. New York: Houghton Mifflin Harcourt, 2018.

Singer, P. W., and Allan Friedman, *Cybersecurity and Cyberwar*. Oxford, UK: Oxford University Press, 2014.

Singer, Peter. *Wired for War: The Robotics Revolution and Conflict in the Twenty-first Century*. New York: Penguin Books, 2009.

Small Arms Survey 2015: Weapons and the World. Graduate Institute of International and Development Studies, Geneva. Cambridge, UK: Cambridge University Press, 2015.

Smith, Merritt Roe. *Harpers Ferry Armory and the New Technology: The Challenge of Change*. Ithaca, NY: Cornell University Press, 1977.

Smith, Merritt Roe. *Military Enterprise and Technological Change: Perspectives on the American Experience*. Cambridge, MA: MIT Press, 1985.

Smith, Rupert. *The Utility of Force: The Art of War in the Modern World*. London: Allen Lane, 2005.

Smithurst, Peter. *The Gatling Gun*. New York: Bloomsbury, 2015.

Smithsonian Institution. *Firearms: An Illustrated History*. New York: DK Publishing, 2014.

Sohlman, Ragner, and Schück, Henrik, *Nobel: Dynamite and Peace*. New York: Cosmopolitan, 1929.

Spencer, C., ed. *The Encyclopedia of North American Indian Wars, 1607–1890*. Oxford, UK: ABC-CLIO, 2011.

Standage, Tom. *The Victorian Internet: The Remarkable Story of the Telegraph and the Nineteenth Century's On-line Pioneers*. New York: Walker, 1998.

Stevens, R. Blake, and Edward C. Ezell, *The Black Rifle: M16 Retrospective*. Cobourg, Ontario: Collector Grade Publications, 2015.

Strachan, Hew. *Carl von Clausewitz's* On War: *A Biography*. London: Atlantic Books, 2008.

Strachan, Hew. *The Direction of War: Contemporary Strategy in Historical Perspective*. Cambridge, UK: Cambridge University Press, 2013.

Strachan, Hew. *European Armies and the Conduct of War*. London: George Allen & Unwin, 1983.

Sutherland, John. *iGuerrilla: Reshaping the Face of War in the 21st Century*. Palisades, NY: History Publishing Company, 2015.

Taylor, Frederick. *The Principles of Scientific Management*. New York: Dover Publications, 1997.

Thompson, Peter G. *Armed Groups: The 21st Century Threat*. Lanham, MD: Rowman & Littlefield, 2014.

Thurman, James R. *Practical Bomb Scene Investigation*. 2nd ed. Boca Raton, FL: Taylor and Francis, 2011.

Toffler, Alvin. *The Third Wave*. New York, William Morrow, 1980.

Trinquier, Roger. *Modern Warfare: A French View of Counterinsurgency*. New York: Praeger, 2006.

Tucker, Jonathan, and Jason Pate. *Toxic Terror: Chemical and Biological Weapons*. Cambridge, MA: The Belfer Center, 2000.

Tufekci, Zeynep. *Twitter and Tear Gas: The Power and Fragility of Networked Protest*. New Haven, CT: Yale University Press, 2017.

Twain, Mark. *Following the Equator: A Journey around the World*. Hartford, CT: The American Publishing Company, 1897; Dover Productions, Reprint Edition, 1989.

Ulam, Adam B. *Russia's Failed Revolutions: From the Decembrists to the Dissidents*. New York: Basic Books, 1981.

Valeriano, Brandon, and Ryan C. Maness. *Cyber War versus Cyber Realities*. Oxford, UK: Oxford University Press, 2015.

Van Creveld, Martin. *Technology and War from 2000 B.C. to the Present*. New York: Free Press, 1989.

Vandervort, Bruce. *Wars of Imperial Conquest in Africa, 1830–1914*. London: Routledge, 2014.

Van Evera, Stephen. *Causes of War: Power and the Roots of Conflict*. Ithaca, NY: Cornell University Press, 1999.

Vanguri, Stav Medzerian, ed. *Rhetorics of Names and Naming*. London: Routledge, 2016.

Venturi, Franco. *The Roots of Revolution: A History of the Populist and Socialist Movements in 19th-Century Russia*. New York: Alfred A. Knopf, 1960.

Volk, S. S. *Narodnaya Volia 1879—1882*. Moscow and Leningrad, 1966.

Wagner, William. *Lightning Bugs and Other Reconnaissance Drones*. Fallbrook, CA: Aero Publishers and Armed Forces Journal International, 1982.

Walter, John. *Rifles of the World*. 3rd ed. Iola, WI: Krause Publications, 2006.

Weber, Max. "Politics as a Vocation" [Politik als Beruf], lecture given in Munich, 28 January 1919. In *Essays in Sociology*, translated and edited by Howard Garth and Cynthia Wright Mills. New York: Free Press, 1946.

Weimann, Gabriel. *Terror on the Internet: The New Arena, the New Challenges*. Washington, DC: US Institute of Peace, 2006.

Weimann, Gabriel. *Terrorism in Cyberspace: The Next Generation*. Washington, DC, and New York: Woodrow Wilson Center Press and Columbia University Press, 2015.

Whelehan, Niall. *The Dynamiters: Irish Nationalism and Political Violence in the Wider World, 1867–1900*. Cambridge, UK: Cambridge University Press, 2015.

White, Lynn, Jr. *Medieval Technology and Social Change*. Oxford, UK: Oxford University Press, 1962.

Whittle, Richard. *Predator: The Secret Origins of the Drone Revolution*. New York: Henry Holt, 2014.

Wilkins, Mira. *Foreign Investment in the United States to 1914*. Cambridge, MA: Harvard University Press, 1989.

Wilkinson, Paul. *Technology and Terrorism*. London: Frank Cass, 1993.

Wilkinson, Paul, and Brian M. Jenkins, eds. *Aviation Terrorism and Security*. London: Frank Cass, 1999.

Wilson, James. *Bureaucracy*. New York: Basic Books, 1991.

Winkler, Johnathan Reed. *Nexus: Strategic Communications and American Security in World War I*. Cambridge, MA: Harvard University Press, 2008.

Wittes, Benjamin, and Gabriella, Blum. *The Future of Violence: Robots and Germs, Hackers and Drones*. New York: Basic Books, 2015.

Woodcock, George. *Anarchism: A History of Libertarian Ideas and Movements*. Cleveland and New York: World Publishing Co., 1962.

Wright, Quincy. *Mandates under the League of Nations*. Chicago: University of Chicago Press, 1930.

Yarmolinksy, Avrahm. *Road to Revolution: A Century of Russian Radicalism*. New York: Macmillan, 1959.

Zadka, Saul. *Blood in Zion: How the Jewish Guerrillas Drove the British out of Palestine*. London: Brassey's, 1995.

Zarzecki, Thomas W. *Arms Diffusion: The Spread of Military Innovations in the International System*. London: Routledge, 2002.

Zisk, Kimberley. *Engaging the Enemy: Organization Theory and Soviet Military Innovation, 1955–1991*. Princeton, NJ: Princeton University Press, 1993.

Zubok, Vladislav M. *A Failed Empire: The Soviet Union in the Cold War from Stalin to Gorbachev*. Chapel Hill: University of North Carolina Press, 2009.

INDEX

Note on index: *For the benefit of digital users, indexed terms that span two pages (e.g., 52–53) may, on occasion, appear on only one of those pages.*

Dragon Runner 10, 204
Dresdner Dynamit-Fabrik, 106
drones
 3D printing used to create, 200,
  226–27, 229, 249
 A1A drone and, 212–13
 A1C drone, 212–13
 Ababil A1B drone, 212–13
 Ababil-T drone, 211–12, 213
 Afghanistan War (2001-) and, 11–
  12, 175, 210, 215
 al-Qaeda as target of attacks by, 31–
  32, 175, 206
 artificial intelligence and, 221, 224
 autonomous flight capability and,
  200, 203, 208, 215, 224,
  236, 247–48
 Awlaki killed in US military attack
  by, 174, 175
 battery power and, 221, 249, 250–51
 Bind and Fly (BNF) drones and, 214
 Blackhawk helicopter's collision
  (2017) with, 224–25
 China and, 210–11, 214–16, 235,
  250, 359n55
 collisions with aircraft and, 224–25
 commercial applications of, 9, 203
 cost of, 201–2, 208–9, 210–11, 218–
  19, 221, 223, 226–27, 230
 countermeasures against, 221, 223–
  24, 230, 265
 cover and concealment maneuvers
  and, 250
 Crimea conflict (2014-) and, 220
 customization of, 205,
  216–17, 258–59
 digital communications capacity
  and, 210
 Eleron-3SV drone and, 220
 encrypted communication and, 212
 expanding commercial market
  for, 214
 global diffusion of, 216

 GPS technology and, 200, 205, 223,
  224, 225–26
 "gray zone conflicts" and, 220
 guidance for missile or aircraft
  attacks provided by, 206
 Harpy loitering munition and, 214,
  234–35, 359n55
 Hezbollah's use of, 212
 Houthi rebels' use of, 213
 Iraq War (2003-12) and, 201–2,
  206, 210, 218–19, 261
 ISIS's use of, 215, 218–19, 261, 262
 Israel and, 210, 212–13, 230, 250
 K-Max drone and, 214
 Lashkar-e-Taiba and, 217
 lethal autonomous weapons (LAWS)
  and, 235
 loitering capabilities of, 250–51
 Mavic Pro and, 214–15
 militaries' role in developing, 1–2, 5,
  6, 201–2, 205, 206, 210–11
 Mirsad-1 surveillance drone
  and, 212
 as missiles, 223
 missiles compared to, 208–9
 missiles fired from, 233
 nano-sized versions of, 201–2
 nineteenth-century predecessors
  of, 206
 package delivery and, 224
 payload capacity of, 216–17
 Perdix drones and, 249
 Plug and Fly (PNF) drones, 214
 precision-strike capability and, 224
 Predator drones and, 11–12, 203,
  206, 209, 210–11
 private military contractors' use
  of, 219–20
 Qasef 2K drone and, 213
 quadcopter drones and, 214–15,
  218–19, 222–23, 225–26, 229,
  233–34, 256, 261
 radar capacity and, 225

gunpowder
  Bacon's introduction to Europe of, 64
  cannons' use of, 65
  challenges of using, 66, 68
  Civil War time bombs created
    from, 80
  construction and mining applications
    of, 67, 69
  dynamite viewed as improvement on,
    61–62, 68
  First World War and, 98
  grenades' use of, 64–65, 97
  Gunpowder Plot of 1605 (England)
    and, 65–66
  improvised explosive devices and, 51
  invention in ninth-century China of,
    30–31, 63–64
  military strategy changed by
    introduction of, 22
  "Orsini bomb" and, 66
  saltpeter and, 64, 68
Guns, Germs, and Steel (Diamond), 57

hacking
  artificial intelligence and, 253
  drone technology and, 223, 238
  encryption as means of
    thwarting, 195
  Internet of Things and, 7, 244, 266
  self-driving vehicles as target
    of, 243–44
  smartphones as target of, 196, 200
Hamas
  cyberwarfare and, 181
  drones and, 211–13, 218, 251–52
  Gaza War (2014) and, 212–13
  Iran and, 42, 211–13
  money transfers and, 186
  targeted killings of leaders
    of, 212–13
Hammami, Seif Allah, 37–38
Hanson, Ole, 92
Haqqani terrorist network, 217

Harpy loitering munition, 214, 234–
  35, 359n55
Harris, Arthur "Bomber," 31–32
Harris, Tristan, 197
Hasan, Nidal, 174
Hawala finance system, 186
Hawking, Stephen, 177, 235,
  238, 239–40
Haymarket Square massacre (Chicago,
  1886), 87–88, 110–11, 118–19
Hazard Powder Company, 106–7
Hearst Publishing Company, 115–16
Heliograf, 195
Hellfire missiles, 233
Henry, Emile, 116
Hezbollah, 1–2, 20, 42–43, 181,
  186, 211–13
Hiroshima atomic bomb blast (1945),
  22, 138
Hitler, Adolf, 32, 135
Hmeimim airbase (Syria), 230
Hodgkinson, James Thomas, 180
Homestead Steel Strike (1892),
  118–19, 131
Homs (Syria), 191
Horowitz, Michael, 29–30, 31, 211,
  292n48
Houthi rebels (Yemen), 211–12, 213,
  250, 361n68
Huawei, 6–7
Hull, Charles, 227
Hundred Years' War (1339-1453),
  24, 65
Hungary, 135, 144–46, 145f, 149, 154
Hussein, Saddam, 32

IBM, 8, 34, 176, 222
Iklé, Fred, 263
improvised explosive devices (IEDs)
  Afghanistan War (2001-) and, 11–
    12, 32, 49, 50–51, 52–53
  Boston Marathon bombing (2013)
    and, 172–73

physical mobilization and, 177
propaganda and, 182
psychological manipulation and,
177, 182–84, 186, 197
social media and, 178, 180, 182
social mobilization and, 177
Model T car, 9–10, 11, 284n12
Molotov cocktails, 144, 185
Monero cryptocurrency, 185–86
Mongol Empire, 247
Monsonego, Miriam, 191–92
Mooney, Thomas J., 81–82
Moore's Law, 176–77
Morgan, J. P., 61, 108
Mosin-Nagant rifles, 148
Most, Johann, 111–13, 182
Mosul (Iraq), 183, 218
Mozambican Liberation Front
(FRELIMO), 159
Mozambican National Resistance
(RENAMO), 159
Mozambique, 159
Mozorov, Nicholas, 84–85
Mujahideen forces (Afghanistan), 43–
44, 159, 188
Mumbai terrorist attacks (India, 2006),
50, 126
Munich Olympics massacre (1972), 15,
44, 50, 158
Musk, Elon, 235, 238–39
Muslim Assassins
(Ismailis-Nazari), 40–41
Mustafa, Faisal, 188
Mutually Assured Destruction
Theory, 23
Myanmar, 96, 359–60
My Disillusionment with Russia
(Goldman), 91

Nagasaki atomic bomb blast
(1945), 138
Napoleon Bonaparte, 33, 66, 294n62
Napoleonic Wars, 180

Napoleon III (emperor of France),
66, 78
Narodnaya Volya (People's Will)
organization, 41–42, 74
National Labor Relations Act (1935),
92–93, 266
National Security Agency (NSA),
175, 196–97
National Union for the Total
Independence of Angola
(UNITA), 159–60
Native Americans, 128–29, 131–32
NATO (North Atlantic Treaty
Organization), 149, 152,
215, 236
Nechayev, Sergei, 41–42
Neo-Nazi National Alliance, 47
Nerekhta mini tank (Russia), 241
The Netherlands, 183
New Left, 42
New York City draft riots (1863), 131
New York Times, 115, 260
New Zealand, 96, 190
Niger, 34
Nigeria, 184, 210
Nimitz, Chester, 21
nitroglycerine
blasting oil developed by Nobel
from, 71–72, 99
discovery of, 69–70
dynamite's use of, 73, 96, 99–100,
102–3, 105, 106
medical applications for, 70
Nitro-Glycerine Act (Great Britain,
1869), 105
Nobel's experiments with,
70–71, 72
volatility of, 69–70, 71, 72, 73
Nixon, Richard, 31–32
Nobel, Alfred
biographical background of, 69
blasting caps invented by, 72
blasting gelatin and, 57, 123–24

quadcopter drones
  3D printing as means of
    creating, 229
  military use of commercially
    developed models of, 218–19
  as most popular model of commercial
    drone, 214–15, 261
  secure areas breached by, 222–23
  terrorists' use of, 233–34, 261
  underground use of, 225–26
Qualcomm, 253
Quarterly Journal of Science, 95–96
Qutb, Sayyid, 171

radar, 25, 225, 233, 234–35
railroads
  dynamite used in the construction of,
    119–20, 265–66
  expansion in nineteenth century
    of, 67–68
  gunpowder used in the construction
    of, 67, 68
  military implications of, 24–25
  role in regulating dynamite, 105–6,
    119–20, 265–66
  US civil war and, 67
Raines, Gabriel, 97–98
Rapoport, David C., 41–42, 43
Rassler, Don, 218
Ravachol (François Claudius
    Koenigstein), 113, 116–18
Raven (RQ-11B) drone, 201–2, 229
Raymond, Henry Jarvis, 131
Razor smart drone, 200–1
"reach" concept, 202–3, 206
Ready to Fly (RTF) drones, 214
Reaper drones, 203, 206, 209, 210–11,
    213, 233
Red Army (Soviet Union), 134, 135–
    36, 139, 145
Red Army Faction (RAF; also Baader-
    Meinhof Group) (West
    Germany), 42

Red Brigades (Italy), 42
Red Scare (United States, 1919-21), 92
religious fundamentalism, 41,
    42–43, 46
Remington, 129
remote control weapons, 217
RepRap 3D printer, 227
Restrepo, Pascual, 253–54
Revolution in Military Affairs (RMA),
    26–27, 238
Revue Générale des Sciences Pures et
    Appliquées, 95–96
Rexilius, Jason, 219–20
ricin, 37, 38, 195–96
Ripley, James, 130–31
robots and robotics. *See also* drones
  algorithms and, 203–4
  autonomous capacity and, 56–57,
    203–4, 241, 258–59
  "centaur warfighting" and, 203–4
  commercially available products and,
    203, 241
  costs of, 204, 221, 230, 241–42, 252
  countermeasures to prevent attacks
    and, 252
  customization options and, 258–59
  diffusion of access to, 251
  explosive ordnance disposal (EOD)
    and, 203–4, 242
  "killer robots" and, 251
  law enforcement's use of, 242
  legal and policy guidelines for
    using, 242
  nanorobots and, 204
  potential disruptions to economy
    and, 253–54
  reconnaissance capacity and,
    204, 242
  sensors and, 242
  terrorism potential of, 240, 241,
    252, 267
  unmanned aerial vehicles (UAVs),
    203, 205, 206

socialism, 41, 42, 43

social media. *See also specific platforms*
  advertising on, 197–98
  Arab Spring uprisings and, 191
  bombmaking information conveyed
    via, 184–85
  bundling of content on, 199
  contagion and, 53–54
  content monitors for, 190
  counterterrorism and, 186
  data collection via, 187, 197–98
  fake accounts on, 193–95
  "fake news" publicized through, 193
  "filter bubbles" and, 198
  government regulation of, 199
  interactive nature of, 187
  jihadists and, 180–81, 184–85
  livestreaming capacity of,
    189–91, 260
  mobilization opportunities and, 178,
    180, 182
  psychological processes activated by
    participating in, 198
  radicalization via, 180
  Syrian civil war and, 191
  targeting by platform companies
    on, 172–98
  terrorist attacks publicized through,
    46, 54, 188, 189, 192, 199
  terrorist organizations' utilization of,
    53, 179–80, 182, 183, 184–85,
    188–90, 198, 199, 260
  traditional media gatekeepers
    bypassed by, 183–84
  unforeseen consequences from
    technologies of, 199
  video capability of, 178
sockpuppets (falsely-attributed social
  media accounts), 194
Solid Concepts Inc., 228, 229
Somalia
  drone warfare in, 175, 206, 210
  improvised explosive devices in, 53

Kalashnikov rifles and, 158–59
private military contractors
  in, 219–20
terrorist recruitment and, 184
US military involvement in, 34
"Sophia" (artificially intelligent
  humanoid robot), 36f
South Africa, 20–21, 152, 159–60
Southampton University Laser Sintered
  Aircraft (SULSA), 200–1
South China Sea, 21
Southern Powder Company, 106–7
South Korea, 234–35, 250
South West Africa, 132
Soviet Union
  Afghanistan War (1979-89) and, 11–
    12, 175, 210, 215
  agricultural collectivization
    of, 134–35
  AK-47 produced by, 127, 136, 138–
    39, 143–44, 146, 147, 148,
    149–50, 162–63, 257
  Central Planning Agency in, 146
  Cheka (counterrevolutionary secret
    police) in, 90–91
  China and, 148–49
  collapse (1991) of, 43–44, 164
  command economy and global
    currency shortages in, 146–47
  Hungarian Uprising (1956)
    and, 144–46
  innovation as a threat to, 5–6
  Military Industrial Commission
    (VPK) in, 147
  New Left terrorist organizations and,
    42, 46
  nuclear arms race with United States
    and, 21–22, 181
  nuclear arsenal of, 138
  Paris Convention for the Protection
    of Industrial Property
    and, 150
  propaganda operations of, 183

ACKNOWLEDGMENTS

THIS BOOK WAS a years-long fixation. Apologies to family, friends, neighbors, students, and colleagues neglected during that time.

The argument's framework emerged during a week-long return to Oxford, where I was hosted by the Changing Character of War (CCW) Programme and did research in the Codrington Library at All Souls College. Thanks to Rob Johnson, Director, and all the CCW associates, present and past, who inspired and improved my thinking. Professor Sir Hew Strachan gave me Codrington access and invaluable feedback on chapter 1. My lingering errors are not his fault.

I am grateful to the Smith Richardson Foundation for financial support that made this book possible.

Many colleagues and students offered wisdom and support. Special thanks to Anne Baker, Jeffrey Chase, Bernard "Bud" Cole, Alexander Falbo, James Goldgeier, Carolyn Just, Ingrid Korsgard, Ricardo Sanchez Ortegon, Jim Pfiffner, Edward Rhodes, Laurie Schintler, and especially John Gudgel, whose loyalty, ideas, and patience were deeply appreciated. A large number of people read drafts or gave other feedback, including Martha Crenshaw, Richard Bach Jensen, Elsa Kania, Laurent Baker, Kenneth Klothe, Richard Bitzinger, Steven Kurth, Elizabeth "Libbie" Prescott, Joshua Rovner, Karen Wilhelm, and others who must remain anonymous.

Dave McBride, George Lucas, my agent, and Emily Loose helped me sharpen the arguments and outgrow some bad academic writing habits. Thank you. I'll never look back. At Oxford University Press I also particularly thank Holly Mitchell, Cheryl Merritt, Cayla DiFabio, and Ashley Noelle.

My career has had unanticipated hurdles along the way. Without my Oxford dissertation supervisor and steady supporter, Adam Roberts, I'd have capitulated long ago. The book is dedicated to him. I also received key encouragement from my beloved and now-departed next-door neighbor, Ira Berlin, and his wife Martha.

Finally, I thank Christopher, Natalie, and especially my patient, optimistic, energetic husband Patrick Cronin, the muse with the news. Your love means everything.